*Arrow Pushing in
Inorganic Chemistry*

Arrow Pushing in Inorganic Chemistry

A Logical Approach to the Chemistry of the Main-Group Elements

Abhik Ghosh

Steffen Berg

WILEY

Published by John Wiley & Sons, Inc., Hoboken, New Jersey
Published simultaneously in Canada

For general information on our other products and services or for technical support, please contact our Customer Care Department within the United States at (800) 762-2974, outside the United States at (317) 572-3993 or fax (317) 572-4002.

Wiley also publishes its books in a variety of electronic formats. Some content that appears in print may not be available in electronic formats. For more information about Wiley products, visit our web site at www.wiley.com.

Library of Congress Cataloging-in-Publication Data is available.

ISBN: 978-1-118-17398-5

Printed in the United States of America

10 9 8 7 6 5 4 3 2 1

Contents

Sections marked with an asterisk (*) may be skipped on first reading.

Foreword

Many years ago George Hammond and I taught a course at Caltech that included discussions of main-group chemistry. We tried to use inorganic textbooks that dealt with the subject, but we were not happy with them, as they paid no attention to reaction mechanisms. Discussions of nucleophilic and electrophilic reagents, associative and dissociative substitutions, reaction energy landscapes, and so on, were nowhere to be found. Faced with this problem, we decided to base our course on reaction mechanisms, but very few instructors adopted this approach in teaching main-group chemistry.

Now, at long last, we have a book on main-group chemistry that students can learn from! They may even read it from cover to cover without going to sleep! The authors, Abhik Ghosh and Steffen Berg, have clearly demonstrated how a mechanistic approach makes the reactions of main-group elements interesting and understandable: Arrow pushing is the key!

There are many parts of the book that I like very much. The treatment of the reactions of nitrogen compounds, largely neglected in inorganic courses, is particularly good. And one of my favorites, the very rich chemistry of high-valent halogen and xenon molecules, is excellent. The bottom line is that arrow pushing is a method that should be used to teach main-group chemistry. As the authors note, their book logically can be used to supplement standard inorganic texts. I urge instructors to try the Ghosh–Berg method when faced with teaching the dreaded "descriptive" section of the inorganic course. Arrow pushing not only is great fun, students who try it may actually learn main-group chemistry!

HARRY B. GRAY
California Institute of Technology
February 2014

Preface

Inorganic chemistry at core consists of a vast array of molecules and chemical reactions. To master the subject, students need to think intelligently about this body of facts, a feat that is seldom accomplished in an introductory course. All too often, young students perceive the field as an amorphous body of information that has to be memorized. We have long been intrigued by the possibility of changing this state of affairs by means of a mechanistic approach, specifically organic-style arrow pushing. We found that such an approach works well for all main-group elements, that is, elements from the s and p blocks of the periodic table. In particular, we found that arrow pushing works well for hypervalent compounds, where the central atom has more than eight electrons in its valence shell in the Lewis structure. Over time, we came to appreciate that full implementation of a mechanistic approach had the potential to transform the teaching of a substantial part of the undergraduate inorganic curriculum. This book is a realization of that vision.

Arrow Pushing in Inorganic Chemistry is designed as a companion to a standard inorganic text. In general, we have devoted one chapter to each group of the main-group elements. Each chapter in this book is designed to supplement the corresponding chapter in a regular inorganic text. A student using this book is expected to have taken general chemistry and a good, introductory course in organic chemistry at the university level. Key prerequisites include elementary structure and bonding theory, a good command of Lewis structures, VSEPR theory, elementary thermodynamics (as usually outlined in general chemistry), simple acid–base calculations, basic organic nomenclature, and a good but elementary understanding of organic mechanisms. Because a basic knowledge of organic chemistry has been assumed, the general level of this book is somewhat higher than that of an undergraduate organic text. The material included in this book (along with related content from a standard inorganic text) has been regularly taught at the University of Tromsø in about 30 h of class time, roughly half of which has been devoted to problem-solving by students. A small number of somewhat specialized topics and review problems have been marked with an asterisk, to indicate that they may be skipped on first reading. We usually take up a few of these at the end of our course and in conjunction with a second or more specialized course.

The approach. Many students are deeply impressed by the logic of organic chemistry. Mechanistic rationales are available for essentially every reaction in the undergraduate (and even graduate) organic curriculum and students learn to write reaction mechanisms right from the beginning of their courses. A survey of current texts shows that a mechanistic approach is universally adopted in introductory organic courses. The situation with inorganic chemistry could not be more different; not one major introductory text adopts a mechanistic approach in presenting descriptive main-group chemistry! In a telling exercise, we went through several textbooks that do an otherwise excellent job of presenting descriptive inorganic chemistry, without finding the words "nucleophile" and "electrophile." Not surprisingly, these texts do not present a single instance of arrow pushing either.

Arrow pushing above all provides a logical way of thinking about reactions, including those as complex as the following:

$$P_4 + 3\,NaOH + 3\,H_2O \rightarrow 3\,NaH_2PO_2 + PH_3$$

$$24\,SCl_2 + 64\,NH_3 \rightarrow 4\,S_4N_4 + S_8 + 48\,NH_4Cl$$

$$2\,HXeO_4^- + 2\,OH^- \rightarrow XeO_6^{4-} + Xe + O_2 + 2\,H_2O$$

These reactions represent important facets of the elements involved but are typically presented as no more than facts. (Why does boiling white phosphorus in alkali lead to hypophosphite and not phosphate?—Current texts make no attempt to address such questions.) Arrow pushing demystifies them and places them on a larger logical scaffolding. The transformative impact of this approach cannot be overstated. Almost to a person, students who have gone through our introductory course say that they cannot imagine how someone today could remain satisfied with a purely descriptive, nonmechanistic exposition of inorganic main-group chemistry.

A mechanistic approach has done wonders for the overall tenor of our classroom—now very much a "flipped classroom," where arrow pushing, instead of videos, have afforded the "flip." Well-designed traditional lectures are still important to us and our students, but they now account for only 50% of total contact hours, with the rest devoted to various types of active learning. Some students solve mechanism problems on their own, others do so in groups, and still others solve them on the blackboard in front of the class. Importantly, such a classroom affords continual feedback from the students so we always have a good idea of their level of understanding and can assist accordingly.

Potential concerns. Given the plethora of advantages of a mechanistic approach, it's worth reflecting why it has never been adopted for introductory inorganic chemistry. A plausible reason is that, in contrast to common organic functional groups, simple p-block compounds such as hydrides, oxides, halides, and so forth, tend to be much more reactive and their vigorous and even violent reactions have been much less thoroughly studied. As good scientists, inorganic chemists may have felt a certain inhibition about emphasizing an approach that has little grounding in experimental fact. This is a legitimate objection, but hardly a dealbreaker, in our opinion, for the following reasons.

Our ideas on main-group element reactivity are not taken out of the blue but are based on parallels with well-studied processes in organic and organoelement chemistry. Second, it no longer necessarily takes a prohibitive amount of resources to test a mechanistic proposal, at least in a preliminary way. Quantum chemical calculations, particularly based on density functional theory (DFT), very often provide an efficient and economical way of evaluating reaction mechanisms. Third, and perhaps most important, it's vastly better to be able to formulate a hypothesis on how a reaction *might* happen than to have no inkling whatsoever about the mechanism.

Content and organization. Chapter 1 attempts to provide a summary of all relevant introductory concepts, paving the way for a full appreciation of the rest of the book. The chapter begins with a discussion of nucleophiles and electrophiles, continues on to present a survey of the major organic reaction types (substitution, elimination, addition, etc.) and of some specifically inorganic reaction types (oxidative addition, reductive elimination, metathesis, migrations, etc.), and concludes with an elementary discussion of hypervalent compounds. The subsequent chapters are organized according to the groups of the periodic table, from left to right. Chapter 2 deals with the s-block elements, providing a combined treatment of hydrogen, the alkali metals, and the alkaline earth metals. For the p block, the chapter number is generally the same as the old group number; thus, the chalcogens are discussed in Chapter 6, the halogens in Chapter 7, and so on. The only exception is group 15, which we have split up into two chapters, 5a and 5b: Chapter 5a is devoted to nitrogen and Chapter 5b to the heavier pnictogens.

As far as any given chapter is concerned, the goal has been not so much to provide a systematic account of a given group of main-group elements (although we believe that we have done so moderately well) as to help students figure out the inner workings of relatively complicated-looking reactions. We have done so by organizing each chapter as a series of vignettes, focusing on reactions that in our opinion are most conducive to sharpening students' arrow-pushing skills. In-chapter review problems are designed to further hone these skills as well as to provide material for in-class discussions and recitation sections. We have refrained from including end-of-chapter problems, in part out of a desire to limit the book to a manageable length. Students in need of additional exercises should find an ample supply of reactions in their regular descriptive inorganic text.

As far as our choice of reactions and topics is concerned, we have attempted to offer a stimulating mix of the traditional and the topical. For the traditional material, we have borrowed freely from introductory and advanced texts with a "descriptive inorganic" emphasis. These books are listed in Appendix 1. The Wikipedia has also been a valuable resource for this purpose. On occasion, we have played science historian and thrown in an anecdote or an amusing quote. The more cutting-edge material has been sourced from the research literature. Examples of such topics include:

- Jones's Mg(I)–Mg(I) reagent
- indium-mediated allylations
- heavy-element carbene, alkene, and alkyne analogs
- the Ruppert–Prakash and Togni reagents
- BrF_3 and higher valent bromine compounds as synthetic reagents
- the recent arsenic-DNA controversy
- the possible role of borate minerals in the origin of life (possibly even on Mars!)

Because this is an introductory text, however, we have cited the original research literature sparingly, often settling for a short list of suggested readings at the end of each chapter.

Stylistic aspects. A few comments on stylistic aspects of the book might be helpful. Perhaps foremost among them is the use of color in our reaction mechanisms, which include blue, black, red, and green. In general, the first nucleophile in a given mechanism is always indicated in blue and the first electrophile in black. Later in the mechanism, if the atoms originating in the initial nucleophile take on a different role, such as that of an electrophile, they are still indicated in blue. Thus, for any given atom or group, its color is maintained the same throughout the mechanism so that its fate can be easily followed throughout the reaction pathway. Curly arrows have throughout been indicated in red; certain atoms "deserving" special attention are also indicated in red. In some cases, where a third reactant is involved, it is indicated in green. In general, the color of a newly formed bond is the same as the color of the lone pair or other electrons from which it may be thought to have originated for bookkeeping purposes.

In this book, curly arrows typically begin from the nucleophilic electron pair and end on the electrophilic atom being attacked. In general, to prevent clutter, we have not shown lone pairs unless they are specifically engaged in a nucleophilic attack.

We have made sparse use of multiple bonds involving higher valent p-block elements. Thus, we have preferred to use the left-hand structures for $POCl_3$ and SO_2Cl_2, as opposed to the multiply bonded structures to the right:

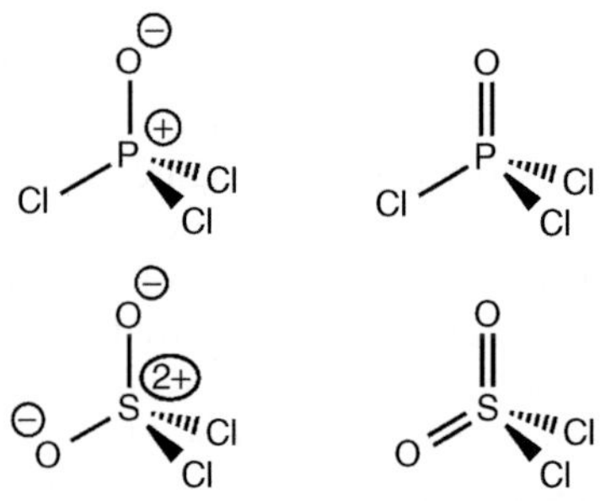

Despite the unrealistic formal charges, we believe that the structures on the left give a clearer sense of the bonding, whereas the multiple bonds shown to the right are harder to appreciate. It is not easy to explain to an undergraduate audience which specific orbitals constitute the double bonds in the right-hand structures. To instructors who would prefer to stick to the more conventional multiply bonded structures, we say: by all means do so; for the vast majority of reactions, arrow pushing will work equally well for both types of structures.

The end of descriptive inorganic chemistry? An interesting question to consider is the following: Does a mechanistic approach, making extensive use of arrow pushing, signal of the end of descriptive inorganic chemistry? The answer, in our opinion, is both yes and no. By emphasizing arrow pushing as a universal tool for rationalizing main-group reactivity, we have placed the field at exactly the same level as organic chemistry. Just as no one speaks of "descriptive organic chemistry," there is no point in treating main-group chemistry as a descriptive subject. That, of course, does not diminish the importance of facts and having an appropriate respect for them. Facts come first, whether it's organic or inorganic chemistry, and mechanisms are primarily useful for understanding and rationalizing them. In that sense, mechanisms can never supplant a descriptive exposition of chemical facts.

ABHIK GHOSH and STEFFEN BERG
The Arctic University of Norway, Tromsø, Norway

Acknowledgments

We are indebted to many friends and colleagues who generously assisted us in the preparation of this book. Prof. Carl Wamser of Portland State University and Dr. David Ware of The University of Auckland read and critiqued the entire manuscript. Our debt to these two loyal friends is immense. Others who read individual chapters and shorter sections include Paul Deck of Virginia Tech (halogens), Penny Brothers of The University of Auckland (Group 13 elements), Barry Rosen of Florida International University (Group 15 elements), Ged Parkin of Columbia University (higher-valent and hypervalent compounds), and Kyle Lancaster of Cornell University (the noble gases). We thank Steven Benner (FFAME, Gainesville, FL; arsenic-DNA), Tristram Chivers (University of Calgary; sulfur nitrides), Harry Gray (Caltech; higher-valent bromine reagents), Roald Hoffmann (Cornell; aspects of halogens), Pekka Pyykkö (University of Helsinki; inert pair effect), and Shlomo Rozen (Tel Aviv University; BrF_3) for helpful advice and correspondence on the topics indicated within parentheses. Our long-time friend and collaborator Prof. Jeanet Conradie of the University of the Free State, South Africa, assisted us with the DFT calculations we needed for a better understanding of certain reactions. Carl Wamser and Penny Brothers also provided wonderful refuges—Portland, Oregon, and Auckland, New Zealand—where one of us (AG) could escape to and write.

The Foreword has been written by Harry Gray, who seemed to us to be uniquely qualified for the purpose. In the 1960s, he and George Hammond tried to adopt a mechanistic approach in teaching aspects of main-group chemistry (see, e.g., *Chemical Dynamics* by J. B. Dence, H. B. Gray, and G. S. Hammond, Benjamin: 1968). Harry's full-throated support of our own approach means a great deal to us.

It is a pleasure to acknowledge Wiley editor Anita Lekhwani for her encouragement and wise counsel throughout the writing process. We are similarly grateful to Sangeetha Parthasarathy of Laserwords Pvt. Ltd. Chennai, India, for the considerable efforts involved in the final production of the book.

Finally, we thank our families and some of our closest friends for their love and encouragement: AG thanks Avroneel, Sheila, Ranjita, Matthew, and Daniel; and SB thanks Kenneth, Andreas, Eirik, Tor Håvard, and above all Cathrine.

Advance praise for *Arrow Pushing in Inorganic Chemistry: A Logical Approach to the Chemistry of the Main-Group Elements*

I tell my organic students to "think like a molecule". What are the molecules doing, and why are they doing that? Since the essence of a chemical reaction is the reorganization of bonds (i.e., electrons), the primary tool for understanding it is arrow pushing. It's a real delight to see that this fundamental approach indeed works beautifully in inorganic chemistry as well. It makes one wonder why it hadn't been "discovered" sooner. Congratulations to the authors for an excellent expository textbook.— Professor Carl C. Wamser, Portland State University

It's great to see a key organic skill, arrow pushing, applied to inorganic chemistry, where there's plenty extra to think about—redox chemistry along with wide variations in atomic size and electronegativity. The strength of the approach is that all this can be taken into account. A powerful new way of thinking for inorganic chemists!—Dr. David Ware and Professor Penny Brothers, University of Auckland, New Zealand

In my Metals in Biology *course, I tell my students the simplest lesson of chemistry: electrons flow from where they are to where they aren't. This is the essence of the 'arrow pushing' formalism, which had its origins in physical organic chemistry. My early training in that field led me to use the arrow pushing language in my own research in bioinorganic chemistry. I am delighted to see this language applied much more generally to inorganic chemistry in this very illuminating and instructive book. Students will learn where electrons want to go and their appreciation of how reactions occur will be greatly enhanced.* — Professor John T. Groves, Princeton University

Nice up-to-date stuff, including frustrated Lewis pairs, Jones's Mg(I) reagent, high-valent bromine and lots more! It would have been easy for the authors to ignore the last twenty years (or fifty) but they didn't do that! — Professor Paul A. Deck, Virginia Tech

I was struck by the sheer amount of innovation, thought, and attention to detail that has gone into the making of this book. In cases where arrow pushing does not immediately indicate a unique mechanism, the authors have even resorted to DFT calculations to resolve the ambiguity. — Professor Jeanet Conradie, University of the Free State, Republic of South Africa

… Valence is an important concept in inorganic chemistry and it's nice to see the authors do full justice to the topic. They carefully distinguish valence and oxidation state, which are often confused, and draw structures with appropriate formal charges that shed light on the bonding. Furthermore, their treatment of the fascinating chemistry of the higher-valent states of p-block elements is superb. — Professor Gerard Parkin, Columbia University

…The marriage between descriptive inorganic chemistry and the language of organic reaction mechanisms is convincingly consummated in this new and most useful contribution. — Professor Peter R. Taylor, University of Melbourne

1

A Collection of Basic Concepts

> *In solving a problem of this sort, the grand thing is to be able to reason backward. That is a very useful accomplishment, and a very easy one, but people do not practise it much. In the everyday affairs of life it is more useful to reason forward, and so the other comes to be neglected.*
>
> Sherlock Holmes in *A Study in Scarlet*, By Sir Arthur Conan Doyle

We assume you've had an introductory course in organic chemistry and hope you found it logical and enjoyable. The logic of organic chemistry is of course key to its charm, and mechanisms are a big part of that logic. In this book, we will present a similar approach for inorganic chemistry, focusing on the main-group elements, that is, the s and p blocks of the periodic table (Figure 1.1). As in organic chemistry, our main tool will be the curly arrows that indicate the movement of electrons, typically electron pairs, but on occasion also unpaired electrons. As we shall see, this approach—arrow pushing—works well in inorganic chemistry, especially for the main-group elements.

We want to get you started with arrow pushing in an inorganic context as quickly as possible, but we'd also like to make sure that you are equipped with the necessary conceptual tools. In this chapter, we'll try to provide you with that background as efficiently as possible. Unavoidably, the concepts form a somewhat disparate bunch but they do follow a certain logic. Sections 1.1–1.6 introduce the idea of nucleophiles and electrophiles, in the context of the S_N2 displacement, and discuss physical concepts such as electronegativity, polarizability, pK_a, redox potentials, and bond energies in relation to chemical reactivity. Armed with these concepts, we'll devote the next several Sections 1.7–1.21 to survey key mechanistic paradigms, focusing on major organic reaction types but also on a few special inorganic ones. Sections 1.22 and 1.23 then present practical tips on arrow pushing, that

Arrow Pushing in Inorganic Chemistry: A Logical Approach to the Chemistry of the Main-Group Elements,
First Edition. Abhik Ghosh and Steffen Berg.
© 2014 John Wiley & Sons, Inc. Published 2014 by John Wiley & Sons, Inc.

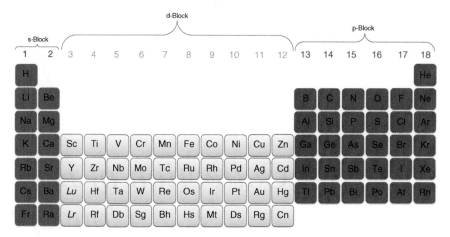

Figure 1.1   *The periodic table: group numbers and the s, p, and d blocks.*

is, how you might approach a given mechanistic problem. In the course of our mechanistic survey, we'll encounter a number of so-called hypervalent p-block compounds, which you may not have encountered until now. These call for a brief discussion of the bonding involved, which we will present in Sections 1.24–1.27. That said, we will not cover some of the more elementary aspects of structure and bonding theory, including the very useful VSEPR (valence shell electron pair repulsion) model; feel free to go back to your general or organic chemistry text for a quick refresher.

1.1   NUCLEOPHILES AND ELECTROPHILES: THE S_N2 PARADIGM

In this book, we will be overwhelmingly concerned with polar or ionic mechanisms. These involve the movement of electron pairs, unlike radical reactions which involve unpaired electrons. The components of a polar mechanism can generally be classified as nucleophiles or electrophiles. A *nucleophile* ("nucleus-lover") is typically an anion or a neutral molecule that uses an electron pair to attack another atom, ion, or molecule. The species being attacked is called an *electrophile* ("electron-lover"). The terms "nucleophile" and "electrophile" often refer to the classic S_N2 reaction of organic chemistry. In the example below (which happens to be a Williamson ether synthesis), the methoxide anion is the nucleophile, methyl iodide is the electrophile, and iodide is the *leaving group*.

$$(1.1)$$

A key feature of the S_N2 reaction is that the nucleophile attacks from the "back side" relative to the leaving group, leading to an umbrella-like inversion of the carbon undergoing

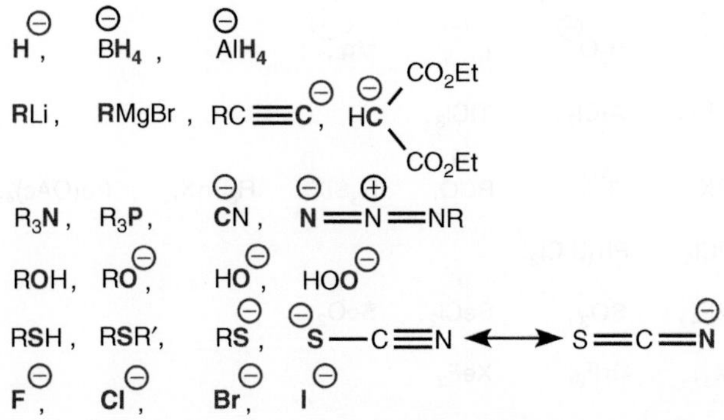

Figure 1.2 *Some common nucleophiles, with the nucleophilic atoms indicated in blue.*

substitution. If this carbon atom is stereogenic,[1] such an inversion of configuration may be discerned experimentally, as in the example below; otherwise the inversion is not detectable, even though it occurs.

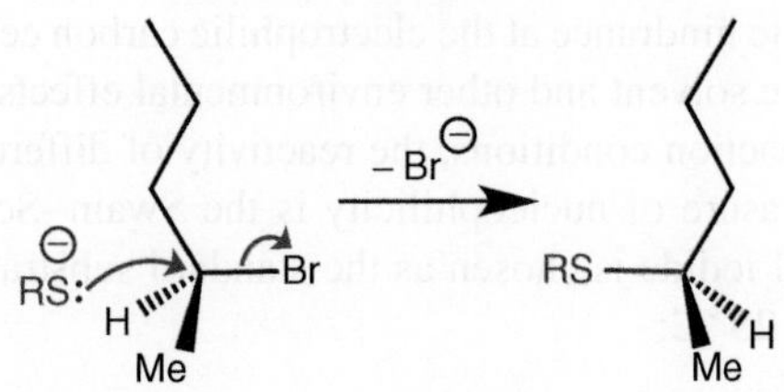

$$(1.2)$$

Several common nucleophiles are depicted in Figure 1.2, where R and R′ denote alkyl groups. Many of them are nitrogen-based, such as ammonia, amines (RNH_2), and azide (N_3^-), or oxygen-based, such as water, alcohols (ROH), and alkoxide (RO^-) and carboxylate (RCO_2^-) anions. Sulfur-based nucleophiles such as thiols (RSH), thiolates (RS^-), and thioethers (RSR′) are also widely used in chemical synthesis. Triphenylphosphine, a phosphorus nucleophile, is an important reagent in organic synthesis, as well as an important transition-metal ligand. Halide ions are widely employed as both nucleophiles and leaving groups. Hydride is used both as a base (typically as NaH or KH) and as a nucleophile (often in complexed forms such as BH_4^- or AlH_4^-).

Carbon nucleophiles play a central role in organic chemistry, as they form the basis of carbon–carbon bond formation. A few are shown in Figure 1.2, including such carbanionic species as organolithiums (RLi), Grignard reagents (typically written as RMgBr), and the cyanide (CN^-) and acetylide ($R-C\equiv C^-$) anions. Other examples such as enolates, enols, and enamines will be briefly discussed in Section 1.15.

Some common electrophiles are shown in Figure 1.3. These include protons and positively charged metal ions, electron-deficient species such as trivalent group 13 compounds (e.g., BF_3, $AlCl_3$), the cationic carbon in carbocations, the halogen-bearing carbon in alkyl

---

[1]A stereogenic center is an atom in a molecule for which interchanging any two of its substituents leads to a different stereoisomer. The term was introduced by Mislow and Siegel in an important foundational paper on modern stereochemical concepts and terminology: Mislow, K.; Siegel, J. *J. Am. Chem. Soc.* **1984**, *106*, 3319–3328.

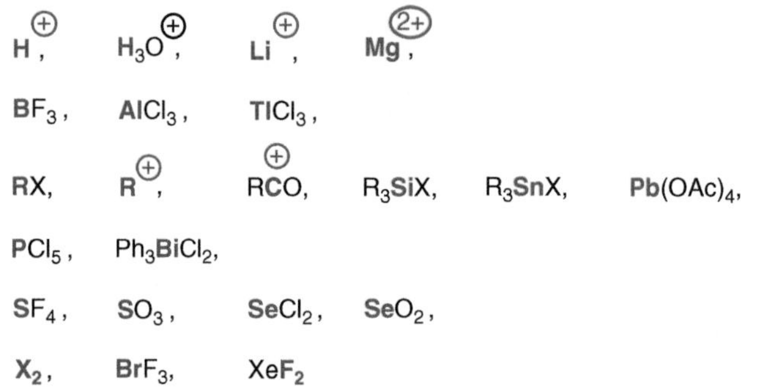

$$\overset{\oplus}{H},\quad H_3O\overset{\oplus}{},\quad \overset{\oplus}{Li},\quad Mg\overset{\text{\textcircled{2+}}}{},$$

$$BF_3,\quad AlCl_3,\quad TlCl_3,$$

$$RX,\quad R\overset{\oplus}{},\quad R\overset{\text{\textcircled{+}}}{C}O,\quad R_3SiX,\quad R_3SnX,\quad Pb(OAc)_4,$$

$$PCl_5,\quad Ph_3BiCl_2,$$

$$SF_4,\quad SO_3,\quad SeCl_2,\quad SeO_2,$$

$$X_2,\quad BrF_3,\quad XeF_2$$

Figure 1.3 *Some common electrophiles; X is a halogen. The electrophilic atoms are indicated in green.*

halides, the Si atom in silyl halides, molecular halogens, and even the fluorine atoms in xenon difluoride (XeF_2).

The ease with which a given S_N2 displacement occurs depends on multiple factors, such as the nucleophilicity of the incoming nucleophile (which depends on both its electronic and steric character), steric hindrance at the electrophilic carbon center, the effectiveness of the leaving group, and the solvent and other environmental effects. By defining a standard substrate and standard reaction conditions, the reactivity of different nucleophiles may be quantified. One such measure of nucleophilicity is the Swain–Scott nucleophilicity constant n, for which methyl iodide is chosen as the standard substrate and reaction rates are measured in methanol at $25\,°C$:

$$n_{CH_3I} = \log \frac{k_{Nu}}{k_{CH_3OH}}$$

where k_{Nu} is the rate constant for the nucleophile of interest (Nu) and k_{CH_3OH} is the rate constant for methanol itself as the nucleophile. Table 1.1 lists n_{CH_3I} values for a number of representative nucleophiles, along with the pK_a values of their conjugate acids (i.e., a measure of the basicity of the nucleophiles). Observe that there is only a very rough correlation between n_{CH_3I} and the conjugate acid pK_a; we'll return to this point in the next section. Table 1.2 presents a more qualitative characterization of some common nucleophiles, classifying them from strong to very weak.

Table 1.1 shows that, for a given electrophile (CH_3I) and standard conditions, the rate constants for common nucleophiles vary by a factor of well over a billion (10^9). This tremendous variation of reactivity of the different nucleophiles might pose a conundrum in relation to their synthetic utility. Note (from either Table 1.1 or 1.2) that alkoxide (RO^-) anions are some 10^3–10^4 times more nucleophilic than neutral alcohols, and the rates for carboxylate anions (RCO_2^-), relative to the un-ionized carboxylic acids, differ by even more: 10^5–10^6. With such low rates, are alcohols and carboxylic acids, in their un-ionized forms, at all useful as nucleophiles? The answer is a clear yes. In acidic media, many of the anionic nucleophiles simply don't exist; they are entirely protonated. Under such conditions, weak nucleophiles such as alcohols and carboxylic acids react effectively with cationic electrophiles such as carbocations. Second, although weaker nucleophiles may not react at a useful rate with alkyl halides, many of them do react at perfectly acceptable rates

TABLE 1.1 Swain–Scott Nucleophilicity Constants and Conjugate Acid pK_a Values of Some Common Nucleophiles

Nucleophile	n_{CH_3I}	Conjugate Acid pK_a
CH_3OH	0.0	−1.7
NO_3^-	1.5	−1.3
F^-	2.7	3.45
$CH_3CO_2^-$	4.3	4.8
Cl^-	4.4	−5.7
R_2S	5.3	−6 to −7
NH_3	5.5	9.25
N_3^-	5.8	4.74
$C_6H_5O^-$	5.8	9.89
Br^-	5.8	−7.7
CH_3O^-	6.3	15.7
HO^-	6.5	15.7
NH_2OH	6.6	5.8
NH_2NH_2	6.6	7.9
$(CH_3CH_2)_3N$	6.7	10.7
CN^-	6.7	9.3
I^-	7.4	−10.7
HO_2^-	7.8	11.75
$(CH_3CH_2)_3P$	8.7	8.7
$C_6H_5S^-$	9.9	6.5
$C_6H_5Se^-$	10.7	5.9

TABLE 1.2 Qualitative Classification of Nucleophiles, Based on the Swain–Scott Nucleophilicity Constants n_{CH_3I}

Nucleophiles	Relative Rate	Characterization
RS^-, HS^-, I^-	>10^5	Strong
N_3^-, CN^-, RO^-, OH^-, Br^-	10^4	Good
NH_3, RCO_2^-, F^-, Cl^-	10^3	Moderate
ROH, H_2O	1	Weak
RCO_2H	10^{-2}	Very weak

with stronger electrophiles such as BF_3 or neutral organosilicon compounds in general. The usefulness of a given nucleophile thus depends enormously on the reaction conditions.

1.2 WHAT MAKES FOR A GOOD NUCLEOPHILE?

Nucleophilicity and electrophilicity are closely related to Lewis basicity and acidity, respectively. Nucleophiles are Lewis bases (electron-pair donors) and electrophiles are Lewis acids (electron-pair acceptors). Now, as discussed previously, nucleophilicity is measured in terms of the *rate* of a nucleophilic attack, so it's a kinetic concept. Basicity, on the other hand, is measured in terms of the *equilibrium constant* for protonation (or for association with some Lewis acid), so it is a thermodynamic concept. Another difference is that, whereas Brønsted basicity refers to the thermodynamic affinity for protons, nucleophilicity in organic chemistry typically refers to the rate of attack on a carbon center. Moreover, in this book, we will talk about nucleophilic attacks on pretty much any p-block element! Understandably, therefore, you should not expect more than a rough correlation between the nucleophilicity of a nucleophile and its basicity.

To better appreciate this point, let us go back to Table 1.1, which lists a number of nucleophiles in increasing order of n_{CH_3I}, an arbitrarily defined measure of nucleophilicity. Observe that the basicities of the nucleophiles, as indicated by the pK_a values of their conjugated acids, increase in a general but highly erratic way with the value of n_{CH_3I}. To illustrate, whereas tertiary phosphines are more nucleophilic than tertiary amines by about two orders of magnitude, the former are *less* basic than the latter by roughly the same factor. What factors then make for a good nucleophile? The following trends have been observed:

- Anions are better nucleophiles than the related neutral molecules. Thus:

$$RO^- > ROH; \quad RS^- > R_2S; \quad NH_2^- > NH_3$$

  where R = H, alkyl, or aryl.
- For analogous species in a given period, nucleophilicity decreases as one progresses to the right of the periodic table:

$$NH_3 > H_2O; \quad R_3P > R_2S$$

  The more electronegative elements hold on to their electrons more tightly.
- Nucleophilicity increases as one goes down a given group of the periodic table. Thus, for example

$$PR_3 > NR_3; \quad PhSe^- > PhS^- > PhO^-$$

  The larger atoms are less electronegative and the anions derived from them are more polarizable, which translates to increasing nucleophilicity as one goes down a group.

Given that electronegativity and size (atomic radius) are the two key determinants of nucleophilicity, it's useful to remind ourselves how the two atomic properties vary across the periodic table. Figure 1.4 presents Pauling electronegativities and Figure 1.5 the atomic radii of the s- and p-block elements. Note that electronegativity increases from left to right along a given period, and decreases down a group. Atomic radii shrink from left to right in a given period and increase down a group.

Against this backdrop, the relative nucleophilicities of the halide anions make for somewhat of a puzzle. The Swain–Scott nucleophilicities (Tables 1.1 and 1.2), based on methanol as solvent, are in the order:

$$I^- > Br^- > Cl^- > F^-$$

The same order is found in other protic solvents. This is also the order expected on the basis of polarizability: the larger and more polarizable anions should be the most nucleophilic. In polar aprotic solvents (e.g., DMSO, DMF, THF, etc.), however, the relative rates are completely reversed:

$$F^- \gg Cl^- > Br^- > I^-$$

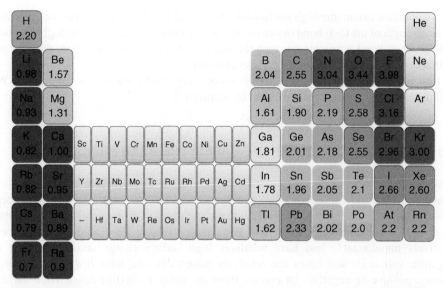

Figure 1.4 *Pauling electronegativities of the main-group elements. A relatively self-explanatory color code has been employed to give a semiquantitative visual indication of the electronegativities.*

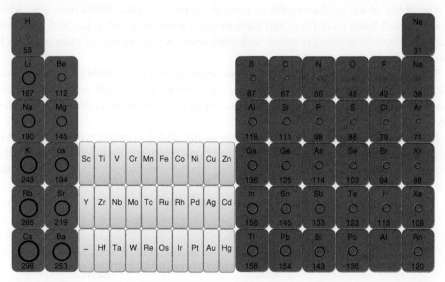

Figure 1.5 *Atomic radii (pm) of s- and p-block elements. (Clementi, E.; Raimond, D. L.; Reinhardt, W. P. J. Chem. Phys. 1967, 47, 1300–1307.)*

This remarkable reversal is due to hydrogen bonding, or the lack thereof in aprotic solvents.

As a powerful hydrogen-bond acceptor, fluoride is understandably a poor nucleophile in protic solvents. Iodide, as the worst hydrogen-bond acceptor, is thus a much more active nucleophile in protic solvents. In the absence of hydrogen-bonding interactions with the solvent, which is the case in dry polar aprotic solvents, fluoride is the strongest nucleophile.

To a significant extent, the high nucleophilicity of "naked" fluoride ions may be attributed to the strength of the C–F bond (more on which in Section 1.6). Because the S_N2 transition state involves bond formation between the incoming nucleophile and carbon, the strength of that bond is a key determinant of nucleophilicity.

Last but not least, steric effects are yet another key determinant of nucleophilicity. We will discuss steric effects to some extent in Section 1.7.

1.3   HARD AND SOFT ACIDS AND BASES: THE HSAB PRINCIPLE

Several of the factors affecting nucleophilicity may be nicely rolled together into the concept of *hard* and *soft* Lewis *acids* and *bases*—HSAB, for short. The HSAB concept was introduced by Ralph Pearson over 50 years ago and was subsequently put on a firmer theoretical foundation by Pearson and Parr, among others. Hard acids and bases are relatively unpolarizable and have relatively high surface charge density, positive or negative; soft acids and bases are relatively polarizable and have low surface charge density, positive or negative. Of course, there are many borderline cases. High surface charge density (hardness) typically results from a high formal charge (FC), positive or negative, and small atomic/ionic size, and the opposite is true for low surface charge density (softness). Examples of hard, borderline, and soft acids and bases are shown in Table 1.3.

The utility of the hardness/softness concept derives from the *HSAB principle*, which states that soft bases react faster and form stronger bonds with soft acids, and hard bases react faster and form stronger bonds with hard acids. A vast amount of chemistry can be rationalized with this principle.

The HSAB concept greatly facilitates our appreciation of nucleophilicity: softer bases often make better nucleophiles. Phosphines, for example, are typically better nucleophiles than the analogous, harder amines, and sulfur compounds are better nucleophiles than their oxygen analogs.

In this book, some of the best illustrations of the HSAB principle will be provided by the so-called ligand exchange or metathesis reactions, which are discussed in more detail in Section 1.19. The principle helps us in deciding whether a metathesis reaction will proceed in a given direction or not:

$$AB + CD \rightarrow AC + BD \tag{1.3}$$

TABLE 1.3   Qualitative Listing of Hard, Intermediate, and Soft Acids and Bases[a]

	Acids	Bases
Hard	H^+, **H–X**, Li^+, Na^+, R_3**SiX** Mg^{2+}, Ca^{2+}, **AlX**$_3$, **SnCl**$_4$, **TiCl**$_4$	**NH**$_3$, R**NH**$_2$ H_2**O**, **HO**$^-$, R**OH**, R**O**$^-$, RCO$_2^-$ **Cl**$^-$, **F**$^-$, **NO**$_3^-$
Intermediate	**CuX**$_2$, **ZnX**$_2$, **SnX**$_2$, **GaX**$_3$, R_3**C**$^+$, R_3**B**	**Br**$^-$, **N**NN$^-$ (azide), Ar**NH**$_2$ pyridine
Soft	RCH$_2$X, R**S**X, R**Se**X, I$_2$, Br$_2$, **Br**F$_3$, **Cu**X, **Ag**$^+$, **Pd**(X/R)$_2$, **Pt**(X/R)$_2$, **Hg**(X/R)$_2$, zero-valent metals	R**SH**, R**S**$^-$, R$_2$**S**, R**Se**$^-$, **I**$^-$, R$_3$**P**, N**C**$^-$, **CO**, R**CH**=**CH**R, benzene

[a]Where warranted, the atom of interest is indicated in bold.

Let us take a couple of concrete examples.

$$SeCl_2 + 2 (CH_3)_3SiBr \rightarrow SeBr_2 + 2 (CH_3)_3SiCl \qquad (1.4)$$

$$PCl_3 + AsF_3 \rightarrow PF_3 + AsCl_3 \qquad (1.5)$$

In reaction 1.4, Se is a softer Lewis acid than Si, and bromide is a softer Lewis base than chloride. It makes sense therefore that Se and Br should link up, as should Si and Cl. In the second example (reaction 1.5), As is a softer Lewis acid center than P, and chloride is a softer Lewis base than fluoride. These ligand exchanges are thus consistent with the HSAB principle.

1.4 pK$_a$ VALUES: WHAT MAKES FOR A GOOD LEAVING GROUP?

Compared with the multitude of factors affecting nucleophilicity, the efficacy of a leaving group is much more easily predictable. In short, a weaker Brønsted base makes a better leaving group. We can simply look up the pK$_a$ of the conjugate acid of a leaving group to arrive at a good idea of its leaving ability. Table 1.4, a short pK$_a$ table, will serve our purposes very well.

Observe that the best leaving groups are conjugate bases of the strongest acids. Thus, iodide and bromide are excellent and popular leaving groups in organic chemistry. The worst leaving groups are very strong bases, such as amide, hydride, and alkyl anions. Hydroxide and alkoxide (RO^-) are also poor leaving groups in organic chemistry. The Williamson ether synthesis mentioned above (reaction 1.1) illustrates this last point well. Like all elementary reactions, the reaction is in principle reversible, but the reverse reaction, I^- displacing a CH_3O^- anion, does not occur for all intents and purposes.

A couple of additional observations are worth making, again with specific reference to organic chemistry.

TABLE 1.4 Common Leaving Groups and the pK$_a$ Values of Their Conjugate Acids

	Leaving Group	Conjugated Acid	pK$_a$
Good	I^-	HI	−10
	Br^-	HBr	−9
	Cl^-	HCl	−8
	HSO_4^-	H_2SO_4	−3
	$p\text{-}CH_3\text{-}C_6H_4\text{-}SO_3^-$	$p\text{-}CH_3\text{-}C_6H_4\text{-}SO_3H$	−3
	H_2O	H_3O^+	−1.7
	F^-	HF	3.2
	CH_3COO^-	CH_3COOH	4.74
	NH_3	NH_4^+	9.25
	HO^-	H_2O	15.74
	CH_3O^-	CH_3OH	15.2
	NH_2^-	NH_3	38
	H^-	H_2	42
Bad	$CH_3\text{-}CH_2\text{-}CH_2\text{-}CH_2^-$	$CH_3\text{-}CH_2\text{-}CH_2\text{-}CH_3$	50

Fluoride and cyanide are very much worse leaving groups than the pK_a values of HF and HCN would imply. This presumably reflects the great strength of the C–F and C–CN bonds.

Sulfonates are better leaving groups than the pK_a values of sulfonic acids suggest. Arenesulfonates ($ArSO_3^-$), especially *p*-toluenesulfonate (also known as *tosylate*, TsO^-), are popular leaving groups in organic chemistry because alkyl tosylates may be readily prepared from the corresponding alcohols. The trifluoromethanesulfonate anion (also known as *triflate*, TfO^-) leaves with even greater alacrity, and even better leaving sulfonate-based leaving groups have been developed:

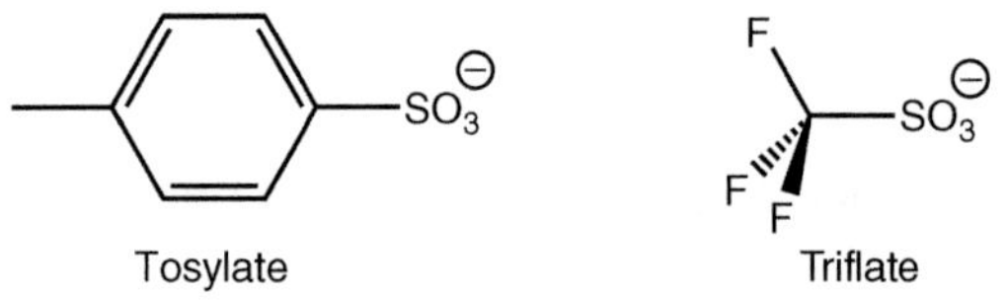

Tosylate   Triflate

A fact that many students struggle with on their first introduction to organic chemistry is the following: the iodide ion is both an excellent nucleophile and an excellent leaving group; by contrast, alkoxide ions (RO^-) are good nucleophiles but lousy leaving groups. What accounts for the difference?

The solution to this conundrum is that, although both nucleophiles and leaving groups are Lewis bases, very different factors control their efficacy. Iodide's nucleophilicity is attributed primarily to its polarizability or softness. The nucleophilicity of alkoxide ions owes more to the hard–hard interaction between O^- and $C^{\delta+}$ and the resulting strength of the C–O bond.

On the other hand, there is a clear inverse correlation between the efficacy of leaving groups and their Brønsted basicity. Thus, iodide is an excellent leaving group because it is a very weak base. Alkoxide anions, being strong bases, are lousy leaving groups.

Protonation greatly enhances the efficacy of leaving groups. For example, the bromide anion by itself (e.g., in the form of NaBr) does not react with an alcohol, OH^- being a notoriously poor leaving group in organic chemistry.

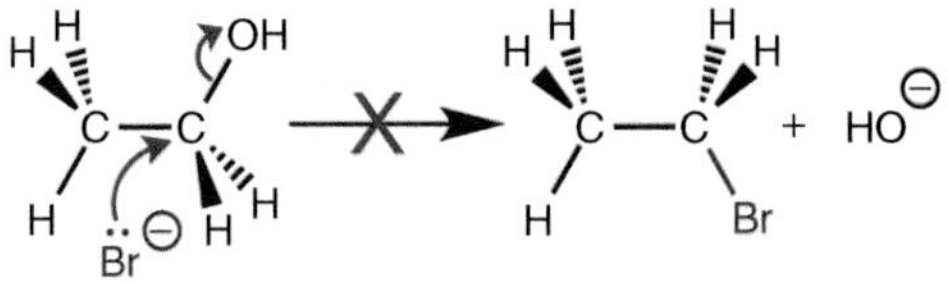

$$(1.6)$$

Protonation of the OH group by concentrated HBr, however, enables the departure of water, a far better leaving group, as shown below:

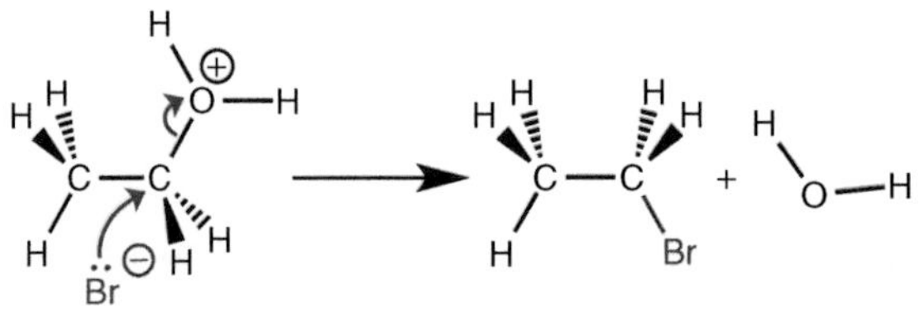

$$(1.7)$$

Concentrated HBr is therefore a suitable reagent for converting simple alcohols to the corresponding alkyl bromides (assuming there are no other acid-sensitive groups in the molecule).

We'd be remiss if we didn't make some amends for presenting a rather "organic-centered" view of leaving groups: a variety of other factors are at play when the electrophilic center is not carbon. Perhaps the most important of these is the fact that single bonds between electronegative elements are typically weak and are easily cleaved. Unlike in organic chemistry, a hydroxide anion is a fair leaving group for a substrate of the form ROOH. Similarly, although thiolates (RS⁻) are hopelessly poor leaving groups in organic chemistry, rings of divalent sulfur atoms are readily broken down by nucleophiles, with S⁻ leaving groups as intermediates (as discussed in Section 6.4).

1.5 REDOX POTENTIALS

Broadly speaking, nucleophilicity correlates with reduction potential. Thus, stronger reducing agents tend to make better nucleophiles, which makes sense because both properties are related to electron donation. The correlation, however, is best limited to a set of structurally closely related nucleophiles. For a broader correlation, Edwards and Ritchie proposed an "oxibase" scale, which afforded a linear correlation of the reactivity of a nucleophile with the reduction potential of its oxidized form and the pK_a of its conjugate acid. Although space doesn't permit a more detailed discussion of this scale, redox potentials are broadly important for the subject matter of this book.

Table 1.5 lists reduction potentials of representative inorganic substances of potential interest in this book. Observe that the half-reactions are all written as reductions, following current convention, as well as to avoid ambiguity. In this convention, the more positive the reduction potential, the more the reduction is favored thermodynamically. Strong oxidants thus exhibit high (i.e., more positive) reduction potentials. A couple of examples should illustrate the utility of this table.

A table such as Table 1.5 provides an indication of whether an oxidant or reductant is suitable for a given redox role. Thus, with a high reduction potential (1.19 V in Table 1.5), ClO_2 is clearly a strong oxidant, which underlies its wide use as a disinfectant.

Observe that several of the reductions involve an oxidant (e.g., O_2, H_2O_2, NO_3^-) in acid solution. This makes sense because protonation is expected to make an oxidant even more powerful, that is, an even more avid acceptor of electrons.

Not much more needs to be said about redox potentials at this point. We will refer back to Table 1.5 once in a while when we make arguments based on redox potentials.

1.6 THERMODYNAMIC CONTROL: BOND DISSOCIATION ENERGIES (BDEs)

Many of the reactions discussed in this book occur under thermodynamic control. In other words, the activation energies for the various pathways are low enough and the temperature is high enough so that the products formed are thermodynamically the most stable ones possible. In contrast, under kinetic control, reaction rates determine the products observed and certain thermodynamically favored products may not predominate because their formation is too slow under the reaction conditions (particularly temperature). Reactions under

TABLE 1.5   Standard Reduction Potentials ($E°$, V) at 25 °C, 1.0 M, 1 atm.

Half-Reaction	$E°$ (V)
$F_2(g) + 2e^- \rightarrow 2F^-(aq)$	+2.87
$XeF_2(aq) + 2H^+ + 2e^- \rightarrow Xe(g) + 2HF(aq)$	+2.32
$O_3(g) + 2H^+(aq) + 2e^- \rightarrow O_2(g) + H_2O(aq)$	+2.07
$H_2O_2(aq) + 2H^+(aq) + 2e^- \rightarrow 2H_2O$	+1.78
$PbO_2(s) + 4H^+(aq) + SO_4^{2-}(aq) + 2e^- \rightarrow PbSO_4(s) + 2H_2O$	+1.70
$HClO_2(aq) + 2H^+(aq) + 2e^- \rightarrow HOCl(aq) + H_2O$	+1.67
$2HOCl(aq) + 2H^+ + 2e^- \rightarrow Cl_2(g) + 2H_2O$	+1.63
$Cl_2(g) + 2e^- \rightarrow 2Cl^-(aq)$	+1.36
$O_2(g) + 4H^+(aq) + 4e^- \rightarrow 2H_2O$	+1.23
$Tl^{3+}(aq) + 2e^- \rightarrow Tl^+(aq)$	+1.23
$ClO_2(g) + H^+(aq) + e^- \rightarrow HClO_2(aq)$	+1.19
$Br_2(aq) + 2e^- \rightarrow 2Br^-(aq)$	+1.09
$NO_3^-(aq) + 4H^+(aq) + 3e^- \rightarrow NO(g) + 2H_2O$	+0.96
$O_2(g) + 2H^+(aq) + 2e^- \rightarrow H_2O_2(aq)$	+0.68
$I_2(s) + 2e^- \rightarrow 2I^-(aq)$	+0.54
$O_2(g) + 2H_2O + 4e^- \rightarrow 4OH^-(aq)$	+0.40
$SO_2(g) + 4H^+(aq) + 4e^- \rightarrow S(s) + 2H_2O$	+0.40
$SO_4^{2-}(aq) + 4H^+ + 2e^- \rightarrow SO_2(g) + 2H_2O$	+0.20
$Sn^{4+}(aq) + 2e^- \rightarrow Sn^{2+}(aq)$	+0.13
$2H^+(aq) + 2e^- \rightarrow H_2(g)$	**0.00**
$Sn^{2+}(aq) + 2e^- \rightarrow Sn(s)$	−0.14
$PbSO_4(s) + 2e^- \rightarrow Pb(s) + SO_4^{2-}(aq)$	−0.31
$2H_2O + 2e^- \rightarrow H_2(g) + 2OH^-(aq)$	−0.83
$Al^{3+}(aq) + 3e^- \rightarrow Al(s)$	−1.66
$Be^{2+}(aq) + 2e^- \rightarrow Be(s)$	−1.85
$Mg^{2+}(aq) + 2e^- \rightarrow Mg(s)$	−2.37
$Na^+(aq) + e^- \rightarrow Na(s)$	−2.71
$Ca^{2+}(aq) + 2e^- \rightarrow Ca(s)$	−2.87
$K^+(aq) + e^- \rightarrow K(s)$	−2.93
$Li^+(aq) + e^- \rightarrow Li(s)$	−3.05

thermodynamic control often lead to the formation of an overall stronger set of bonds; a general idea of bond dissociation energies (BDEs) is therefore quite useful. Table 1.6 lists some typical single BDEs.

A few examples of BDE considerations are as follows:

Bonds between highly electronegative elements are weak and easily broken, for example, O–O, N–O, X–X, N–X, O–X, and so on, where X is a halogen.

TABLE 1.6   Typical Single Bond Dissociation Energies (kJ/mol, in black) and Bond Distances (pm, in green) for Selected p-Block Elements

	H	C	N	O	F	Si	P	S	Cl	Br	I
H	436	414	389	464	569	323	318	339	431	368	297
	74	110	98	94	92	145	138	123	127	142	161
C		347	293	351	439	289	264	259	330	276	238
		154	147	143	141	184	187	181	176	191	210
N			159	201	272		209		201	243	
			140	136	134	187	180	174	169	184	203
O				138	184	368	351		205		201
				132	130	183	176	170	165	180	199
F					159	540	490	285	255	197	
					128	181	174	168	163	178	197
Si						176	213	226	360	289	
						234	227	221	216	231	250
P							213	230	331	272	213
							220	214	209	224	243
S								213	251	213	
								208	203	218	237
Cl									243	218	209
									200	213	232
Br										192	180
										228	247
I											151
											266

Carbon, silicon, and phosphorus form strong bonds with O and F. In addition, the C=O (~899 kJ/mol) and P^+–O^- (~544 kJ/mol) BDEs are very high and these bonds tend to form easily under hydrolysis. These elements are thus said to be strongly oxophilic and fluorophilic.

The HF bond is also extremely strong (BDE 569 kJ/mol), which in part explains why, unlike the other hydrohalic acids, HF is a weak acid. Thus, a number of p-block element fluorides react with proton sources to yield HF.

The bond distances listed in Table 1.6 do not warrant much comment. By all means, browse them briefly; they are simply meant to give you a sense of the comparative dimensions of different bonds. Observe that there is no particular correlation between bond distances and BDEs. For bonds between a given pair of elements, however, a longer bond does correspond to a lower BDE.

♦ ♦ ♦

Our discussion until now has centered around the S_N2 displacement. With some of the key physical concepts in place, we are now in a good position to survey a number of other fundamental mechanisms. These are arranged in somewhat arbitrary order as follows:

- The E2 mechanism (Section 1.7)
- Proton transfers (PTs) (Section 1.8)

- Associative and dissociative processes (Section 1.9)
- S_N2-Si (Section 1.10)
- Two-step mechanisms: the S_N1 and E1 pathways (Section 1.11)
- Electrophilic addition to C–C multiple bonds (Section 1.12)
- Electrophilic substitution on aromatics: Addition–elimination (Section 1.13)
- Nucleophilic addition to C–heteroatom multiple bonds (Section 1.14)
- Carbanions and ylides (Section 1.15)
- Carbenes (Section 1.16)
- Oxidative addition and reductive elimination (Section 1.17)
- Migrations (Section 1.18)
- Ligand exchanges (Section 1.19)
- Radical reactions (Section 1.20)
- Pericyclic reactions (Section 1.21)

◆ ◆ ◆

1.7   BIMOLECULAR β-ELIMINATION (E2)

A process that often competes with S_N2 displacements for organic systems is E2 (which stands for "elimination, bimolecular"). When there is one or more hydrogens β with respect to the leaving group (see reaction 1.8) and the incoming nucleophile is a strong enough base, bimolecular elimination occurs, often in competition with nucleophilic displacement, as shown below:

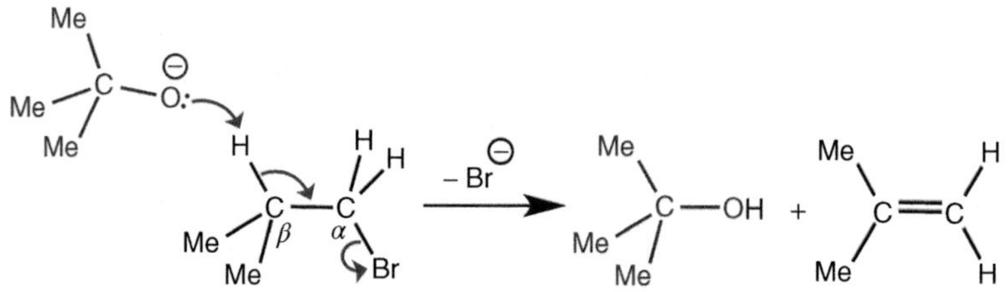

(1.8)

An important point is that, although an E2 reaction involves movement of three electron pairs, it all happens as a concerted one-step process.

A sterically hindered carbon center, such as the tertiary carbon in a *t*-butyl halide, is generally not conducive to an S_N2 displacement.

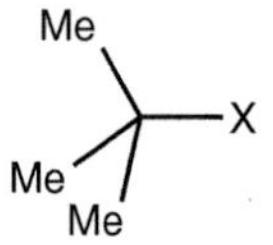

t-Butyl halide

An E2 elimination is often the favored pathway in such cases. Similarly, a strong, sterically hindered base also tends to favor an E2 over an S_N2 pathway. A good example is potassium *t*-butoxide, which is used in reaction 1.8. Examples of sterically hindered nitrogen bases include *N,N*-diisopropylethylamine (DIPEA, also known as *Hünig's base*) and the much stronger lithium diisopropylamide (LDA). The pK_a of the conjugate acid of DIPEA is about 9.0. Amidines, which are all-nitrogen analogs of carboxylic acids, tend to be significantly more basic than tertiary amines. Two moderately sterically hindered bicyclic amidines—1,8-diazabicyclo[5.4.0]undec-7-ene (DBU) and 1,5-diazabicyclo[4.3.0]non-5-ene (DBN)—are fairly widely used as bases promoting E2 eliminations.

DBU   DBN

The pK_a values of the conjugate acids of DBU and DBN have been estimated to be about 2–3 pK_a units higher than those of tertiary amines such as triethylamine or DIPEA.

We will conclude this section with a couple of interesting tidbits.

The first is a piece of information: DBU is not only a synthetic base but also a natural product, an alkaloid isolated from the Caribbean (Cuban) marine sponge *Niphates digitalis*.

The second is a question for you to consider. Observe that DBU and DBN have two nitrogens each. Which is the more basic nitrogen? Why?

1.8   PROTON TRANSFERS (PTs)

PTs between electronegative atoms are fast and reversible. They are enormously important, not only in organic and inorganic chemistry but also in biology. In terms of arrow pushing, PTs look very much like S_N2 displacements:

$$\text{Nu:}^{\ominus} \quad\longrightarrow\quad \text{H—Nu} \quad\longrightarrow\quad \text{Nu—H} \;+\; \text{Nu}^{\ominus} \tag{1.9}$$

Sometimes the proton donor is obvious: in an aqueous solution it is typically the H_3O^+ ion; under nonaqueous conditions it may be a strong acid such as H_2SO_4 or HCl. In such cases, the proton donor may not be explicitly shown but be simply indicated as H^+:

$$\text{Nu}^{\ominus} \quad\overset{\overset{\oplus}{H}}{\longrightarrow}\quad \text{Nu—H} \tag{1.10}$$

Some important concrete examples of PTs are as follows: carbonyl and nitrile groups undergo protonation:

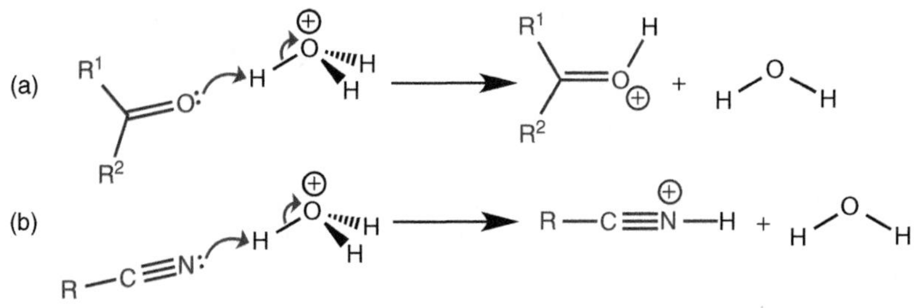

(a)

(b)

$$(1.11)$$

The significance of these protonations is that the cations that result are much stronger electrophiles than the uncharged functional groups. In the same vein, alcohols and thiols may be deprotonated, as shown below, and the resulting anions are much stronger nucleophiles than their uncharged precursors:

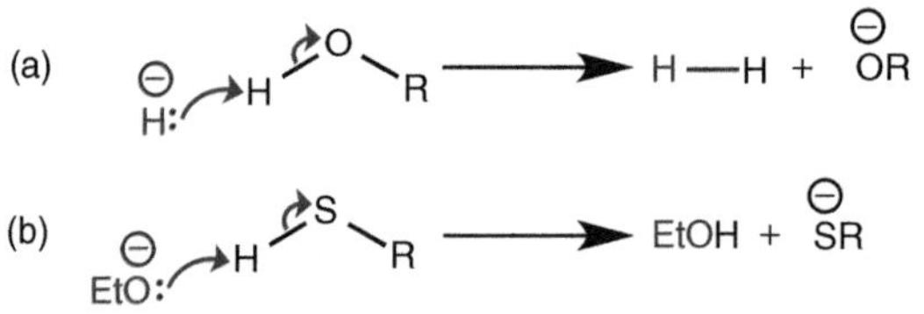

(a)

(b)

$$(1.12)$$

Being fast and reversible, PTs are typically thermodynamically controlled. In other words, for complete PT to occur, the proton donor must be a stronger acid than the protonated proton acceptor. Stated differently, a proton wants to be bonded to the strongest base around. Tables of pK_a values are thus well suited for predicting the direction of PT reactions.

1.9 ELEMENTARY ASSOCIATIVE AND DISSOCIATIVE PROCESSES (A AND D)

Two other elementary polar reactions are worth mentioning right away. These are the polar association (A) and dissociation (D) reactions. An A reaction is simply a bond formation between a nucleophile (Nu or Nu^-) or Lewis base and an electrophile (E or E^+) or Lewis acid, and the product may be positively or negatively charged or neutral:

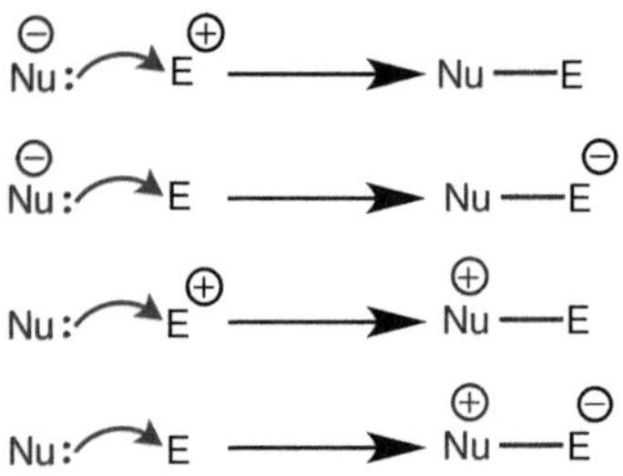

$$(1.13)$$

An A reaction is one of the most common elementary reactions (i.e., a one-step process) we will encounter in this book. Whenever you encounter a reaction of the type "A + B → products," there is a good chance that the first step is either an S_N2 or an A reaction. Indeed, for elements below period 2 of the periodic table, A reactions are at least as common as S_N2 reactions. Several examples of A reactions involving various p-block elements are shown below:

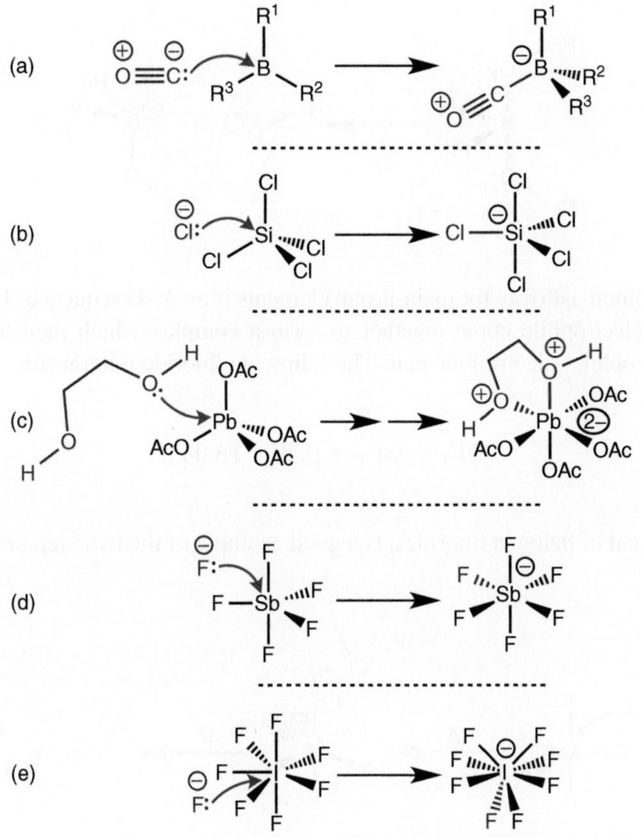

$$(1.14)$$

A D reaction is the reverse of an A reaction; in other words, a D reaction consists of the heterolytic cleavage of a molecule or ion into a nucleophile–electrophile pair, as shown below:

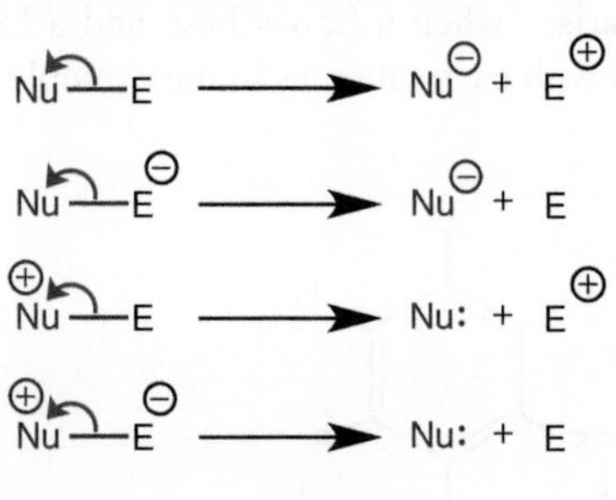

$$(1.15)$$

A classic example of a D reaction is the ionization of *t*-butyl tosylate in a polar solvent:

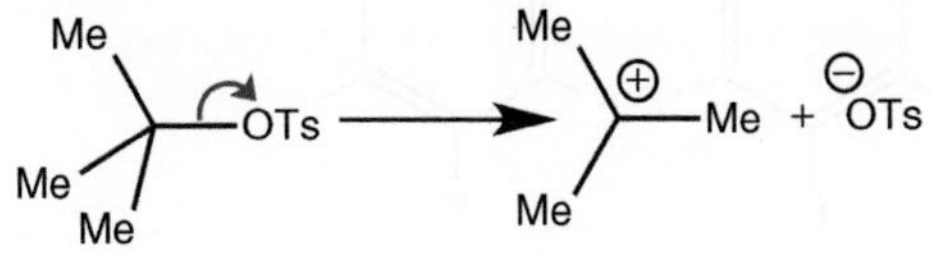

$$(1.16)$$

A more "inorganic" example might be the ionization of Martin sulfurane, a rather fancy tetravalent-sulfur-based dehydrating agent (discussed in Section 6.13):

$$(1.17)$$

A very common pathway for main-group elements is an A–D sequence. Thus, a nucleophile and an electrophile come together to form a complex which then falls apart to a different nucleophile–electrophile pair. The following fluoride ion transfer,

$$ClF_3 + AsF_5 \rightarrow [ClF_2]^+[AsF_6]^- \qquad (1.18)$$

a reaction typical of halogen fluorides, is a good example of the two-step process:

$$(1.19)$$

The D–A sequence is also fairly common. Possibly the best known example of such a sequence is the S_N1 pathway of organic chemistry, briefly described in the next section.

A fascinating situation arises when a Lewis base and a Lewis acid are too sterically encumbered to form a bond with each other, as in the example below:

No reaction $\qquad (1.20)$

Very recently, such "frustrated Lewis acid–base pairs (FLPs)," as they are called, have been found to exhibit unique reactivity, including activation of molecular hydrogen (see Section 2.6 for additional details). The following example shows associative (A) reactions between a P/B-based FLP and CO_2.

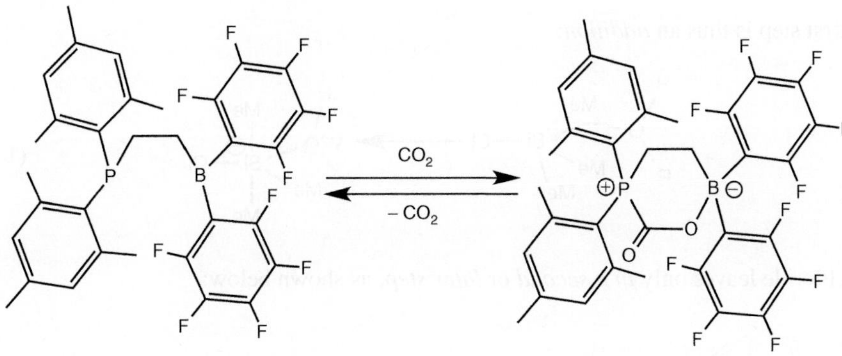

$$(1.21)$$

The mechanism presumably consists of two successive association processes, as shown below:

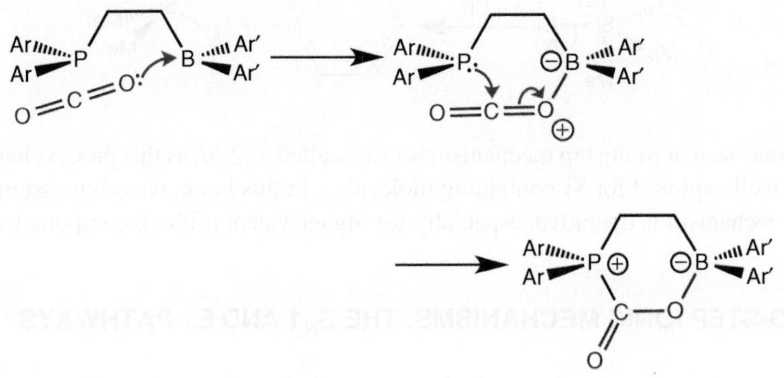

$$(1.22)$$

Sulfur dioxide also reacts in much the same way with a P,B-based FLP. Feel free to sketch it out.

1.10   TWO-STEP IONIC MECHANISMS: THE S$_N$2-Si PATHWAY

Unlike in organic chemistry, nucleophilic attack on a main-group center with a noble gas configuration does not necessarily lead to immediate bond breakage. This is because of the tendency of p-block elements in period 3 and below to form hypervalent molecules ("expanded octets"). In other words, *direct* S$_N$2 displacements (such as the Williamson ether synthesis in reaction 1.1) are far from being the norm for heavier main-group centers. Instead, an incoming nucleophile often *adds* first, and the leaving group departs *only in the next step*—in essence, an A–D sequence, to use the terminology of the last section.

Silylation of an alcohol, a common method for OH group protection in organic chemistry, typically involves such a mechanism:

$$ROH + Me_3SiCl \xrightarrow{\text{Pyridine}} ROSiMe_3 \tag{1.23}$$

The first step is thus an *addition*:

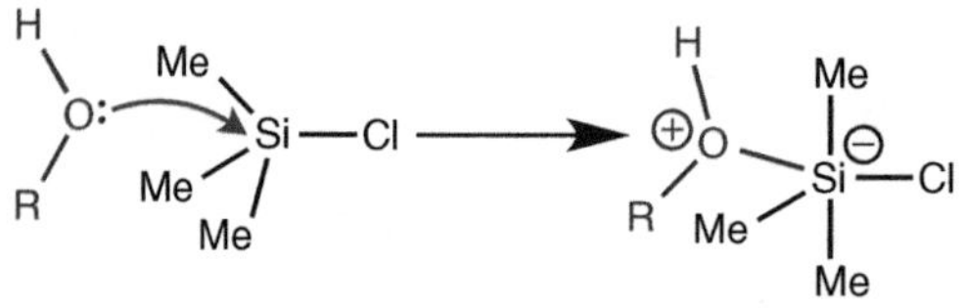

$$(1.24)$$

The chloride leaves only *in a second* or *later step*, as shown below:

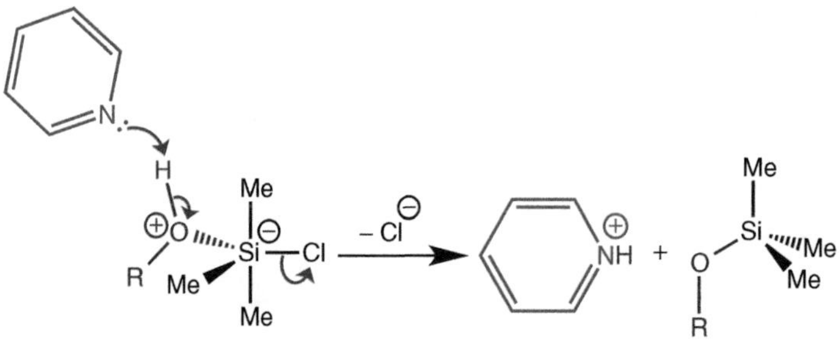

$$(1.25)$$

The overall two-step or multistep mechanism is often called S_N2-Si, as this process has been particularly well explored for Si-containing molecules. In this book, we often assume that the S_N2-Si mechanism is operative, especially for higher-valent p-block compounds.

1.11   TWO-STEP IONIC MECHANISMS: THE S_N1 AND E1 PATHWAYS

Certain carbon–heteroatom bonds may ionize, especially in polar protic media (sometimes referred to as *solvolytic conditions*), to generate carbocations. This is shown below for *t*-butyl tosylate in acetic acid:

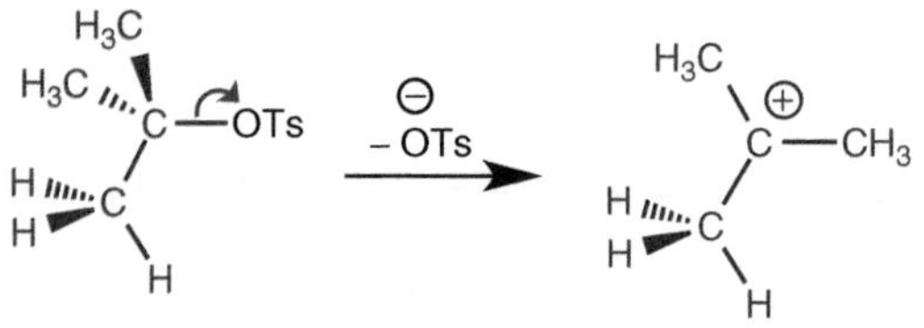

$$(1.26)$$

This rate-determining D step forms the first step of both the S_N1 and E1 pathways. In S_N1 (substitution, nucleophilic, unimolecular), the carbocation combines with a nucleophile, often derived from the solvent. In E1, a base, again often derived from the solvent, deprotonates the carbocation and generates an alkene. In this case, the nucleophile or base is the acetate anion (AcO^-):

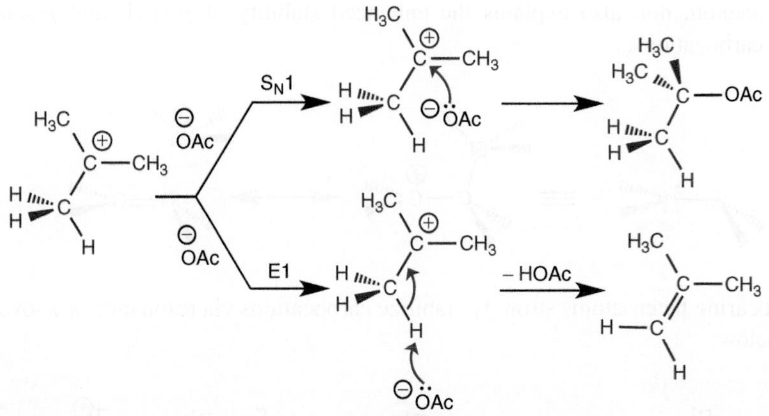

(1.27)

Observe that an S_N1 reaction consists of a two-step D–A sequence, whereas an E1 reaction is a D–PT sequence. Because the carbocation carbon is planar (sp^2), the A reaction with the nucleophile (in this case, AcO^-) can occur via both the front and back sides. Unlike an S_N2 reaction, therefore, an S_N1 reaction is not stereospecific.

We will encounter carbocations again several times in this book, so this is a good place for a brief refresher on carbocation stability. Carbocations are described variously as methyl, primary (1°), secondary (2°), or tertiary (3°), depending on the number of alkyl (or aryl) groups attached to the cationic center. The stability of these cations increases in the following order (3° most stable):

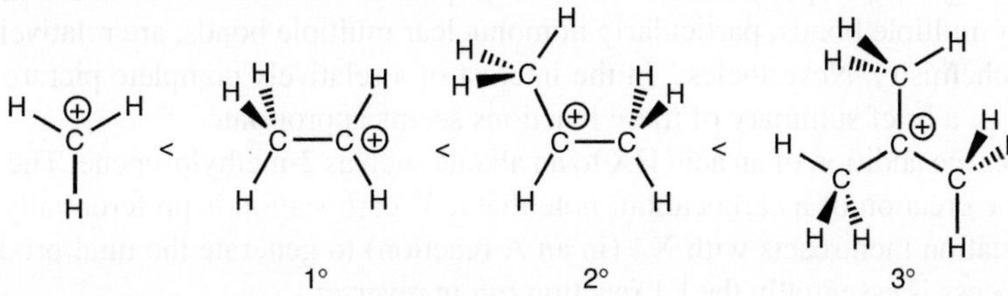

This stability order is typically explained in terms of hyperconjugation, which refers to the overlap of neighboring C–H bonds with the empty p orbital at the cationic center, as shown below:

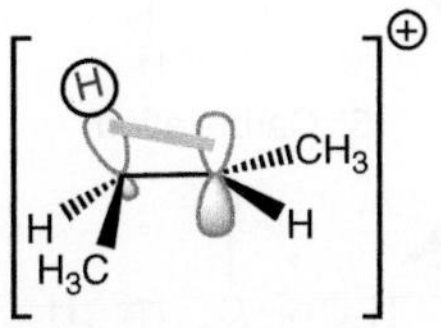

The phenomenon is also often depicted in terms of "no-bond" resonance structures:

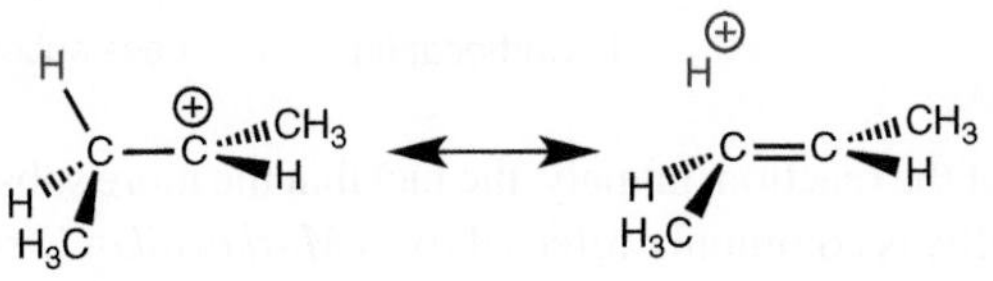

The same phenomenon also explains the enhanced stability of β-silyl- and β-stannyl-substituted carbocations:

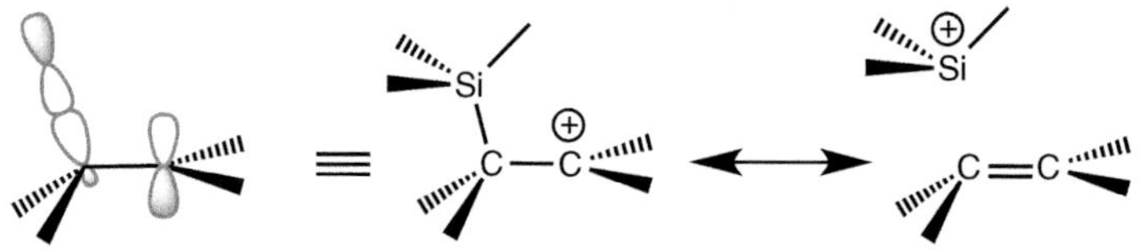

Lone-pair-bearing heteroatoms strongly stabilize carbocations via resonance or π-overlap, as shown below:

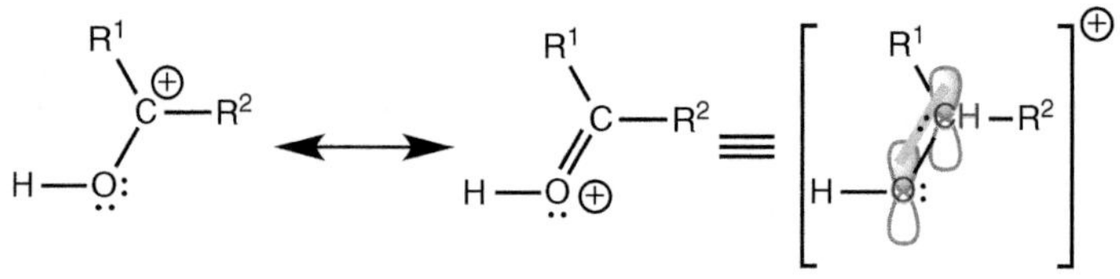

1.12   ELECTROPHILIC ADDITION TO CARBON–CARBON MULTIPLE BONDS

From an inorganic perspective, this is not an important class of reactions for the simple reason that multiple bonds, particularly homonuclear multiple bonds, are relatively rare in inorganic chemistry. Nevertheless, in the interest of a relatively complete picture of polar mechanisms, a brief summary of these reactions seems appropriate.

   Consider the addition of an acid HX to an alkene such as 2-methylpropene. The first step involves the creation of a carbocation; note that a 3° carbocation is preferentially formed. The carbocation then reacts with X⁻ (in an A reaction) to generate the final product. The overall process is essentially the E1 reaction run in reverse:

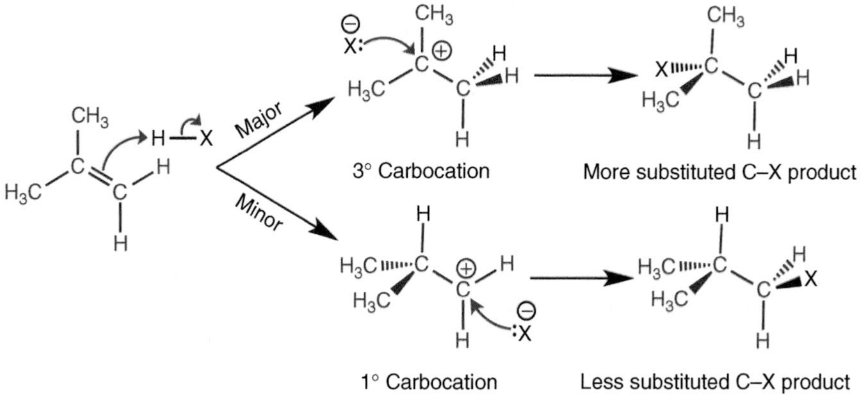

$$(1.28)$$

The regioselectivity of the reaction, namely, the fact that the more substituted addition product forms preferentially, is commonly referred to as *Markovnikov's rule*.

Addition of halogen molecules to double bonds is often stereospecific, and the two halogen atoms typically add to *opposite* faces of the double bond (so-called *anti* addition). This is explained by the intermediacy of cyclic halonium ions, as shown below:

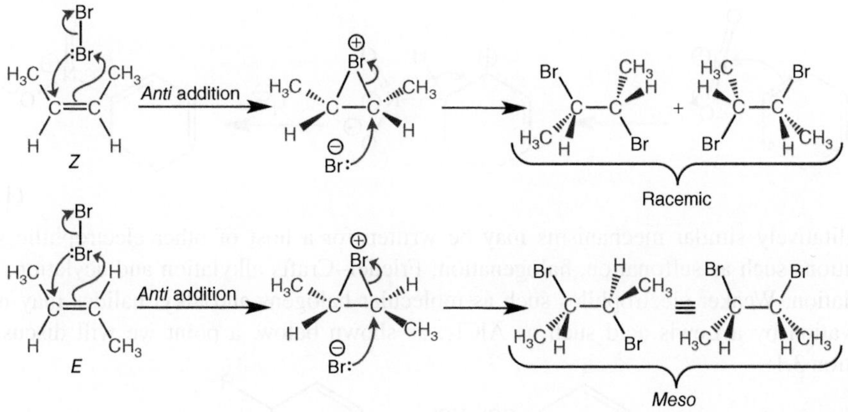

(1.29)

Recall that a racemic substance is an equal mixture of two enantiomers of a chiral molecule. A *meso* compound, on the other hand, has a symmetry element such as a mirror plane or an inversion center that prevents it from being chiral. Analysis of product stereochemistry thus played a crucial role in establishing these mechanisms.

In other cases, addition takes place across a given face of a carbon–carbon double bond (so-called *syn* addition). Epoxidation by a peroxyacid is a good example:

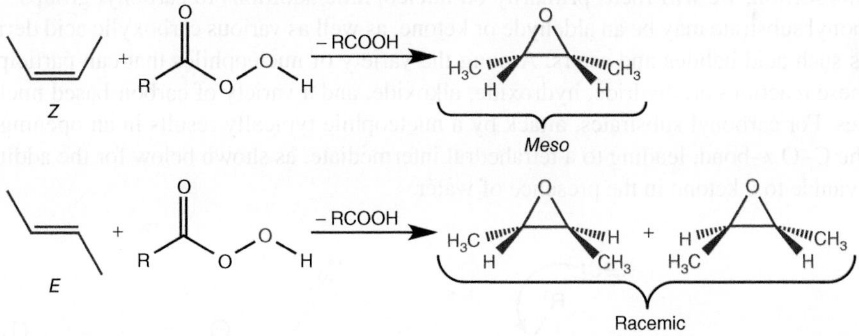

(1.30)

Once again, product stereochemistry provides key clues to the mechanism. Another important example of *syn* addition is osmium tetroxide-mediated *cis*-dihydroxylation of alkenes. Because of their limited relevance in this book, however, we won't discuss the mechanistic details of these reactions, but do consult your organic text if you are curious.

1.13 ELECTROPHILIC SUBSTITUTION ON AROMATICS: ADDITION–ELIMINATION

Electrophiles may add to aromatics, generating transient cationic intermediates. The driving force to regain aromaticity, however, is typically too strong for subsequent addition of a nucleophile. Instead, a proton is eliminated, resulting in a substituted aromatic as the final product. The overall two-step process is depicted below for the nitration of benzene, with the nitronium ion (NO_2^+) as the electrophile:

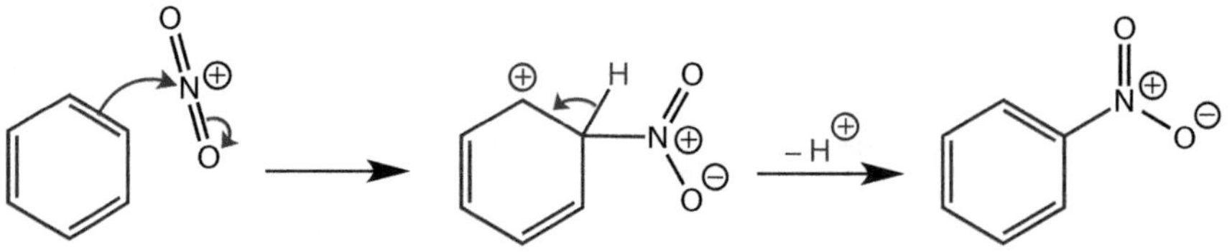

$$(1.31)$$

Qualitatively similar mechanisms may be written for a host of other electrophilic substitutions such as sulfonation, halogenation, Friedel–Crafts alkylation and acylation, and thallation. Weaker electrophiles such as molecular halogens and alkyl halides may need activation by a Lewis acid such as $AlCl_3$, as shown below, a point we will discuss in Section 3.1:

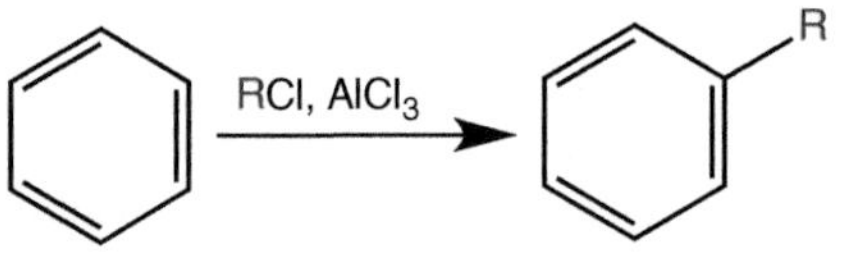

$$(1.32)$$

1.14 NUCLEOPHILIC ADDITION TO CARBON–HETEROATOM MULTIPLE BONDS

In this section, we will focus primarily on nucleophilic additions to carbonyl groups. The carbonyl substrate may be an aldehyde or ketone, as well as various carboxylic acid derivatives such acid halides and esters. Among the variety of nucleophiles that can participate in these reactions are hydride, hydroxide, alkoxide, and a variety of carbon-based nucleophiles. For carbonyl substrates, attack by a nucleophile typically results in an opening up of the C–O π-bond, leading to a tetrahedral intermediate, as shown below for the addition of cyanide to a ketone in the presence of water.

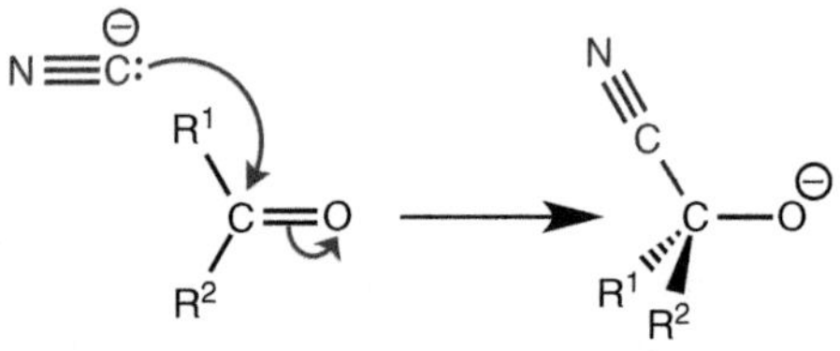

$$(1.33)$$

The anionic alkoxide intermediate picks up a proton from water to generate the "cyanohydrin" product:

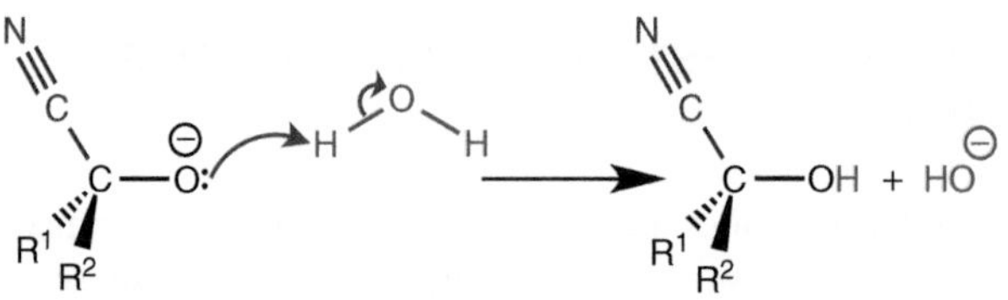

$$(1.34)$$

A similar reaction may also be accomplished under nonaqueous conditions with trimethylsilyl cyanide. Though exceedingly unstable by themselves, trimethylsilyl cations in complexed form may be profitably viewed as "fat protons" and are excellent Lewis acids:

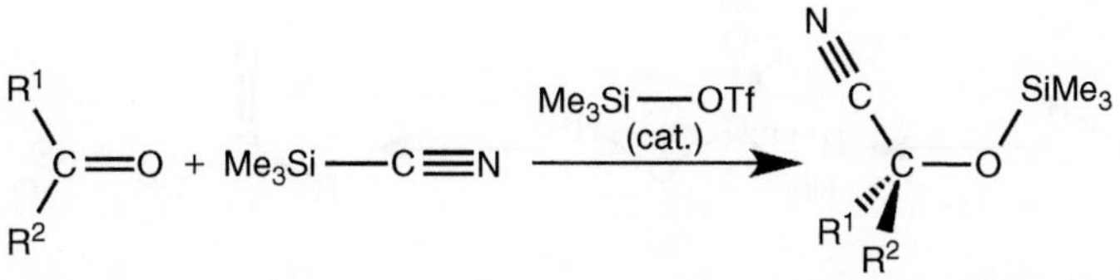

$$\text{(1.35)}$$

The interaction between the carbonyl compound and the catalyst, trimethylsilyl triflate, may be envisioned as an A–D sequence, as shown below:

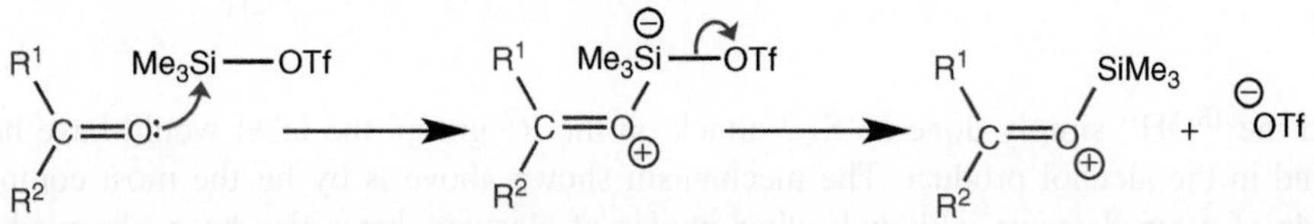

$$\text{(1.36)}$$

A second A–D sequence involving the triflate anion and trimethylsilyl cyanide may then produce free cyanide:

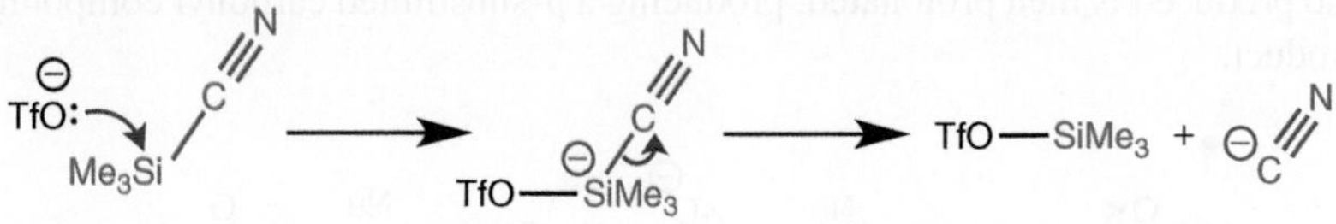

$$\text{(1.37)}$$

The cyanide would then attack the activated carbonyl group to yield a silylated cyanohydrin as the final product, as shown below:

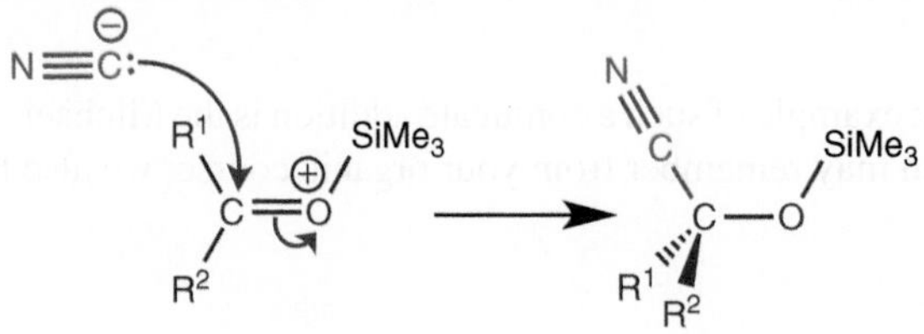

$$\text{(1.38)}$$

It's also conceivable that the actual nucleophile is the five-coordinate species $[Me_3Si(CN)(OTf)]^-$, as opposed to free cyanide.

Ester hydrolysis, an important organic reaction that is also of great biological importance, may be either acid- or base-catalyzed. The base-mediated process may be represented as follows:

$$R^1COOR^2 + NaOH \longrightarrow R^1COONa + R^2OH \qquad \text{(1.39)}$$

The various oxygens have been colored so you may track how they start from a given starting material and end up in a given product. In practice, the color corresponds to the possible use of an isotopic label, typically ^{18}O. Thus, use of $Na^{18}OH/H_2{}^{18}O$ leads to the incorporation *one atom* of ^{18}O into the carboxylic acid product, suggesting that ester cleavage occurs across the $R^1CO–OR^2$ bond, as shown below:

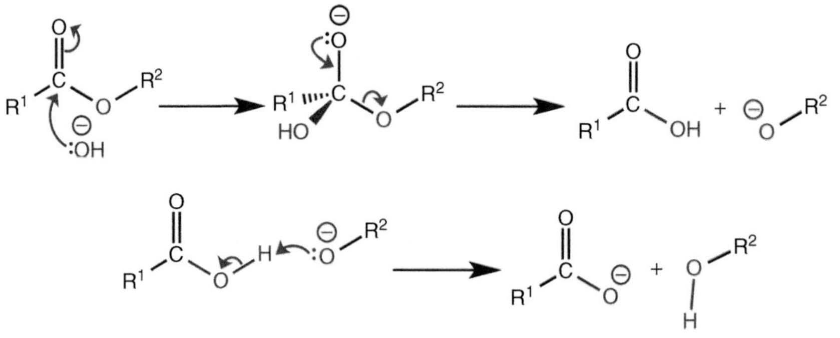

$$(1.40)$$

Had the $^{18}OH^-$ simply done an S_N2 attack on the R^2 group, the label would have been found in the alcohol product. The mechanism shown above is by far the most common mode of ester cleavage, although other modes of cleavage have also been observed for certain substrates and under certain reaction conditions.

An important variant of nucleophilic additions to carbonyl compounds is the "conjugate addition" of a nucleophile to the β-position of an α,β-unsaturated carbonyl compound. The "enolate" so produced is then protonated, producing a β-substituted carbonyl compound as the final product.

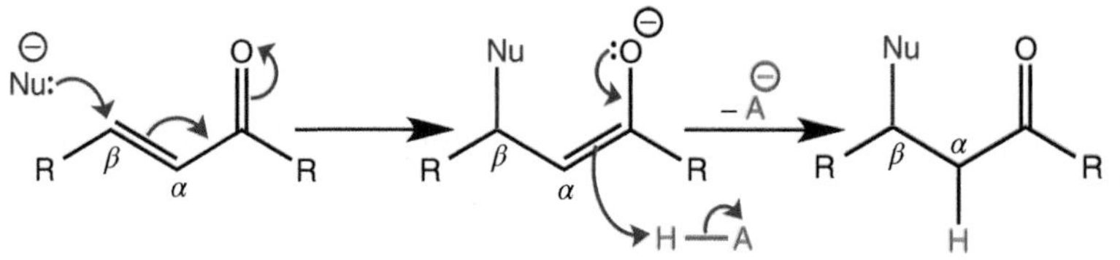

$$(1.41)$$

The most important example of such a conjugate addition is the Michael reaction or Michael addition, which you may remember from your organic course; we also briefly discuss it in the next section.

1.15 CARBANIONS AND RELATED SYNTHETIC INTERMEDIATES

In this section, we'll take a break from our survey of reaction mechanisms and focus instead on a class of intermediates, namely, carbanions. We will also discuss carbanion cognates such as *enols*, *enolates*, *enamines*, and *ylides*. As classic nucleophiles, carbanions react in highly characteristic ways, particularly via S_N2 displacements, as well as via other pathways (e.g., carbonyl addition and conjugate addition) we have discussed above. The material in this section will thus help you flesh out your understanding of what we have discussed so far.

The best known among unstabilized carbanionic derivatives are the Grignard reagents (organomagnesium compounds) and organolithiums:

$$\text{RMgBr} \qquad\qquad \text{RLi}$$

Alkylsodiums and alkylpotassiums are considerably more reactive and are less commonly used. Of particular importance are stabilized carbanions derived from deprotonation of C–H bonds adjacent to carbonyl groups; such anions are often referred to as *enolates* because the negative charge is localized more on the carbonyl oxygen than on the carbanionic carbon:

(1.42)

A strong base such as an alkali metal amide ($pK_a \sim 35$–40) is typically required to quantitatively convert a simple aldehyde or ketone to an enolate. On the other hand, an alkoxide base ($pK_a \sim 16$) is sufficient for deprotonating a β-diketone or a β-ketoester, since the resulting enolate is much more stabilized by resonance:

(1.43)

Stabilized carbanions may also be derived from nitriles:

(1.44)

Enamines are an important class of uncharged synthetic intermediates that exhibit carbanion-like reactivity. They are typically prepared from ketones with one or more α-hydrogens and a secondary amine under acid catalysis:

(1.45)

For reasons of space, we won't go through the mechanism of this reaction, but do look it up in your organic text, if you wish to.

Ylides are an important class of carbanion analogs, which we will encounter several times in this book. Generally, they are not anionic, but 1,2-dipolar compounds in which a carbanionic (or other anionic) center is stabilized by an adjacent cationic p-block center, where both centers have full octets of electrons. The best known ylides are phosphonium and sulfonium ylides, the following being prototypical examples:

Triphenylphosphonium methylide

Dimethylsulfonium methylide

Carbanions and related intermediates react with a variety of carbon electrophiles, such as alkyl halides and carbonyl compounds, forming carbon–carbon bonds. In this capacity, these intermediates are a cornerstone of organic synthesis. A classic example of a carbanion reaction is the aldol reaction. In this, an enolate reacts with a carbonyl compound to yield a β-hydroxycarbonyl compound, as shown in the example below:

Enolate

β-Hydroxycarbonyl

$$(1.46)$$

The conjugate addition of an enolate to an α,β-unsaturated carbonyl compound is called the *Michael reaction* or *Michael addition*. A good example is the following, where an enolate derived from diethyl malonate reacts with methyl vinyl ketone.

Methyl vinyl ketone

Diethyl malonate

$$(1.47)$$

We won't go through the mechanism of this reaction, since it's essentially the same as that depicted for a generic conjugate addition in Section 1.14.

1.16 CARBENES

Divalent carbon species, or carbenes, are another classic group of organic intermediates. Not only are they important in organic chemistry but they are also isoelectronic with other divalent group 14 molecules (based on Ge, Sn, and Pb) and hence of considerable relevance to our further discussions. In addition, nitrogen-stabilized, so-called *N*-heterocyclic carbenes (NHCs) have gained enormous popularity as transition-metal ligands; many of the resulting complexes mediate unique transformations that define a frontier area of contemporary chemistry.

Carbenes occur in one of two low energy states, triplet or singlet. These terms refer to the spin multiplicity of the molecule. The divalent carbon in a triplet carbene has two unpaired electrons, whereas in a singlet carbene the two electrons are paired. Lone-pair-bearing α-heteroatom substituents strongly stabilize the singlet state, as shown by the π-type orbital interaction below:

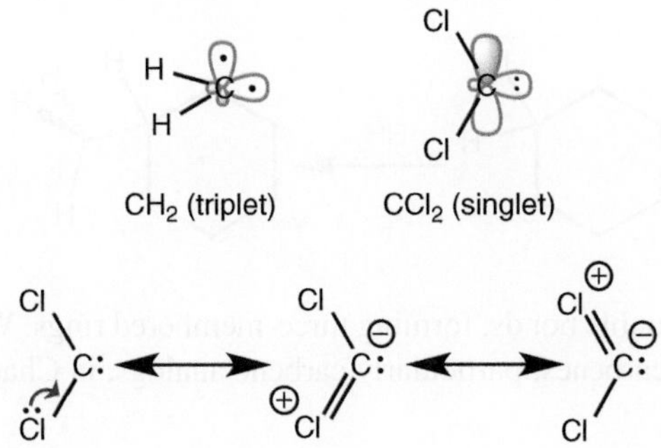

CH$_2$ (triplet)   CCl$_2$ (singlet)

(1.48)

Two heteroatom substituents can result in highly stable "bottle-able" singlet carbenes, two examples of which are shown below:

Additional ionic resonance structures, which can also be drawn for these two carbenes (feel free to draw them out), provide a rationale for their stability.

Carbenes may be accessed via a variety of routes, of which three common ones are shown below:

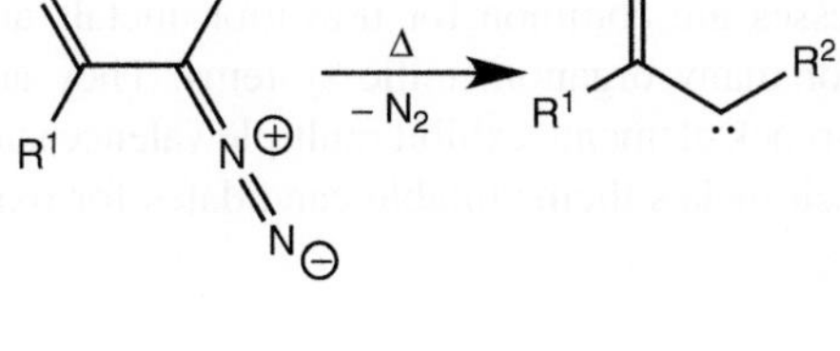

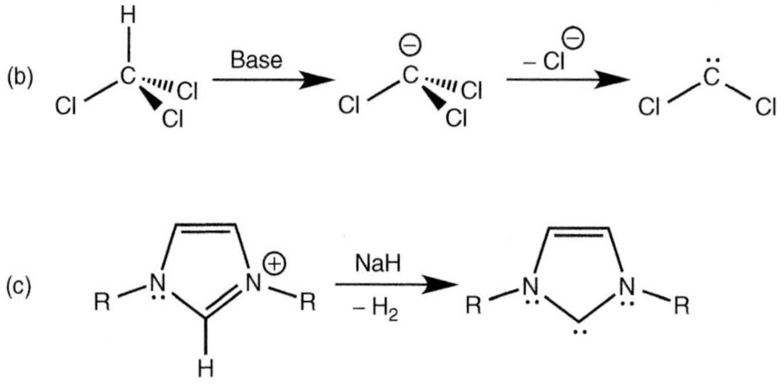

(b)

(1.49)

(c)

Observe that the last of these reactions (1.49c) results in the formation of an NHC.

A highly characteristic reaction of carbenes is insertion, whereby a carbene inserts itself into a C–H or C–C bond, as shown below:

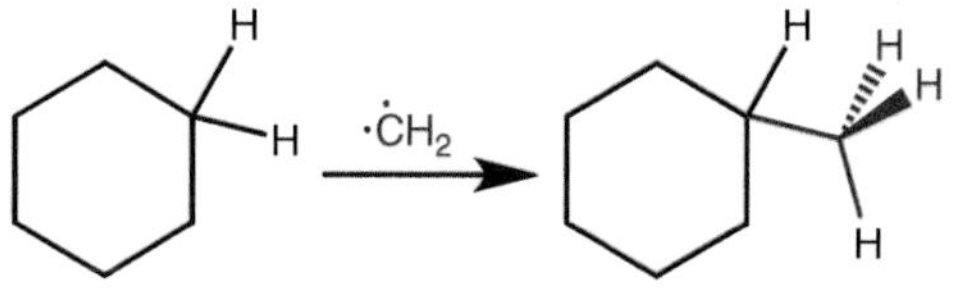

(1.50)

Carbenes also add to double bonds, forming three-membered rings. We'll have more to say about the reactions of carbenes, particularly carbene analogs, in Chapter 4.

1.17   OXIDATIVE ADDITIONS AND REDUCTIVE ELIMINATIONS

We finally come to a pair of reactions that may be described as typically "inorganic" — oxidative addition and reductive elimination. The two processes are the reverse of each other:

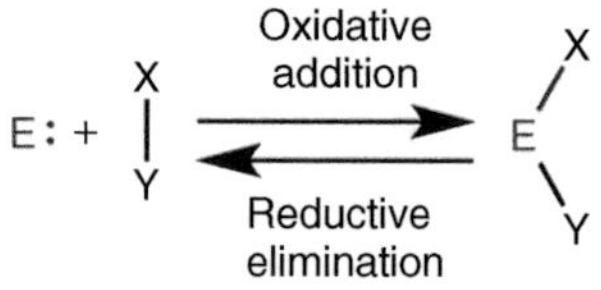

$$E: + \begin{matrix} X \\ | \\ Y \end{matrix} \quad \underset{\substack{\text{Reductive} \\ \text{elimination}}}{\overset{\substack{\text{Oxidative} \\ \text{addition}}}{\rightleftarrows}} \quad E \overset{X}{\underset{Y}{\diagdown}}$$

(1.51)

Observe that oxidative addition results in an increase in the valence of the element E by two units, while reductive elimination results in the reverse. Thus, oxidative addition is a 1,1-addition, that is, addition of two groups to the same atom. Typical organic additions, by contrast, are 1,2-additions or, less commonly, 1,4-additions. It's in this sense that oxidative addition and reductive elimination are characteristically inorganic processes.

The two processes are common for transition metals and have been studied in considerable depth for many organometallic systems. They are also important for p-block elements. Most p-block elements exhibit multiple valence states, with the valence differing by two units, which makes them suitable candidates for oxidative addition and reductive elimination.

Quite a variety of mechanisms are possible for the two processes, including radical pathways. We'll focus here, however, on just a couple of polar mechanisms. A concerted, one-step mechanism may be written as follows:

$$
\text{E:} \quad \overset{\text{X}}{\underset{\text{Y}}{\overset{|}{\text{I}}}} \longrightarrow \text{E} \overset{\text{X}}{\underset{\text{Y}}{\diagdown}} \tag{1.52}
$$

Another common pathway involves a two-step polar addition. As written below, the overall process may be viewed as a tandem S_N2–A process:

$$
\text{E:} \overset{}{\frown} \text{X} \underset{\text{Y}}{\overset{}{\frown}} \text{Y} \xrightarrow{S_N2} \overset{\oplus}{\underset{\text{Y:}^{\ominus}}{\text{E}}}\text{—X} \xrightarrow{A} \text{E} \overset{\text{X}}{\underset{\text{Y}}{\diagdown}} \tag{1.53}
$$

The following chlorination reactions are good examples of oxidative addition involving main-group elements:

$$
PCl_3 + Cl_2 \rightarrow PCl_5 \tag{1.54}
$$

$$
ICl + Cl_2 \rightarrow ICl_3 \tag{1.55}
$$

As in the case of many main-group reactions, the mechanisms have not been investigated for these two reactions. Given that Cl_2 is a good electrophile, either of the above mechanisms seems quite reasonable.

For reductive elimination, we may envision the reverse of the above two pathways. A concerted, one-step reductive elimination may be depicted as follows:

$$
\overset{\text{X}}{\underset{\text{Y}}{\text{E}}} \longrightarrow \text{E:} + \overset{\text{X}}{\underset{\text{Y}}{\text{I}}} \tag{1.56}
$$

Alternatively, we may envision a D–S_N2 sequence:

$$
\text{E} \overset{\text{X}}{\underset{\text{Y}}{\diagdown}} \xrightarrow{D} \overset{\oplus}{\text{E}}\text{—X} \overset{:Y^{\ominus}}{\frown} \xrightarrow{S_N2} \text{E:} + \text{X—Y} \tag{1.57}
$$

Good examples of reductive eliminations involving main-group elements include the following:

$$(1.58)$$

1.18 MIGRATIONS

Migrations are a somewhat quirky process in which an atom or a group tears itself from its site of origin and jumps to a neighboring atom along with its pair of bonding electrons. Thus, typically, the migrating group acts as a nucleophile and the migration terminus as an electrophile. The process is typically intramolecular, that is, occurring within a single covalently bound entity.

Some of the best studied examples of migration are provided by carbocation rearrangements. Generally, less stable cations (e.g., 2°) rearrange to more stable ones (3°) by undergoing hydride or alkyl shifts, as shown in the example below:

$$(1.59)$$

Indeed, observation of rearranged products is often seen as "proof" of the involvement of carbocations and hence of an $S_N1/E1$-type reaction pathway; concerted $S_N2/E2$ pathways are not expected to result in a rearrangement.

A typical migration involved in a carbocation rearrangement is called a *1,2-shift* because the origin and terminus of the migration are (typically, but not invariably) adjacent to each other. We will encounter a few examples 1,2-shifts involving main-group elements. A common reaction pattern is the following:

$$(1.60)$$

The migrating group R is often an alkyl or aryl group; the E–L bond is generally weak, which makes L a good leaving group and E a good migration terminus. The negative charge on the migration origin M enhances the migratory aptitude of the R group. Typically, M is a relatively electropositive p-block element such as boron or silicon, or a higher-valent state of an electronegative element such as a halogen. A good example of this type of a 1,2-shift is provided by hydroperoxide-mediated oxidation of organoboranes.

A standard method of work-up for organoboranes obtained with the hydroboration reaction involves reaction with alkaline hydrogen peroxide. Mechanistically, the first step is an A reaction between the organoborane and a hydroperoxide anion that is present in solution:

$$(1.61)$$

The anionic boron center then acts as a launchpad for a migrating R group:

$$(1.62)$$

Note that the leaving group here is hydroxide, normally a lousy one in organic chemistry. Here, however, the electrophilic site is an oxygen and an O–O bond is a weak one that is easily cleaved. We will encounter several examples of similar 1,2-shifts as we progress through the book.

1.19 LIGAND EXCHANGE REACTIONS

Main-group element chemistry is replete with ligand exchange reactions or metatheses. We will encounter a fair number of such reactions in this book, a few examples being as follows:

$$SnCl_4 + SnR_4 \rightarrow 2\ SnR_2Cl_2 \qquad (1.63)$$

$$SO_2 + PCl_5 \rightarrow SOCl_2 + POCl_3 \qquad (1.64)$$

$$2\ (CH_3)_3SiBr + SeCl_2 \rightarrow 2\ (CH_3)_3SiCl + SeBr_2 \qquad (1.65)$$

$$IF_7 + POF_3 \rightarrow IOF_5 + PF_5 \qquad (1.66)$$

For a mechanistic discussion, we will choose the last reaction. A reaction pathway may not be obvious at this point, based on the mechanistic paradigms we have discussed so far. Clearly, a series of simple D and A reactions, whereby oxide and fluoride ligands detach from one central atom and reattach to another one, are unreasonable, given the covalent nature of the molecules involved. An A reaction, however, is still a promising starting point:

$$(1.67)$$

The oxo-bridged P^+–O–I^- intermediate may now react in a number of different ways. A fluoride could depart from the anionic iodine and reattach to the cationic phosphorus center, as shown below:

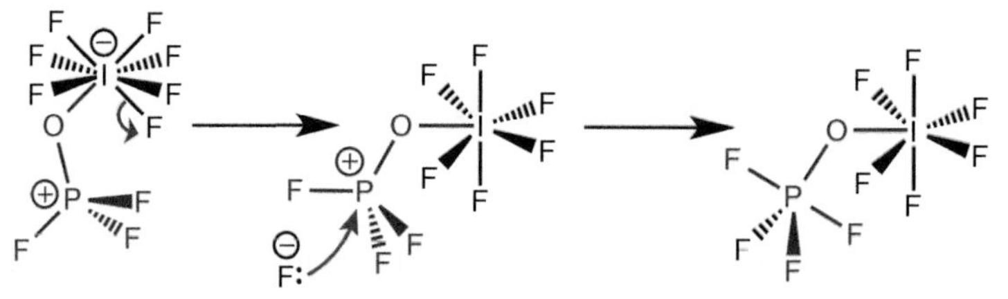

$$(1.68)$$

A second fluoride could then attack the neutral phosphorus center, leading to the final products PF_5 and IOF_5:

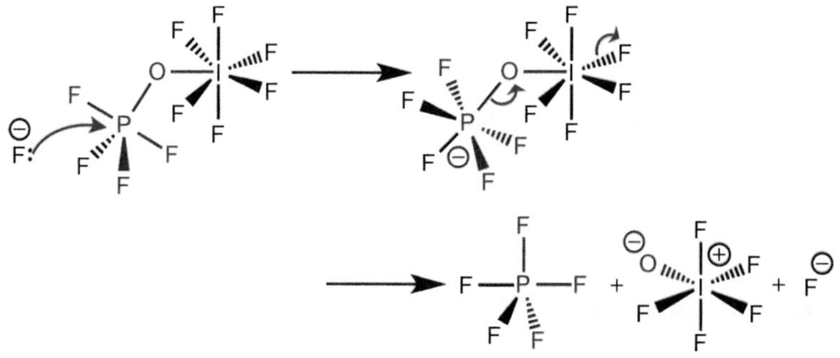

$$(1.69)$$

Other pathways are also conceivable for the initially formed oxo-bridged pathway. For example, a 1,3-shift of a fluoride provides "quick" access to the neutral intermediate F_4P–O–IF_6:

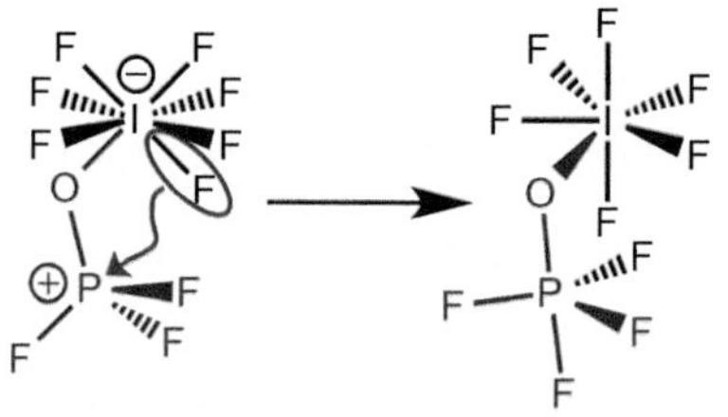

$$(1.70)$$

A second fluoride 1,3-shift, along with cleavage of the I–O bond, could then lead to the final observed products.

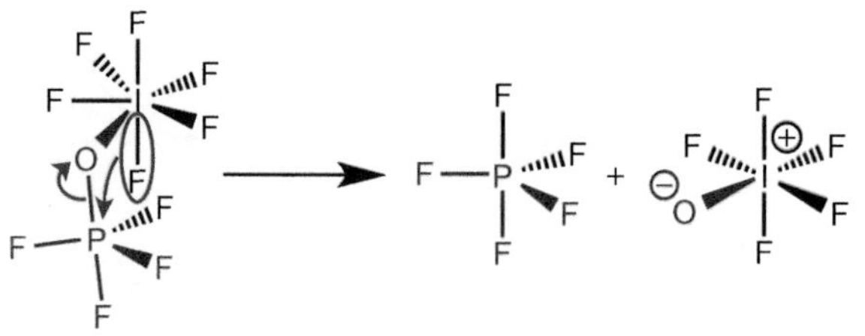

$$(1.71)$$

At present, we do not know whether fluoride 1,3-shifts provide a low energy pathway or not. Therefore, we cannot state categorically which of the above two pathways, or for that matter a different one, is the one that operates in reality. Note that both pathways require the formation of an oxo-bridged intermediate. That appears to be a general feature of ligand exchange reactions of this type. When dealing with such reactions, simply join up the two reactants via a lone pair on one of the migrating groups; subsequent D and A reactions, or ligand 1,3-shifts, would then lead to the final products.

A word is in order on the thermodynamic driving forces underlying the above reaction. Relief of steric strain at the 7-coordinate iodine is a possible factor, but the main driving force is undoubtedly the formation of two highly stable P–F bonds, whose combined BDEs ($\sim$490 kJ/mol each) more than outweigh that of one P^+–O^- unit ($\sim$544 kJ/mol).

1.20 RADICAL REACTIONS

Although our focus is clearly on polar or ionic mechanisms, we should not and will not ignore radical pathways altogether. A very brief introduction is therefore provided here. Observe that, in the discussion below, single-headed fishhook arrows indicate "movement" of unpaired electrons.

Radicals are typically produced by thermal or photochemical homolytic cleavage of a weak single bond:

$$\text{Homolysis:} \qquad A\text{---}B \longrightarrow A\cdot + B\cdot \qquad (1.72)$$

Homolysis refers to the separation of a bonding electron pair into two unpaired electrons, that is, radicals. Heterolytic mechanisms, by contrast, are characterized by a bond-breaking step where the electron pair constituting the bond leaves with one of the fragments, as shown below:

$$\text{Heterolysis:} \qquad \begin{array}{c} A\text{---}B \longrightarrow A^{\oplus} + B^{\ominus} \\[6pt] A\text{---}B \longrightarrow A^{\ominus} + B^{\oplus} \end{array} \qquad (1.73)$$

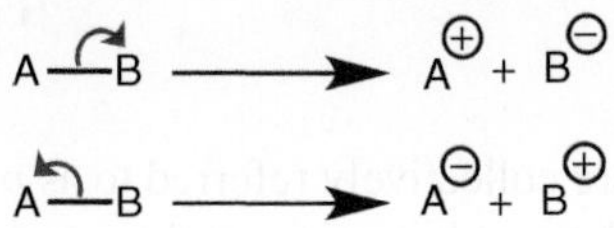

The term "heterolytic mechanism" is thus more or less synonymous with a polar or ionic mechanism. Some classic radical-generating reactions are as follows:

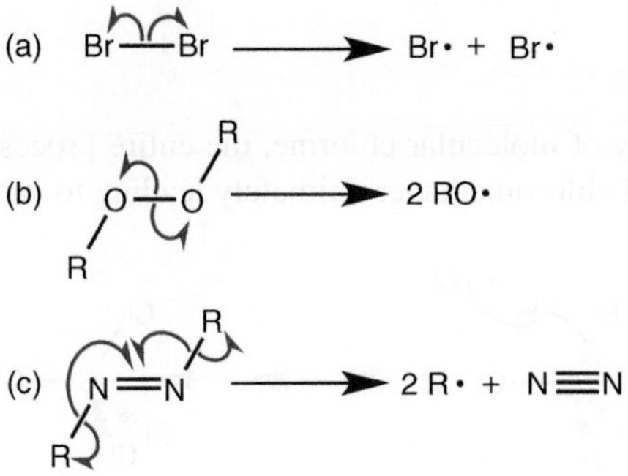

$$\text{(a)} \quad Br\text{---}Br \longrightarrow Br\cdot + Br\cdot$$

$$\text{(b)} \quad \underset{R}{RO}\text{---}OR \longrightarrow 2\,RO\cdot \qquad (1.74)$$

$$\text{(c)} \quad N\text{==}N \longrightarrow 2\,R\cdot + N\text{≡}N$$

Let us work our way through a radical chain reaction. A good example is the photo-chemical chlorination of alkanes, shown below for methane:

$$CH_4 \xrightarrow[-HCl]{Cl_2, h\nu} CH_3Cl \xrightarrow[-HCl]{Cl_2} CH_2Cl_2$$
$$\xrightarrow[-HCl]{Cl_2} CHCl_3 \xrightarrow[-HCl]{Cl_2} CCl_4 \tag{1.75}$$

The reaction begins with the light-induced splitting of molecular chlorine into chlorine atoms; this is called the *initiation step*:

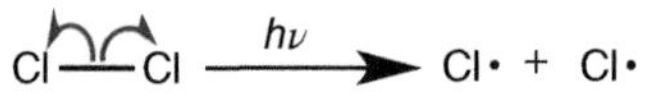

$$Cl \overset{h\nu}{\underset{}{\longrightarrow}} Cl\cdot + Cl\cdot \tag{1.76}$$

A chlorine atom can then abstract a hydrogen from methane:

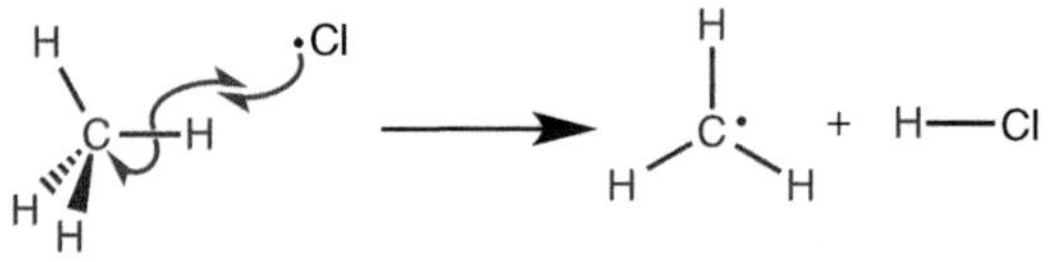

$$\tag{1.77}$$

The methyl radical then abstracts a chlorine atom from molecular chlorine, forming chloromethane and another chlorine atom, thus perpetuating the presence of chlorine radicals in the system.

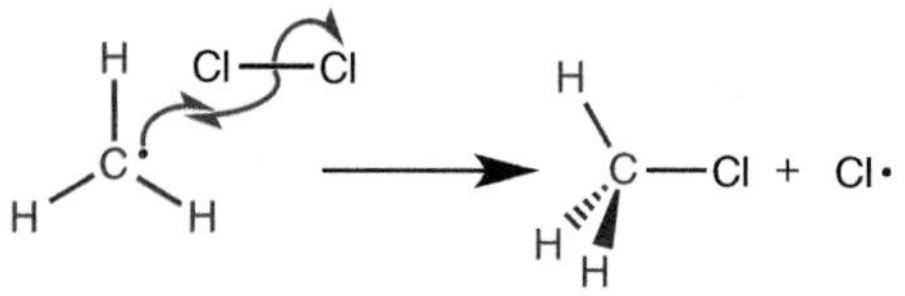

$$\tag{1.78}$$

The above two steps are collectively referred to as *propagation steps*. A methyl radical and a chlorine atom may also combine to form chloromethane, as shown below, in what is called a *termination step*:

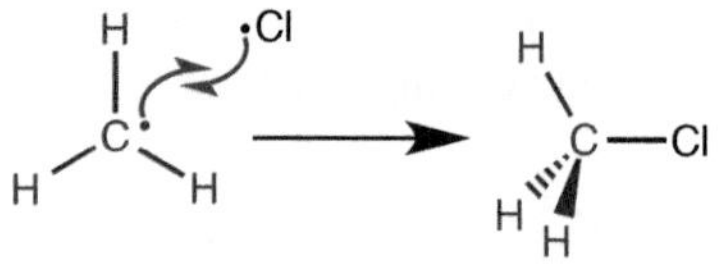

$$\tag{1.79}$$

Depending on the supply of molecular chlorine, the entire process may continue, leading to further chlorination of chloromethane, ultimately leading to carbon tetrachloride:

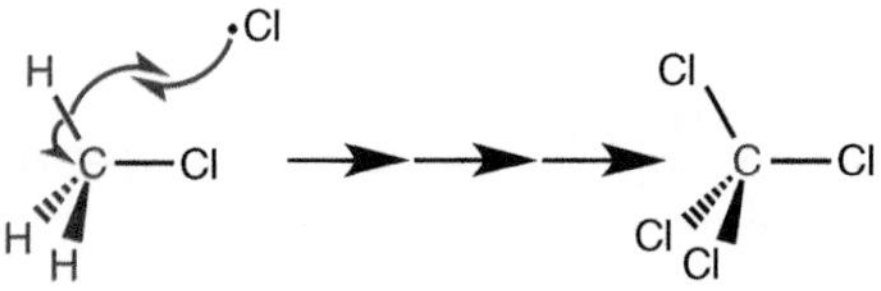

$$\tag{1.80}$$

Although much less commonly used in chemical synthesis relative to polar reactions, radical reactions nonetheless form a distinct "genre" of synthetic reactions. Creatively orchestrated, they can lead to a variety of complex structures with a surprising degree of efficiency.

1.21 PERICYCLIC REACTIONS

Pericyclic reactions, most notably the Diels–Alder reaction, other cycloadditions, and certain sigmatropic rearrangements in which two or more electron pairs move in a more or less concerted manner along a cyclic pathway are a cornerstone of organic synthesis. Much of their importance derives from the efficiency with which they create two or more bonds in one step and also in a stereospecific manner. Some examples are as follows:

$$(1.81)$$

Pericyclic reactions provide some of the most elegant examples of the importance of orbital symmetry in chemical reactions. Unlike in organic chemistry, however, pericyclic reactions are not of great importance in inorganic chemistry. That said, we will encounter a few significant examples in this book, including the reduction of carbon–carbon double bonds by diimide (Section 5.7a) and certain selenium dioxide oxidations (Section 6.16).

◆ ◆ ◆

That concludes our survey of the major reaction types that we are likely to encounter in this book. We are therefore in a position now to think in somewhat more general terms about arrow pushing. This we do in the next two sections.

◆ ◆ ◆

1.22 ARROW PUSHING: ORGANIC PARADIGMS

Organic chemists have codified a number of rules and guidelines as aids to arrow pushing. We'll summarize a few of them here as a starting point for exploring the more diverse world of inorganic mechanisms.

1. Stable molecules, the final products of reactions, typically have noble gas configurations on all atoms, that is, octets on all atoms B through F.

2. Nucleophilic attack on a center with an inert gas configuration (an octet) must lead to bond breakage, that is, departure of a leaving group.

3. Stable organic molecules strive toward charge neutrality for all atoms, although charged species such as carbocations may arise as intermediates. Like charges on two adjacent atoms are a taboo in organic chemistry!

4. Good leaving groups are relatively nonnucleophilic and nonbasic. Thus, N_2 is a superb leaving group, and water (H_2O) and alcohols (ROH) are very good leaving groups as well. By contrast, OH^- and RO^- are lousy leaving groups and H^- (hydride) and R^- (alkyl anions) are far worse.

For us, the key question is: how well do these guidelines carry over to inorganic chemistry? Not too well, it turns out, as we discuss below!

1.23 INORGANIC ARROW PUSHING: THINKING LIKE A LONE PAIR

Generally speaking, the mechanisms of p-block element reactions are not particularly consistent with the rules outlined above. The reason for this boils down to the so-called first-row anomaly, where both first- and second-period elements (H–Ne) are all somewhat unreasonably lumped together as first row. The expression means that the chemical properties of first-row elements are anomalous relative to those of their heavier congeners. Let us go through the above four rules one by one and see how well they hold up in a main-group inorganic context.

1. The octet rule breaks down routinely as soon as one goes down to period 3. Main-group centers with more than eight electrons in their valence shells abound for period 3 and below. Molecules containing such centers are called *hypervalent*. Well-known examples include SiF_6^{2-}, PF_5, PF_6^-, SF_4, SF_6, BrF_3, IF_5, IF_7, XeF_2, XeF_4, XeF_6, and XeF_8^{2-}. Even this short list of paradigmatic examples should show with complete clarity that hypervalent molecules are anything but unusual; they are ubiquitous for p-block elements in period 3 and below. Not surprisingly, therefore, the octet rule is essentially irrelevant for these elements. (*Note*: The bonding in the hypervalent molecules listed above might seem puzzling at first sight, but we will address that issue over the next few sections.)

2. As mentioned in Section 1.9, nucleophilic attack on a heavier main-group center with a noble gas configuration does not necessarily lead to immediate bond breakage. Instead, the first step may be an A reaction leading to a hypervalent intermediate with an expanded octet. The leaving group would them depart in a subsequent D process. As mentioned, this two-step process is known as S_N2-Si. In this book, we have

often tacitly assumed that the S_N2-Si mechanism is operative. Although there is no question that it is pervasive, whether it applies near-universally to all heavier p-block scenarios has yet to be settled.

3. Compared with organic chemistry, charged centers are much more common in main-group element chemistry. Thus, monatomic anions such as O^{2-}, S^{2-}, and N^{3-} are all stable species when stabilized by cations in a solid lattice. Ammonium, phosphonium, and sulfonium ions are an important part of the chemistry of the elements in question. Oxoanions are an important part of the chemistry of the great majority of p-block elements. Thus, rule no. 3 (concerning charge-neutral reactants and products) applies rather more weakly in inorganic chemistry than in organic chemistry; there are many exceptions.

4. We mentioned that strong bases such as OH^- and RO^- generally make poor leaving groups. These, however, are "poor leaving groups" only from an organic perspective, that is, where the reaction center is carbon. Non-carbon p-block centers vary hugely in their electronegativity from near-metallic or metalloidal B, Al, Sn, and so on, to highly electronegative elements such as N, O, and F. Indeed, for N, O, or halogen reaction centers, OH^- and RO^- are rather good leaving groups. The reason for this is that bonds between two highly electronegative atoms are weak and are readily broken, both via S_N2 displacements as well as homolytically. Rule no. 4 too thus has limited application in inorganic chemistry.

Organic paradigms accordingly are not very helpful in providing rules of thumb for arrow pushing in main-group inorganic chemistry, especially for elements below period 2. Yet we are far from helpless. Simple pattern recognition skills, along with some basic ideas about nucleophiles, electrophiles, and leaving groups, go a long way in helping us arrive at reasonable mechanisms for reactions involving main-group elements. Our approach may be summarized as follows:

1. Look at the product structure(s) carefully and determine what bonds have been broken in the course of the reaction and what new bonds have been formed.

2. Identify the nucleophile and the electrophilic site of attack.

3. Apply steps (1) and (2) iteratively until you arrive at the product structures (assuming they are known).

Step (1) consists of pattern recognition, somewhat similar to the logic involved in putting together a puzzle. Note that in the quote at the beginning of the chapter, Sherlock Holmes describes this ability to "reason backward" as easy! Based on many years of experience, we would echo the same assessment. Step (2) is where your chemistry knowledge comes in handy: Apply your knowledge of nucleophilicity, electrophilicity, leaving groups, bond strengths (e.g., bonds between two electronegative atoms are easily cleaved), and so on. Beyond that, we do not advocate an overly algorithmic approach. Several of the reactions discussed in this book are too complex for that. To us, a semi-intuitive approach is what makes inorganic arrow pushing both challenging and fun. Follow your nose! Or, to use our favorite metaphor: think like a lone pair! Where would you attack if you were a lone pair?

♦ ♦ ♦

It's time now to think about hypervalent compounds. You have encountered a few of them already, as products of A reactions and as intermediates in S_N2-Si mechanisms. But what is special about such compounds? Is the term "hypervalent" synonymous with higher-valent? (No.) To better understand these issues, we'll take a step back in Section 1.24 and remind ourselves what the term "valence" *exactly* means and how it differs from related concepts such as coordination number (CN), FC, and oxidation state (OS). Confusion between these terms and incorrect usage are widespread in both textbooks and the research literature. From there we'll proceed on to some related topics such as an elementary molecular orbital description of hypervalent bonding (Section 1.25). We'll conclude this chapter with a brief discussion of the inert pair effect, an important aspect of the variable valence of the heaviest (sixth-period) p-block elements.

◆ ◆ ◆

1.24   DEFINITIONS: VALENCE, OXIDATION STATE, FORMAL CHARGE, AND COORDINATION NUMBER

Table 1.7 presents compact definitions for all four concepts.

Valence is most simply defined as the number of electrons an atom uses in bonding. More accurately, *valence may be defined as the number of valence electrons in an atom minus the number of nonbonding electrons on the atom in the molecule in question.* Using either of these definitions, we may readily see that the valence of sulfur is 2, 4, and 6 in H_2S, SF_4, and SF_6, respectively:

Hydrogen sulfide

Sulfur tetrafluoride

Sulfur hexafluoride

Valence is often confused with OS (also called oxidation number, ON) and less frequently with CN. Not infrequently, valence equals OS. For example, the valence and OS of carbon in CF_4 are each 4; the same holds for CO_2. Consider, however, CH_2F_2 and H_2CO; carbon

TABLE 1.7   Definitions of Valence, Oxidation State, Formal Charge, and Coordination Number

Term	Definition
Valence (V)	Number of electrons that an atom uses in bonding
Oxidation state (OS)	The charge remaining on an atom when all ligands have been removed heterolytically, with the electrons being transferred to the more electronegative partner; homonuclear bonds (i.e., bonds between atoms of the same element) do not contribute to the OS
Formal charge (FC)	The charge remaining on an atom when all ligands have been removed homolytically
Coordination number (CN)	The number of atoms bonded to the atom of interest

This table is adapted from: Parkin, G. *J. Chem. Educ.* **2006**, *83*, 791–799.

is still tetravalent in both compounds but its OS is 0. In the same manner, CN often equals valence. While these equalities can be understood on a case-by-case basis, it's probably best to view them as coincidental.

Another very important concept in this connection is that of FC. *FC is the charge remaining on an atom when all the ligands have been removed homolytically.* FCs are the charges that are commonly shown in structural formulas and reaction mechanisms. Thus the FC on nitrogen in NH_4^+ is $+1$ and that on boron in BH_4^- is -1.

The valence of atoms bearing a nonzero FC can be a bit of a tricky affair. Let's consider the ions NH_4^+ and NH_2^- and compare them with NH_3. The nitrogen in ammonia is clearly trivalent. In NH_4^+ and NH_2^-, nitrogen uses four and two electrons, respectively, to form bonds with hydrogen atoms. In addition, the nitrogen has lost one of its original valence electrons in NH_4^+; in NH_2^-, the nitrogen has gained an extra electron relative to its free atomic state. To account for these excess charges, it is helpful to recognize that valence is *the number of electrons an atom uses in forming bonds plus the net number of electrons it has lost in forming the molecule/ion in question.* This is not a new definition of valence; it's perfectly equivalent to the "more accurate" definition that we have presented above. Thus, nitrogen in NH_4^+ is pentavalent: four valence electrons used in bonding plus one lost. Similarly, nitrogen in NH_2^- is monovalent: two valence electrons used in bonding and one electron *gained* relative the free, neutral atom, that is, $2 + (-1) = 1$. Thus, a very useful relation is the following:

$$\text{valence} = \text{no. of bonds} + \text{formal charge}$$

Some of the consequences of this definition can seem a bit mind-bending, until you get used to them. Thus, applying this definition, we get an oxygen valence of 0 for OH^- and 4 for H_3O^+! After a while, these results won't seem quite as bizarre as they might do now.

1.25 ELEMENTS OF BONDING IN HYPERVALENT COMPOUNDS

A hypervalent molecule is characterized by a main-group element atom with more than eight electrons in its valence shell, according to the molecule's Lewis structure. Figure 1.6 presents the Lewis structures of a selection of hypervalent molecules and ions, with different central atoms but all with fluoride as the terminal ligands. To assist with electron bookkeeping, we have indicated the valence electrons of the central atoms as dark blue dots, those of fluorine as crosses, and those corresponding to any excess negative charge as red crosses.

The main question we will try to address in this section concerns how the central atom in a hypervalent molecule accommodates more than eight valence electrons. For a long time, such "expanded octets" were thought to reflect participation by the d orbitals of the central atom. That explanation is now believed to be incorrect. Instead, a perfectly straightforward explanation is available from standard molecular orbital theory, which we will discuss below using the trigonal bipyramidal molecule PF_5 as an example.

We may view the equatorial atoms of PF_5 as a planar PF_3 unit with an sp^2-hybridized P atom. The lone pair on the P then might be thought of as occupying an unhybridized p orbital, which we will call the p_z orbital. This lone pair interacts with the p_z orbitals of two F atoms, each of which is assumed to have an unpaired electron, forming two apical P–F bonds. As shown in Figure 1.7, the three p_z orbitals form bonding, nonbonding, and

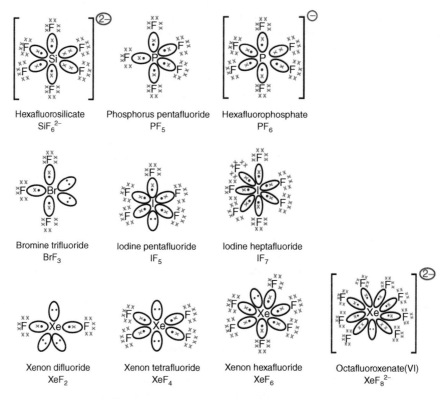

Figure 1.6   *Examples of hypervalent molecules.*

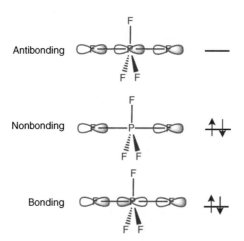

Figure 1.7   *F–P–F σ-bonding along the trigonal axis of PF$_5$.*

antibonding molecular orbitals. The first two of these molecular orbitals accommodate the four electrons in question, two from the P lone pair and one from each F.

   This electron bookkeeping parallels that for allyl species, except that for allyl the orbital overlaps are not head-on or σ but of the sideways or π type, as shown in Figure 1.8.

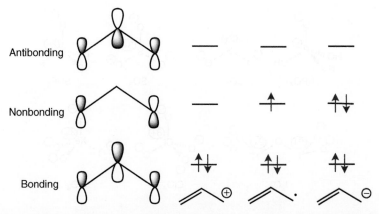

Figure 1.8 π-*Molecular orbitals of the allyl system. From left to right, the electron occupancies are for the cation, radical and anion, which have two, three, and four π-electrons, respectively.*

Arguments of a similar nature also provide a qualitative explanation of the bonding in SF_6, BrF_3, IF_5, IF_7, and the various xenon fluorides. This book is not an appropriate place for a discussion of each of these cases. The key point is that, for the purposes of arrow pushing, all the element–fluorine bonds may be viewed as normal two-electron bonds.

We are now in a position to discuss the question of multiple bonds between higher-valent main-group elements and typical multivalent ligands such as O and N. For this, consider the following molecules/ions:

$$H_3PO_4, SO_2, SO_3, SNF_3, I_2O_5, ClO_4{}^-, \text{ and } XeO_4$$

Drawn traditionally, the structural formulas all have a hypervalent central atom and one or more multiple bonds:

H_3PO_4 SO_2 SO_3 SNF_3

I_2O_5 $ClO_4{}^-$ XeO_4

The structural formulas immediately raise questions about the nature of the multiple bonds in them. Specifically, are there π-bonds involved? The answer to this question is: to a degree, yes, but not in the conventional sense. Thus, in SNF_3, any S–N π-bonding involves lone pairs on the nitrogen and S–F σ-*antibonding orbitals*. This is a subtle issue that, in our opinion, should not preoccupy us, nor for that matter dominate day-to-day mechanistic reasoning. Consequently, we prefer the ionic formulations shown below, which also *happen to be* nonhypervalent (i.e., the central atom in each case has an octet of electrons):

H_3PO_4 SO_2 SO_3 SNF_3

I_2O_5 ClO_4^- XeO_4

The great advantage of this style is that the basic nature of the bonding is immediately obvious. For the central atom, no more than a set of s and p orbitals are involved. Importantly, we must add that rewriting multiply bonded structural formulas in ionic form does not *necessarily* make them nonhypervalent. A good example is periodic acid, H_5IO_6, where both the traditional and newer ionic formulas are hypervalent:

"Traditional"   "Ionic"

Let's now consider whether the representation chosen—the traditional one with multiple bonds versus the newer one—affects the valence of the central atom. Let us do so with a couple of examples involving higher-valent sulfur, say, $SOCl_2$ and SO_2Cl_2. If we consider the traditional structural formulas with double bonds, the S is clearly tetra- and hexa-valent in the two molecules, respectively. Simply count the electrons, denoted by blue dots, that the sulfur has used to form bonds.

Thionyl chloride

Sulfuryl chloride

However, if we consider the alternative ionic structures, we see that $SOCl_2$ and SO_2Cl_2 contain three and four bonds to sulfur, respectively. In these cases, it's useful to recall that the sulfur shares one of its valence electron pairs with oxygen in $SOCl_2$ and two of its valence electron pairs with the oxygens in SO_2Cl_2, as follows:

Thionyl chloride

Sulfuryl chloride

Applying the above definition of valence (= no. of bonds + FC), we find

$$V(SOCl_2) = 3 + 1 = 4$$

$$V(SO_2Cl_2) = 4 + 2 = 6$$

where the symbol $V(SOCl_2)$ refers to the valence of sulfur in $SOCl_2$. Thus, the valence of sulfur in $SOCl_2$ or SO_2Cl_2 does not depend on the Lewis structures we choose to draw, that is, whether we use double bonds or not.

Your inorganic textbook, in all likelihood, adopts the traditional structural formulas, involving a greater number of multiple bonds. As mentioned, for the purpose of arrow pushing, this should not make a difference, at least not in the vast majority of cases. Both are equivalent methods for electron bookkeeping. A valid criticism of the new style is that it results in unusually high FCs on certain atoms. But then FCs and OSs rarely provide a realistic description of the electrostatic character of atoms. Also, note that the unusually high charges are always balanced by opposite charges on adjacent atoms. On balance, therefore, we have decided to adopt the new style in this book.

As a final example of a hypervalent species, let us consider SiF_6^{2-}, whose Lewis structure is shown in Figure 1.6. As shown, silicon uses all four of its valence electrons in bonding in SiF_6^{2-} and is therefore tetravalent. Indeed, the species may be represented by a "no-bond resonance form" consisting of SiF_4 and two closed-shell F^- anions; the latter do not contribute to the valence of the silicon. This is a general point: coordination of closed-shell Lewis bases has no effect on the valence of an atom. For SiF_6^{2-}, we can also determine the valence of silicon more mechanically:

$$V(\text{Si in } SiF_6^{2-}) = \text{number of bonds} + \text{formal charge} = 6 + (-2) = 4$$

Thus, although the silicon in SiF_6^{2-} has a perfectly normal valence of 4, the fact that it has 12 valence electrons around itself in the Lewis structure of SiF_6^{2-} makes it hypervalent. The term "hypervalent" thus does *not* indicate an unusually high valence.

1.26 THE λ CONVENTION

Given that multiple valence states are a common feature of p-block elements, it's useful to familiarize oneself with the λ convention, which is useful for describing nonstandard valence states of p-block elements. The superscript in this notation equals the valence of the atom in question, as defined in Table 1.7. Thus, tri- and penta-valent phosphorus compounds are called λ^3- and λ^5-*phosphoranes*, respectively; tetra- and hexa-valent sulfur compounds are called λ^4- and λ^6-*sulfuranes*, and so on. The system is particularly useful for organoiodine compounds, where the iodine can exhibit a variety of different valence states. Thus, the hypothetical parent compounds H_3I, H_5I, and H_7I would be known as λ^3-, λ^5-, and λ^7-*iodane*, respectively. To take a couple of slightly more complicated examples, a phosphonium ylide such as $Ph_3P^+CH_2^-$ is called λ^5-*(methylene)triphenylphosphorane* in this system; similarly, the sulfur ylide $(CH_3)_2S^+CH_2^-$ is called λ^4-*(methylene)dimethylsulfurane*. We will use this nomenclature from time to time as we progress through the p-block elements.

1.27   THE INERT PAIR EFFECT

The inert pair effect is another important concept to be aware of, especially in connection with the heavier main-group elements. It refers to the increasing tendency of the outermost s electrons to remain un-ionized or unshared for the heaviest p-block elements. The origin of the effect is rather complex, involving differences in nuclear screening and relativistic stabilization, as experienced by the outermost s versus p electrons, and is not greatly relevant to our discussion. The chemical consequences of the inert pair effect, on the other hand, are striking and particularly important for groups 13 (In, Tl), 14 (Sn, Pb), and 15 (Sb, Bi); the following examples should illustrate the point.

While the chemistry of aluminum is largely that of the trivalent state, the monovalent state becomes increasingly stable as one goes down group 13; for thallium, the monovalent state is the common state. Trivalent Tl salts are reactive and used as oxidants in organic chemistry, as shown below for the thallation of benzene:

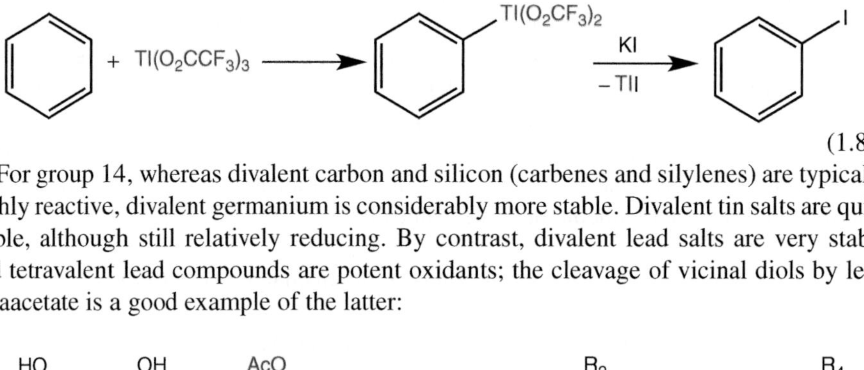

$$(1.82)$$

For group 14, whereas divalent carbon and silicon (carbenes and silylenes) are typically highly reactive, divalent germanium is considerably more stable. Divalent tin salts are quite stable, although still relatively reducing. By contrast, divalent lead salts are very stable and tetravalent lead compounds are potent oxidants; the cleavage of vicinal diols by lead tetraacetate is a good example of the latter:

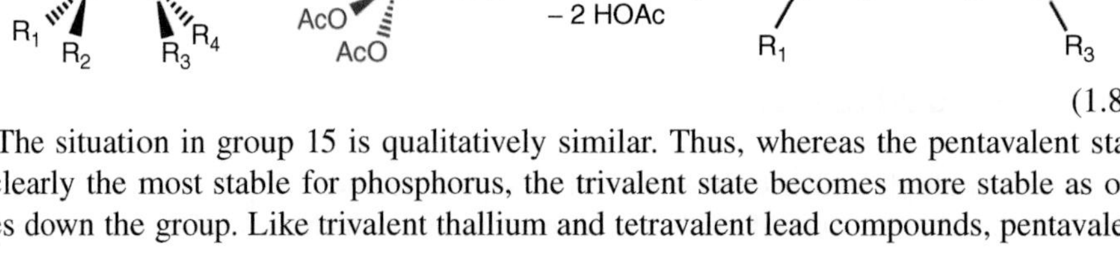

$$(1.83)$$

The situation in group 15 is qualitatively similar. Thus, whereas the pentavalent state is clearly the most stable for phosphorus, the trivalent state becomes more stable as one goes down the group. Like trivalent thallium and tetravalent lead compounds, pentavalent bismuth compounds (such as triphenylbismuth carbonate) are useful oxidants in organic chemistry, as shown in the example below:

$$(1.84)$$

As we explore p-block chemistry, we will encounter a wide variety of hypervalent molecules used as reagents in organic synthesis; in quite a few of these, the inert pair effect is believed to play a role.

1.28 SUMMARY

We have covered a fair amount of ground in this chapter, attempting to provide a summary of the background knowledge you'll need as you explore the chemistry of the main-group elements over the next eight chapters. Some of the highlights of our discussion are the following:

1. The S_N2 paradigm provided a starting point for our discussion, from which the following key points are worth remembering:
   - Soft bases tend to make better nucleophiles.
   - The HSAB principle provides a good description of many nucleophile–electrophile interactions: soft–soft and hard–hard interactions are both thermodynamically and kinetically favored.
   - Leaving group ability correlates with the pK_a of the conjugate acid of the group; that is, weaker bases are better leaving groups.
   - Bonds between two electronegative atoms break easily.
2. We then moved on to a wide-ranging survey of different reaction types, with some emphasis on reactions you have previously encountered in organic chemistry. Fortunately, only a relatively small number of fundamental reaction types are truly important for us in this book.
   - As in organic chemistry, the S_N2 reaction is important for other period 2 reaction centers.
   - Simple association (A) and dissociation (D) reactions are important reaction pathways in p-block chemistry.
   - For period 3 and below, nucleophilic displacement often involves a sequential A–D process, that is, addition of the incoming nucleophile, followed by departure of the leaving group in a second step. The overall process is often referred to as S_N2-Si.
   - PTs are important in protic media.

- Oxidative additions and reductive eliminations are much more important in inorganic chemistry than in organic chemistry. Many higher-valent period-6 compounds are prone to reductive elimination because of the inert pair effect; several such compounds are employed as oxidants in organic synthesis.
- Ligand exchange or metathesis reactions are common in inorganic chemistry. They are generally dictated by the HSAB principle. Mechanistically, they generally involve bridged intermediates.

3. Many A reactions and oxidative additions lead to so-called hypervalent products. S_N2-Si intermediates are also generally hypervalent; that is, they contain main-group centers with more than eight valence electrons around them in a Lewis structure. As we saw from a case study (PF_5), the bonding in such systems can be readily understood in terms of simple molecular orbital considerations. No special considerations are involved in applying arrow pushing to hypervalent compounds.

4. A two-pronged approach is suggested for coming up with reasonable mechanistic proposals:

- Pattern recognition goes a long way toward identifying a reasonable mechanistic pathway: simply look at the reactant and product structures and determine what bonds have been broken and what new bonds have formed.
- Draw on your chemistry knowledge (nucleophilicity, pK_a values, HSAB, bond energies, etc.) to identify plausible nucleophiles and electrophiles.

Apply the two strategies iteratively and, with a bit of practice, nice, well-crafted mechanisms should emerge.

FURTHER READING

Much of the material presented in this chapter is well covered in the organic and inorganic texts listed in Appendices 1 and 2. Among the references listed below, the first four are excellent resources for arrow pushing in organic chemistry.

1. Weeks, D. P. *Pushing Electrons: A Guide for Students of Organic Chemistry,* 4th ed.; Brooks/Cole: Belmont, CA, 2013; 224 pp. *This is arguably the simplest and gentlest book on the subject. It's a good place to start if you feel if you need some extra help.*
2. Levy, D. E. *Arrow Pushing in Organic Chemistry: An Easy Approach to Understanding Reaction Mechanisms*; John Wiley & Sons, Inc.: Hoboken, NJ, 2008; 320 pp. *A companion text not unlike this one but meant for organic chemistry. The choice of subject matter is more advanced in this book, since we have assumed previous exposure to organic chemistry.*
3. Grossman, R. B. *The Art of Writing Reasonable Organic Reaction Mechanisms,* 2nd ed.; Springer: New York, 2010; 360 pp. *Another excellent companion to a standard organic text.*
4. Scudder, P. H. *Electron Flow in Organic Chemistry: A Decision-Based Guide to Organic Mechanisms*; John Wiley & Sons, Inc.: Hoboken, NJ, 2013; 448 pp. *This book presents a highly algorithmic approach to organic arrow pushing. Applying such an approach to the more diverse world of inorganic mechanisms, however, seems unwieldy from a teaching perspective.*

5. Parkin, G. "Valence, Oxidation Number, and Formal Charge: Three Related but Fundamentally Different Concepts," *J. Chem. Educ.* **2006**, *83*, 791–799. *Don't be sloppy and misuse these terms, as so many people do! Read the relevant sections in this chapter with care and, for a little more, read this little gem of a paper!*

6. Akiba, K.-y. *Organo Main Group Chemistry*; John Wiley & Sons, Inc.: Hoboken, NJ, **2011**; 288 pp. *A treasure trove of main-group chemistry applied to organic synthesis, presented by a leading practitioner.*

2

The s-Block Elements: Alkali and Alkaline Earth Metals

Sir Humphry Davy
Abominated gravy.
He lived in the odium
Of having discovered sodium.

E. Clerihew Bentley, in *Biography for Beginners* (1905)

The s-block metals, admittedly, are not the most fertile ground for arrow pushing because their overwhelming tendency is to lose their outermost s electrons and generate the corresponding cations: +1 for alkali metals and +2 for alkaline earth metals. The elements never occur in their free state in nature, and even their discovery in the early 1800s by the English chemist and inventor Sir Humphrey Davy relied on the then brand-new technique of electrolysis. Yet we wouldn't dream of glossing over them in this book: they are to chemistry what water and salt—the best known s-block compounds—are to life, if we may wax poetic for a moment. *Water and Salt*, incidentally, is also an Italian fairy tale where a princess tells her father, "Oh father, I love thee as meat loves salt," which indirectly inspired Shakespeare's *King Lear*. Life processes depend on Na^+, K^+, Mg^{2+}, and Ca^{2+} ions in countless ways. An important example is the adenosine triphosphate (ATP) powered Na–K pump, which sits astride cell membranes (as schematically depicted in Figure 2.1) and ensures a low intracellular concentration of Na^+ and a high concentration of K^+. Add to that lithium batteries, hydrogen-powered cars, and the prospect of a hydrogen economy, and there can be little doubt about the all-round importance of the s-block elements.

Arrow Pushing in Inorganic Chemistry: A Logical Approach to the Chemistry of the Main-Group Elements,
First Edition. Abhik Ghosh and Steffen Berg.
© 2014 John Wiley & Sons, Inc. Published 2014 by John Wiley & Sons, Inc.

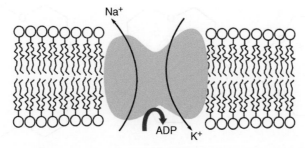

Figure 2.1 *Schematic diagram of the Na–K pump sitting astride a lipid bilayer. (This graphic has been obtained from the public domain source Wikimedia Commons.)*

It would be wrong, however, to give the impression that the s-block elements do not have much of a covalent chemistry at all. Thus:

1. The chemistry of hydrogen is largely covalent. Low temperature activation of dihydrogen (H_2), traditionally the preserve of transition-metal catalysts, has recently been accomplished with main-group element reagents and catalysts.

2. The smaller, least electronegative s-block atoms Li, Be, and Mg also form many covalent compounds, most notably with carbon. (Have a look at Figure 1.4 to get a sense of the Pauling electronegativities of these elements.) Organolithium and organo-magnesium compounds are cornerstones of organic synthesis, where they form the basis of some of the most reliable methods of forming carbon–carbon bonds.

3. By donating single electrons, the s-block metals promote a variety of radical reactions, including a number of reductive coupling reactions.

With those introductory remarks, we are ready to explore some of the details of s-block chemistry.

2.1 SOLUBILITY

One of the key roles of s-block metals in chemistry is to act as counterions for anionic nucleophiles. Some basic ideas about the solubility of s-block metal salts should therefore allow more informed discussions as we progress through the book.

As ionic substances, most alkali metal salts and many alkaline earth metal salts dissolve in water. Certain di- and tri-valent anions such as carbonate, sulfate, and phosphate form insoluble salts with alkaline earth metals; examples of such insoluble salts include $CaCO_3$, $BaSO_4$, and $Ca_3(PO_4)_2$.

For salts of strong acids, the aqueous solutions are typically neutral, for example, NaCl, $Ba(NO_3)_2$, and so on. The analogous salts of Be and Mg, however, give rise to acidic solutions. The small sizes of these cations and the resulting high surface charge densities lead to ionization of a coordinated water, as shown below:

$$[Be(H_2O)_4]^{2+} + H_2O \rightleftharpoons [Be(H_2O)_3(OH)]^+ + H_3O^+ \tag{2.1}$$

$$[Mg(H_2O)_6]^{2+} + H_2O \rightleftharpoons [Mg(H_2O)_5(OH)]^+ + H_3O^+ \tag{2.2}$$

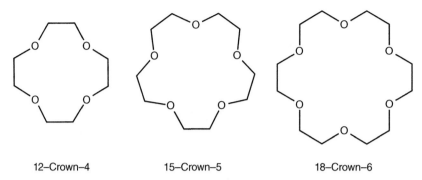

12–Crown–4   15–Crown–5   18–Crown–6

Figure 2.2 *Examples of crown ethers.*

As expected for ionic compounds, alkali and alkaline earth metal salts generally do not dissolve in highly nonpolar solvents that are commonly used in organic synthesis. Polar cation-solvating solvents such as alcohols, tetrahydrofuran (THF), dimethoxyethane (DME), dimethylformamide (DMF), and dimethyl sulfoxide (DMSO), however, often do a reasonable job of dissolving s-block metal salts. Should a highly nonpolar solvent prove essential, macrocyclic ligands such as crown ethers and cryptands may be employed, which solubilize alkali metal salts by encircling the cations. The anions of the salts then act as highly reactive "bare" nucleophiles, which are of considerable synthetic use. Depending on the size of their cavities, the crown ethers exhibit significant selectivity for different alkali metal cations. Thus, 12-crown-4 exhibits high affinity for Li^+, 15-crown-5 for Na^+, and 18-crown-6 for K^+ (Figure 2.2).

Certain salts are unexpectedly soluble in organic solvents. Thus, lithium perchlorate is highly soluble in ether, where it's widely used as a Lewis catalyst for Diels–Alder reactions (see Section 1.21). The small size of the lithium ion should lead to a degree of covalent character in the Li^+–ClO_4^- interaction, which may enhance solubility in organic solvents. Potassium carbonate and cesium carbonate are widely employed as bases in organic chemistry. The latter, though more expensive, is much more soluble and is often preferred. The solubility of Cs_2CO_3 probably reflects a size mismatch between the large Cs^+ cations and the carbonate anion, which may lead to a less stable crystal packing.

2.2 THE s-BLOCK METALS AS REDUCING AGENTS

The s-block metals are among the strongest reductants known and reduce most materials capable of reduction. The reaction with water is shown below for a generic alkali metal (M):

$$M^\bullet \quad \overset{\curvearrowright}{\underset{H}{\diagdown}} \overset{O}{\diagup} \underset{H}{\diagdown} \quad \overset{\ominus}{-OH} \quad \longrightarrow \quad \overset{\oplus}{M} + H^\bullet \tag{2.3}$$

Note carefully the nature of the two arrows, a "fishhook" to indicate movement of a single electron and a normal double-headed arrow to indicate the movement of an electron pair.

The hydrogen atom produced in the above reaction then combines with another one to form hydrogen gas:

$$\text{H} \cdot \quad \cdot \text{H} \longrightarrow \text{H—H} \tag{2.4}$$

The overall balanced equation is thus

$$2\,\text{M} + 2\,\text{H}_2\text{O} \rightarrow 2\,\text{M}^+\text{OH}^- + \text{H}_2 \tag{2.5}$$

An analogous reaction also happens with alcohols (ROH).

The metals reduce halogens and other nonmetals to yield binary salts, as illustrated below for chlorine:

$$2\,\text{M} + \text{Cl}_2 \rightarrow 2\,\text{MCl} \tag{2.6}$$

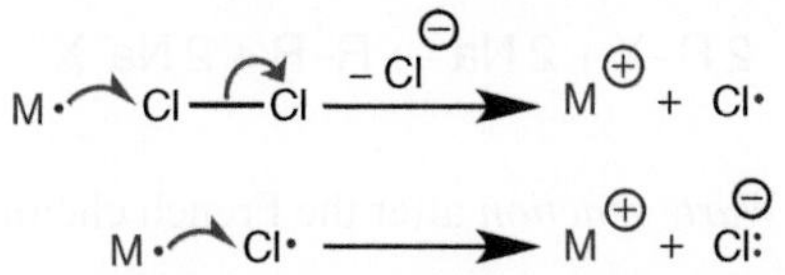

$$\text{M} \cdot \quad \text{Cl—Cl} \xrightarrow{-\overset{\ominus}{\text{Cl}}} \text{M}^{\oplus} + \text{Cl} \cdot \tag{2.7}$$

$$\text{M} \cdot \quad \text{Cl} \cdot \longrightarrow \text{M}^{\oplus} + \overset{\ominus}{\text{Cl}}{:}$$

Under suitable conditions, the metals even reduce dihydrogen. Sodium hydride (NaH), for example, is produced by the interaction of hydrogen gas and liquid sodium:

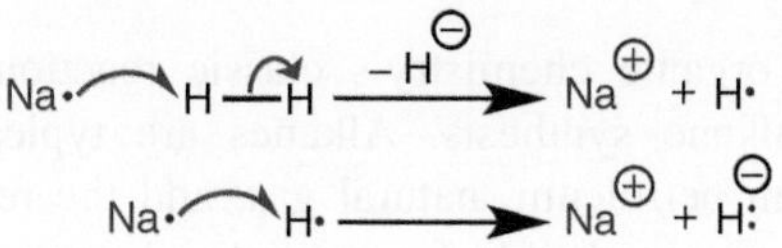

$$\text{Na} \cdot \quad \text{H—H} \xrightarrow{-\overset{\ominus}{\text{H}}} \text{Na}^{\oplus} + \text{H} \cdot \tag{2.8}$$

$$\text{Na} \cdot \quad \text{H} \cdot \longrightarrow \text{Na}^{\oplus} + \overset{\ominus}{\text{H}}{:}$$

Sodium hydride is widely used as a strong base in organic chemistry, whereas calcium hydride is used as a drying agent for solvents.

REVIEW PROBLEM 2.1

Lithium metal is unique in terms of its ability to activate N_2 at room temperature and ordinary pressure:

$$3\,\text{Li} + \text{N}_2 \rightarrow 2\,\text{Li}_3\text{N}$$

Draw a mechanism using fishhooks and regular arrows. Can you rationalize the unique reactivity of Li, relative to the heavier alkali metals, which do not react with N_2 under ordinary conditions?

2.3 REDUCTIVE COUPLINGS

The strong tendency of s-block metals to give up their valence electrons is the basis of a variety of useful coupling reactions in organic chemistry. To moderate the potentially violent nature of the reactions, the metals are often used in amalgam form (i.e., as solid solutions in mercury). With alkyl halides, alkyl radicals are generated:

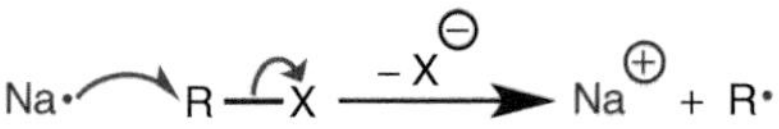

$$\text{Na·} \frown \text{R} \frown \text{X} \xrightarrow{-\text{X}^{\ominus}} \text{Na}^{\oplus} + \text{R·} \tag{2.9}$$

The radicals then dimerize to form alkanes:

$$\text{R·} \frown \text{·R} \longrightarrow \text{R} \text{—R} \tag{2.10}$$

The overall reaction then is

$$2\,\text{R–X} + 2\,\text{Na} \rightarrow \text{R–R} + 2\,\text{Na}^+\text{X}^- \tag{2.11}$$

This reaction is called the *Wurtz reaction* after the French chemist Charles-Adolphe Wurtz (1817–1884).

REVIEW PROBLEM 2.2

Although it's one of organic chemistry's classic reactions, the Wurtz reaction is seldom used for alkane synthesis. Alkanes are typically much more conveniently obtained from petroleum, natural gas, and the reduction of fatty acids. The Wurtz reaction is most valuable for specialized applications, particularly for closing cyclopropane rings, as in the synthesis of bicyclo[1.1.0]butane, shown below:

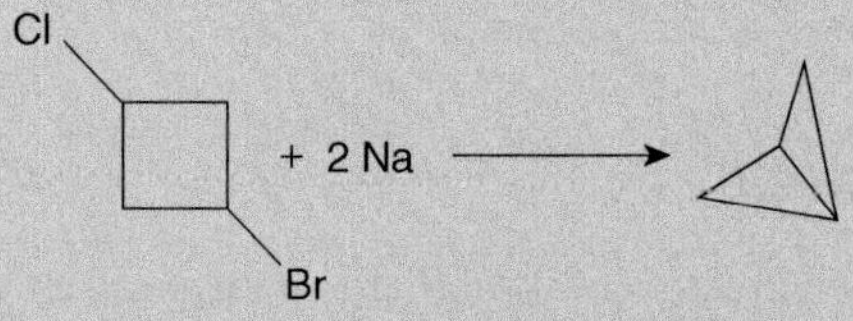

Draw a mechanism for this reaction.

A much more widely used reductive coupling is the pinacol coupling of carbonyl compounds, depicted below for acetone:

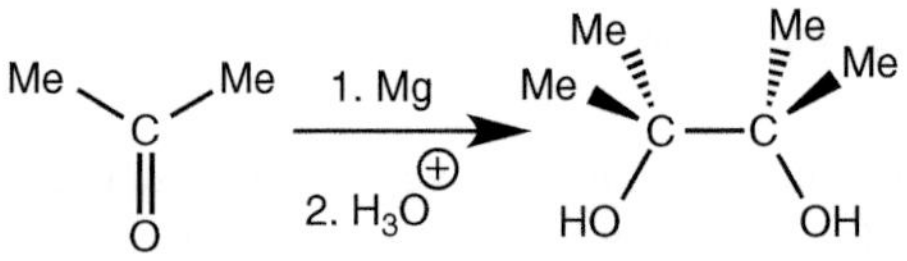

$$\begin{array}{c} \text{Me} \diagdown \underset{\underset{\text{O}}{\overset{\|}{\text{C}}}}{} \diagup \text{Me} \quad \xrightarrow[\text{2. H}_3\text{O}^{\oplus}]{\text{1. Mg}} \quad \overset{\text{Me}}{\underset{\text{HO}}{\text{Me}\diagdown \text{C}}} \text{—} \overset{\text{Me}}{\underset{\text{OH}}{\text{C} \diagup \text{Me}}} \end{array} \tag{2.12}$$

The metal presumably donates a single electron to the ketone to form a ketyl radical:

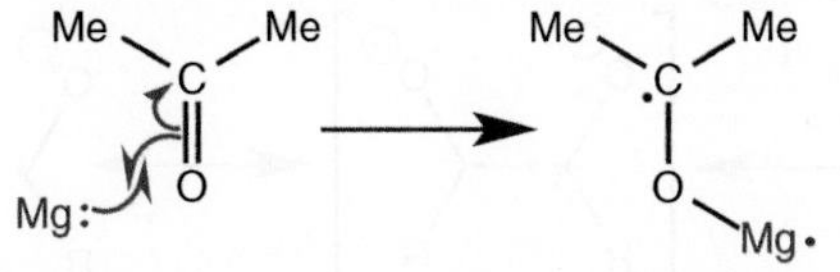

$$(2.13)$$

Two ketyl radicals then couple to form a magnesium dialkoxide:

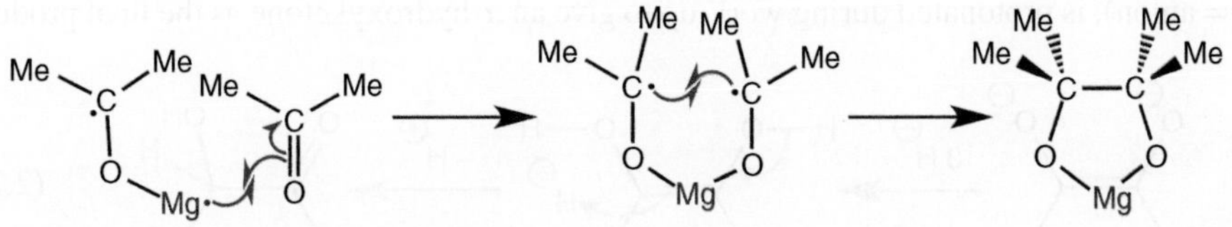

$$(2.14)$$

On aqueous work-up, the dialkoxide forms the expected 1,2-diol; the specific one shown below is called *pinacol*, after which this reaction is named:

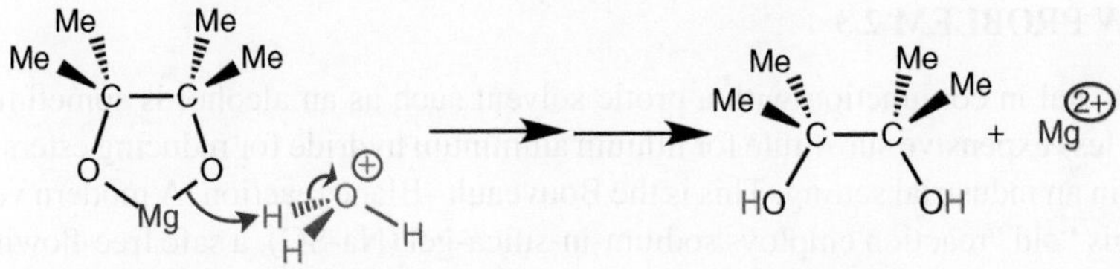

$$(2.15)$$

Yet a third example (there are many!) of reductive coupling is the acyloin condensation, which involves the coupling of two esters:

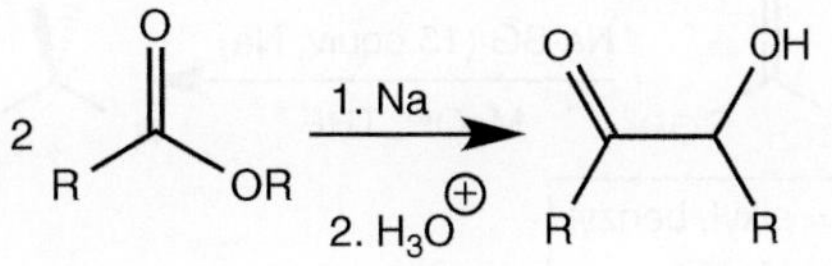

$$(2.16)$$

The first step is similar to that of pinacol coupling:

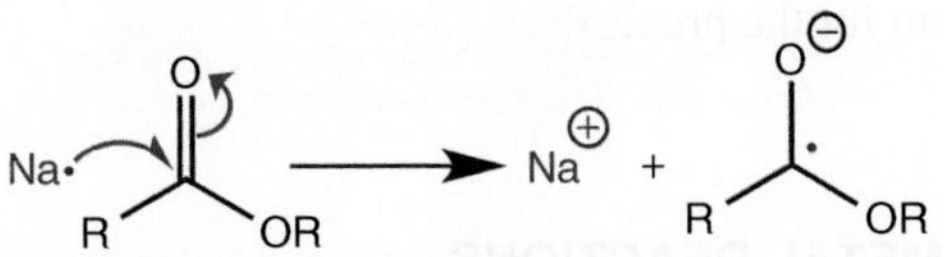

$$(2.17)$$

The radicals then couple, and the coupling product eliminates alkoxide ions (RO⁻) to yield a diketone, as shown below:

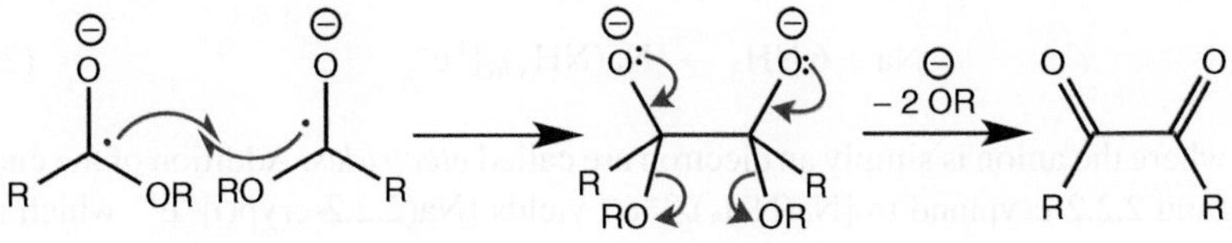

$$(2.18)$$

The diketone undergoes further reduction:

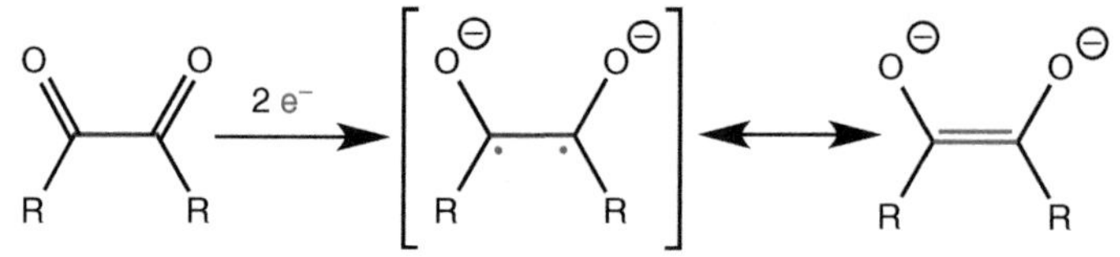

$$(2.19)$$

The resulting dianion, known as an *enediolate* (ene = double bond, diol = double alcohol, ate = anion), is protonated during work-up to give an α-hydroxyketone as the final product:

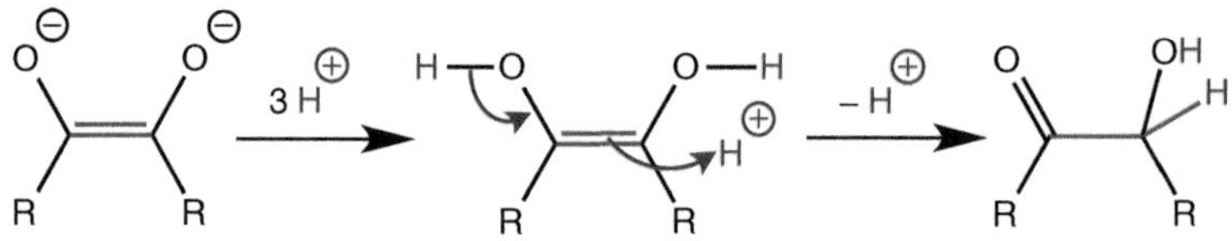

$$(2.20)$$

REVIEW PROBLEM 2.3

Sodium metal in conjunction with a protic solvent such as an alcohol is sometimes used as a less expensive substitute for lithium aluminum hydride for reducing esters to alcohols in an industrial setting. This is the Bouveault–Blanc reaction. A modern version of this "old" reaction employs sodium-in-silica-gel (Na-SG), a safe free-flowing powder, instead of the bulk metal (B. S. Bodnar, P. F. Vogt, *J. Org. Chem.* **2009**, *74*, 2598–2600):

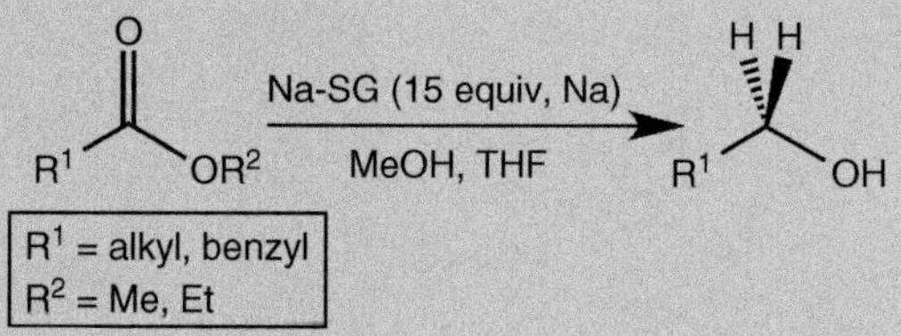

Suggest a mechanism for the process.

2.4 DISSOLVING METAL REACTIONS

Alkali metals dissolve in liquid ammonia to yield solvated cations and solvated electrons, as shown below for sodium:

$$Na + 6\,NH_3 \rightarrow [Na(NH_3)_6]^+e^- \qquad (2.21)$$

Such salts where the anion is simply an electron are called *electrides*. Addition of the encapsulating ligand 2,2,2-cryptand to $[Na(NH_3)_6]^+e^-$ yields $[Na(2,2,2\text{-crypt})]^+e^-$, which may be isolated as a blue-black paramagnetic salt after the evaporation of the ammonia:

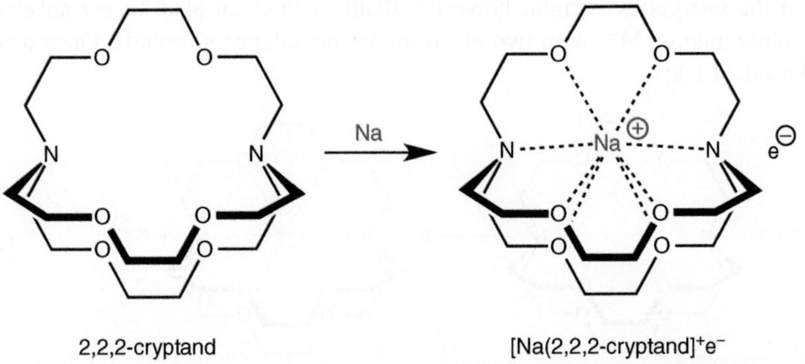

2,2,2-cryptand   [Na(2,2,2-cryptand]⁺e⁻

(2.22)

Electrides are synthetically useful and form the basis of so-called dissolving metal reactions, of which the Birch reduction of aromatic compounds is the paradigm, shown below for benzene and naphthalene:

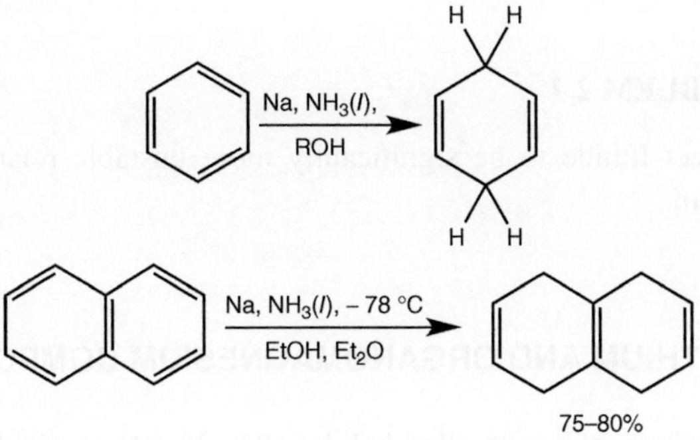

(2.23)

The mechanism, shown only for benzene, involves a succession of electron and proton additions:

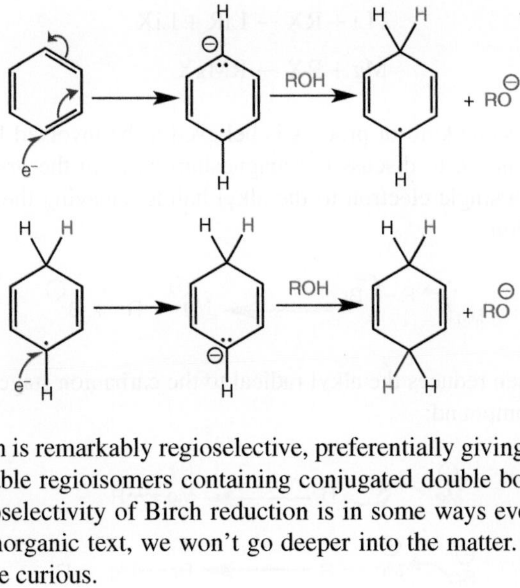

(2.24)

The Birch reduction is remarkably regioselective, preferentially giving the products shown instead of more stable regioisomers containing conjugated double bonds. For substituted benzenes, the regioselectivity of Birch reduction is in some ways even more fascinating, but this being an inorganic text, we won't go deeper into the matter. Do read up a bit on your own, if you are curious.

We have emphasized the s-block metals' overriding tendency to lose their valence electrons. When the energetics is right, however, alkali metals can also *accept* an electron to yield alkalide anions, M^-, with two electrons in the valence s orbitals. Once again, a cryptand does the trick:

$$2\,\text{Na} \; + \qquad\qquad\qquad\qquad \longrightarrow \qquad\qquad\qquad\qquad + \; \text{Na}^{\ominus} \qquad (2.25)$$

Although we won't comment on the point, we urge you to pause a moment and appreciate the remarkable role of the cryptand. Sodide or natride (Na^-), potasside or kalide (K^-), rubidide (Rb^-), and ceside (Cs^-) are all known species; lithide (Li^-), however, is not.

REVIEW PROBLEM 2.4

Would you expect lithide to be significantly more unstable relative to the other alkalides? Explain.

2.5 ORGANOLITHIUM AND ORGANOMAGNESIUM COMPOUNDS

Lithium and magnesium react with alkyl halides (RX, $X \neq F$) to yield alkyllithium (RLi) and alkylmagnesium (RMgX) compounds, respectively:

$$2\,\text{Li} + \text{RX} \rightarrow \text{LiR} + \text{LiX} \qquad (2.26)$$

$$\text{Mg} + \text{RX} \rightarrow \text{RMgX} \qquad (2.27)$$

Mechanistically, the same kind of process is believed to be involved for both metals; but here we arbitrarily choose to discuss the magnesium case. In the first step, the metal is believed to transfer a single electron to the alkyl halide, cleaving the latter into an alkyl radical and a halide ion:

$$\qquad (2.28)$$

A second electron then reduces the alkyl radical to the carbanion stage, that is, effectively an organometallic compound:

$$\qquad (2.29)$$

There is good evidence for the existence of radical intermediates in this process. Treatment of the Mg/RX reaction mixture with the stable organic radical 2,2,6,6-*te*tramethyl*p*iperidinyl-1-*o*xy (TEMPO, see structure below) results in high yields of a TEMPO-R adduct, strongly indicating the intermediacy of alkyl radicals:

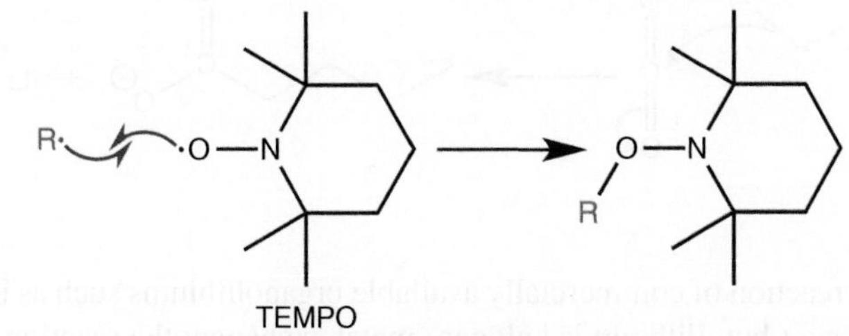

(2.30)

TEMPO

A reagent such as TEMPO that specifically reacts with radicals is called a *radical trap*; such compounds are clearly very useful as mechanistic probes.

The organomagnesium compounds described above are known as *Grignard reagents*, after the French chemist François Auguste Victor Grignard (1871–1935; Nobel Prize winner in chemistry in 1912), who correctly identified them as organometallic compounds and recognized their importance to organic chemistry. Although the Grignard reagents are commonly written as RMgX, there is strong evidence for the following equilibrium in ether, the solvent in which Grignard reagents are typically generated:

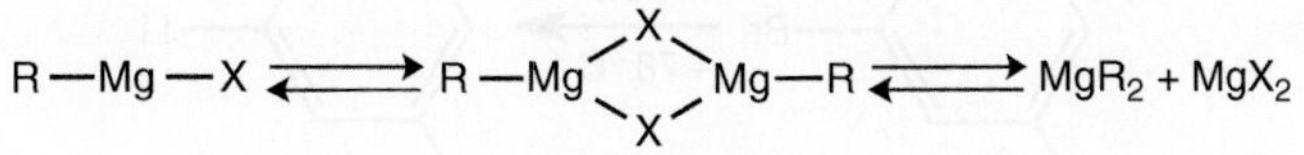

(2.31)

A variety of oligomeric structures are also known for organolithiums.

Organolithium and organomagnesium reagents represent a reasonable approximation to carbanions; indeed, they are frequently referred to as such. As highly basic compounds, they are moisture sensitive (as well as air-sensitive) and need to be handled accordingly. The hydrolysis of *n*-butyllithium may be depicted as follows:

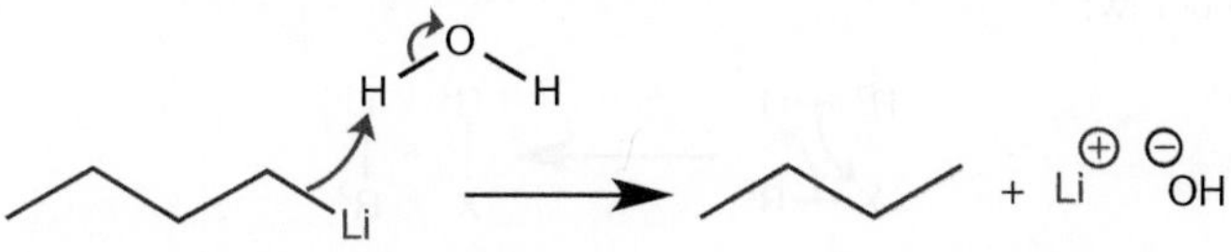

(2.32)

n-Butyllithium and other organolithiums are most commonly employed as strong *bases*, for example, to deprotonate carbonyl compounds and thereby generate enolate anions, as mentioned in Section 1.15. Grignard reagents, on the other hand, are most often employed as *nucleophiles* toward carbonyl compounds, resulting in, after aqueous work-up (not shown explicitly below), alcohols with carbon frameworks larger than those of the starting materials:

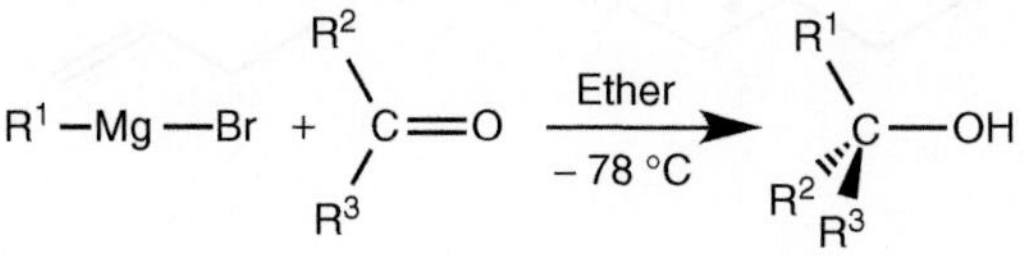

(2.33)

Both organolithiums and Grignard reagents react with carbon dioxide to yield carboxylic acids, as illustrated below for *n*-butyllithium:

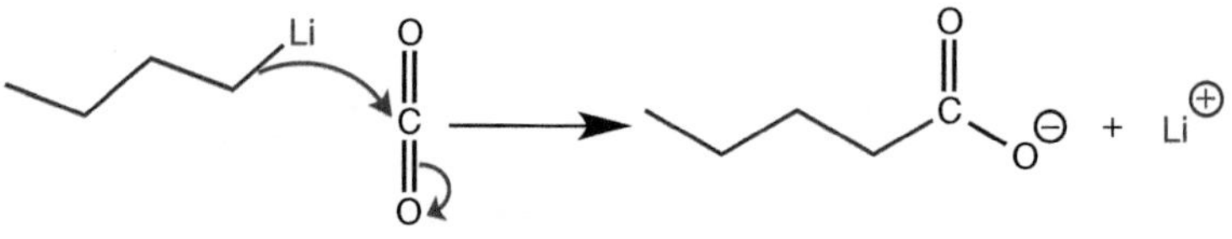

$$(2.34)$$

A very useful reaction of commercially available organolithiums such as butyllithium or the even more basic *t*-butyllithium is halogen–metal exchange; the reaction is particularly suitable for alkenyl or aryl halides:

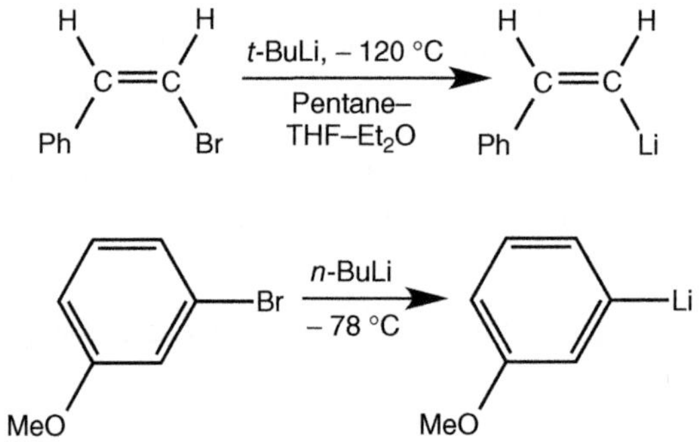

$$(2.35)$$

The reaction is fast and has to be conducted at low temperature. Overall, the reaction proceeds in the direction that converts a more strongly basic organolithium to one that is less basic. Naively, a simple four-center cyclic pathway would account for the observed products, as shown below:

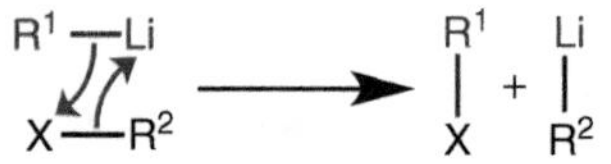

$$(2.36)$$

What happens in reality may well involve more complex assemblages of the reactants.

Finally, alkyllithiums are thermally unstable and undergo hydride elimination as shown below:

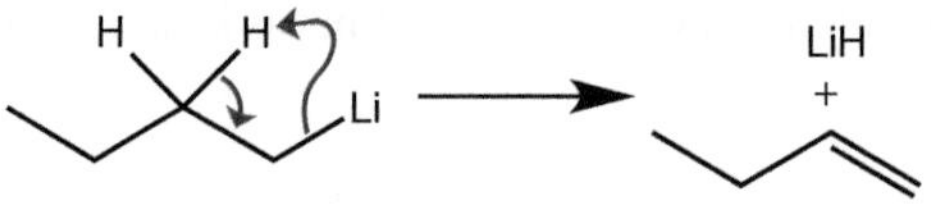

$$(2.37)$$

REVIEW PROBLEM 2.5

n-Butyllithium is widely used as a polymerization initiator in the synthesis of poly-butadiene or styrene–butadiene–styrene (SBS), both of which are used in the production of tires, hoses, shoe heels and soles, golf balls, and even chewing gum. The first step may be depicted as follows:

$$C_4H_9Li + CH_2=CH-CH=CH_2 \rightarrow C_4H_9-CH_2-CH=CH-CH_2Li$$

Using arrow pushing, show how the polymers might now arise.

2.6 DIHYDROGEN ACTIVATION BY FRUSTRATED LEWIS PAIRS (FLPs)

Recall from Section 1.9 that the Lewis acid and base in a frustrated pair are too bulky or otherwise sterically constrained to bond with each other but the cleft between them acts like an enzyme active site and promotes remarkable chemical reactions, perhaps none more so than the activation of dihydrogen at or below room temperature. Two representative reactions mediated by P,B-based FLPs are as follows:

(2.38)

Observe that each of these reactions accomplishes a net heterolytic cleavage of an H_2 molecule, an incredibly weak acid ($pK_a = 42$), with the P picking up a proton and the B a hydride. How might this occur? Neither a naked proton nor a naked hydride anion is plausible as an intermediate. A reaction path in which both the P and the B act in concert to break the H–H bond heterolytically is envisioned below, even though explicit experimental evidence for such a pathway is still lacking:

(2.39)

Even more excitingly, the above zwitterion catalyzes the hydrogenation of polar carbon–carbon double bonds, traditionally a bastion of transition-metal catalysts. The following example involves an enamine, an uncharged nitrogen equivalent of an enolate anion (RT = room temperature):

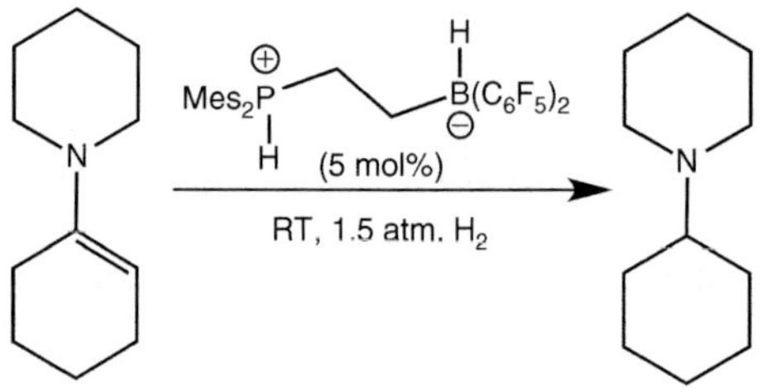

(2.40)

The mechanism is likely to be straightforward. The phosphonium part of the catalyst should protonate the enamine; the hydroborate part of the catalyst should then transfer a hydride to the cationic intermediate produced, thereby generating the "FLP form" of the catalyst.

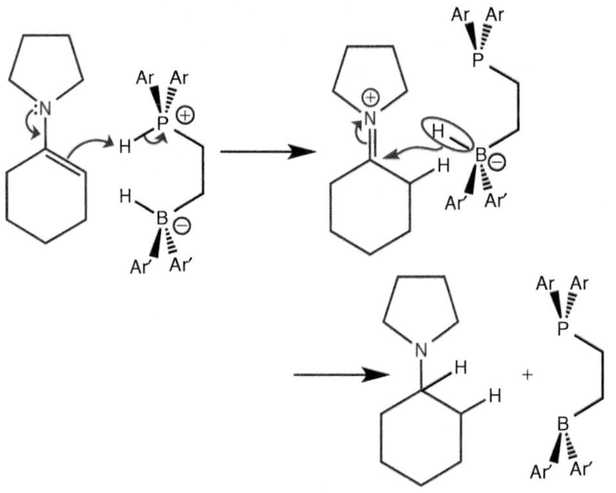

(2.41)

Since the reaction is carried out under dihydrogen, the zwitterionic dihydro form of the catalyst is then regenerated.

Inspired by this remarkable FLP-mediated chemistry, other researchers have sought additional dihydrogen-activating main-group reagents. Stable diaminocarbenes (see Section 1.16), including the N-heterocyclic carbenes (NHCs), seemed promising but didn't quite work. A sterically hindered monoaminocarbene, however, proved equal to the task:

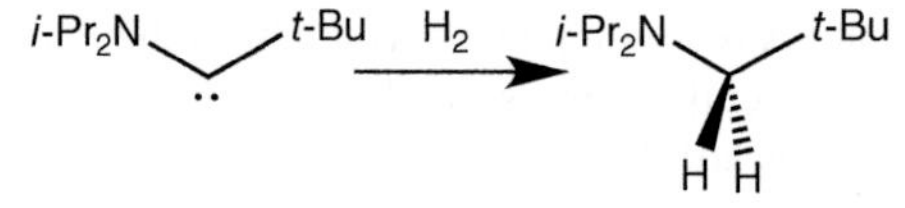

(2.42)

Unlike the overly nucleophilic diaminocarbenes, a monoaminocarbene seems to have the perfect balance of electrophilicity and nucleophilicity that seems to be required for cleaving H_2.

> **REVIEW PROBLEM 2.6**
>
> Little mechanistic information is available for the carbene-mediated dihydrogen activation shown above. In the absence of such information, a concerted oxidative addition seems as plausible as anything else. Sketch out such a pathway using conventional arrow pushing.

2.7   A Mg^I–Mg^I BOND

Recently, chemists in Australia have reductively coupled magnesium(II) complexes with bulky capping ligands to create an unprecedented Mg^I–Mg^I bond:

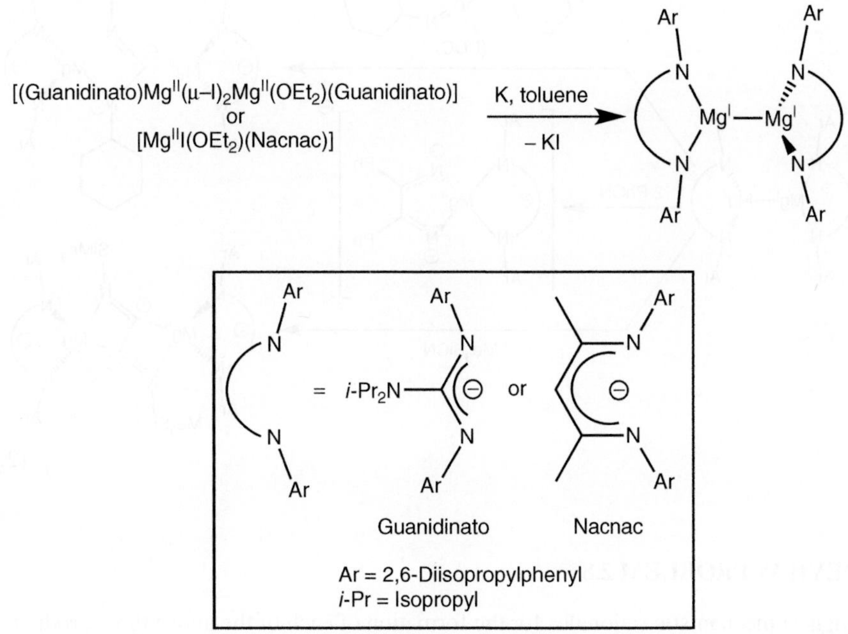

$$(2.43)$$

In case the bulky ligands are too distracting, here's the essence of what's going on:

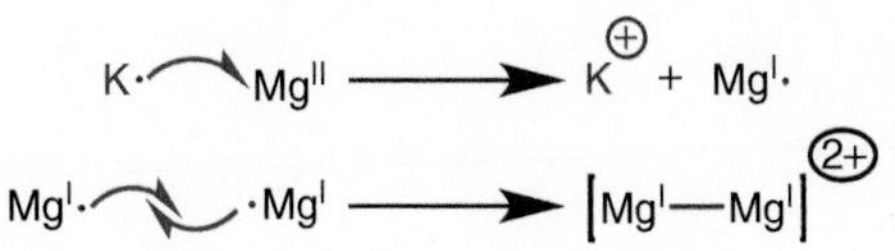

$$(2.44)$$

The remarkable feature of this synthesis is that, under normal circumstances, it's impossible to stop alkaline earth metals from losing their second valence electron. The bulky designer ligands allowed the researchers to accomplish just that. The air- and moisture-sensitive product, which is often referred to as *Jones's Mg^I–Mg^I reagent* (after Cameron Jones, the senior investigator), is remarkably thermally stable.

REVIEW PROBLEM 2.7

Comment on the magnesium oxidation state, valence, and coordination number in Jones's reagent.

Although only a few years old at the time this book is being published, the Jones reagent is already finding applications as a synthetic reagent, a few of which are shown below. Observe that with benzonitrile (PhCN) and trimethylsilyl cyanide, it accomplishes a reductive coupling, not unlike a pinacol coupling.

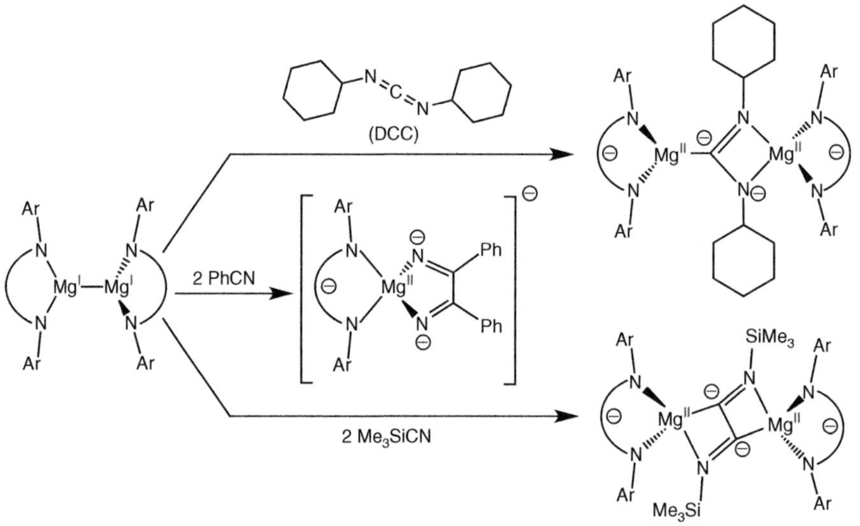

(2.45)

REVIEW PROBLEM 2.8*

Suggest mechanistic rationales for the formation of each of the above three products. What final products would you expect after aqueous work-up of the products shown? (*Note*: Don't be overly concerned that the reaction with Me_3SiCN has involved an $SiCN \rightarrow SiNC$ isomerization. For now, simply write Me_3SiCN as Me_3SiNC!)

2.8 SUMMARY

The following are some of the key points we discussed in this chapter.

1. While the chemistry of hydrogen is predominantly covalent, the chemistry of s-block metals is dominated by their overwhelming tendency to lose their valence electrons. The metals are thus best known as *counterions* in a variety of salts.

2. The metals act as strong reducing agents. The metals promote a variety of reductive processes, including the Wurtz, pinacol, and acyloin couplings, as well as the Birch reduction.

3. The metals exhibit an extensive covalent chemistry involving carbon, of which Grignard and organolithium reagents deserve particular mention.

4. The hydride anion is well known in salts. Much more exotic are the alkalide anions, which, with the exception of Li^-, have all been synthesized under specialized conditions.

5. Dihydrogen activation, traditionally the preserve of transition-metal catalysts, has recently been accomplished with p-block reagents, including a nucleophilic carbene and FLPs.

6. A unique metal–metal-bonded Mg(I)–Mg(I) molecule has recently been synthesized with the help of bulky capping ligands. The molecule, often referred to as the Jones's Mg(I)–Mg(I) reagent, has already been applied to synthetic problems.

FURTHER READING

Besides the general references listed in Appendices 1 and 2, the following provide additional information on the recent developments mentioned in this chapter.

1. Several review articles on frustrated Lewis pairs have been assembled in (a) Erker, G.; Stephan, D. W., eds. "Frustrated Lewis Pairs I: Uncovering and Understanding," *Top. Curr. Chem.* **2013**, *332*, 1–350. (b) Erker, G.; Stephan, D. W., eds. "Frustrated Lewis Pairs II: Expanding the scope," *Top. Curr. Chem.* **2013**, *334*, 1–317. *These two volumes provide comprehensive coverage of this remarkable class of molecules.*

2. (a) Green, S. P.; Stasch, A.; Jones, C. "Stable Magnesium(I) Compounds with Mg–Mg Bonds," *Science* **2007**, *318*, 1754–1757. (b) Ma, M. T.; Stasch, A.; Jones, C. "Magnesium(I) Dimers as Reagents for the Reductive Coupling of Isonitriles and Nitriles," *Chem. Eur. J.* **2012**, *18*, 10669–10676. *These are two key original papers on Jones's reagent.*

3

Group 13 Elements

Why did I decide to undertake my doctorate research in the exotic field of boron hydrides? As it happened, my girlfriend, Sarah Baylen, soon to become my wife, presented me with a graduation gift, Alfred Stock's book, The Hydrides of Boron and Silicon. *I read this book and became interested in the subject. How did it happen that she selected this particular book? This was the time of the Depression. None of us had much money. It appears she selected as her gift the most economical chemistry book ($2.06) available in the University of Chicago bookstore. Such are the developments that can shape a career.*

Herbert C. Brown.

In "*From Little Acorns Through to Tall Oaks: From Boranes Through Organoboranes*,"

Nobel Lecture, December 8, 1979.

With these elements, we begin our introduction to the p block, home of all nonmetals (and of a few metals) and of classic covalent chemistry. The elements are chemically rather diverse: boron is a moderately electropositive nonmetal, whereas aluminum, gallium, indium, and thallium are all metals. The covalent chemistry we will encounter in this chapter, therefore, will still be somewhat limited, compared to what is to follow in the next chapters. Some general remarks are in order:

- The elements are generally trivalent, although the monovalent state is also well established for all the elements except boron. For thallium, the monovalent state is the normal stable state, whereas for the lighter metals the monovalent state is reducing and accessible only with considerable care and difficulty.
- With only six electrons in their valence shells, the trivalent compounds are generally potent Lewis acids, forming a variety of adducts with bases (e.g., $H_3N \rightarrow BF_3$).

Arrow Pushing in Inorganic Chemistry: A Logical Approach to the Chemistry of the Main-Group Elements,
First Edition. Abhik Ghosh and Steffen Berg.
© 2014 John Wiley & Sons, Inc. Published 2014 by John Wiley & Sons, Inc.

- Like beryllium and magnesium halides, aluminum and gallium halides undergo partial hydrolysis in aqueous solutions, resulting in acidic solutions.
- To satisfy their hunger for electron pairs, many of the trivalent compounds exist as dimers such as diborane (B_2H_6) and Al_2Cl_6.

The bonding in diborane is rather special and may be understood in terms of the following simple molecular orbital (MO) picture. Each B–H–B unit is associated with a three-center doubly occupied MO, the bonding MO in the diagram below. This MO results from two boron sp^3 atomic orbitals, one from each B, and a hydrogen 1s orbital (Figure 3.1).

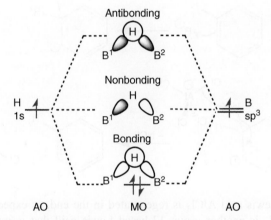

Figure 3.1 *Schematic MO energy level diagram for B–H–B three-center bonding.*

- Boron exhibits a great deal of unique covalent chemistry. We'll focus here on the hydroboration reaction and on further elaboration of the resulting organoboranes.
- With thallium, we will again encounter some unique behavior, resulting from the inert pair effect (see Section 1.27); we will see that thallium(III) salts are somewhat unstable and prone to reduction to Tl(I). This is a phenomenon we will come across repeatedly throughout p-block chemistry; it forms the basis of the use of higher-valent compounds of the heaviest p-block elements as oxidizing agents in organic chemistry.

3.1   GROUP 13 COMPOUNDS AS LEWIS ACIDS

Group 13 halides are widely used as Lewis acid catalysts in organic reactions, of which the Friedel–Crafts reaction is an excellent example. The reaction can bring about both alkylation and acylation of aromatic compounds, as shown below for benzene:

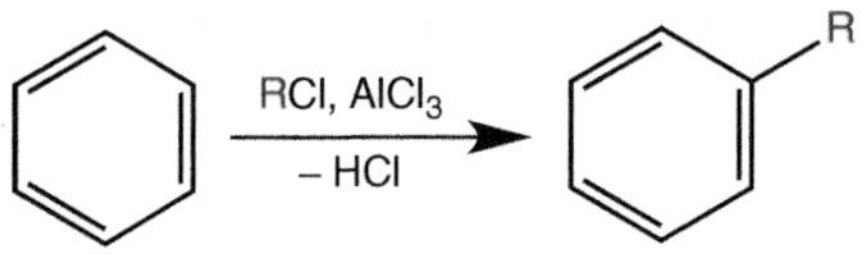

(3.1)

By coordinating to the Cl in RCl, the Lewis acid assists in the heterolysis of the R–Cl bond. If R^+ is a relatively stable carbocation, it may form as a discrete intermediate before the reaction with the aromatic ring, as shown below:

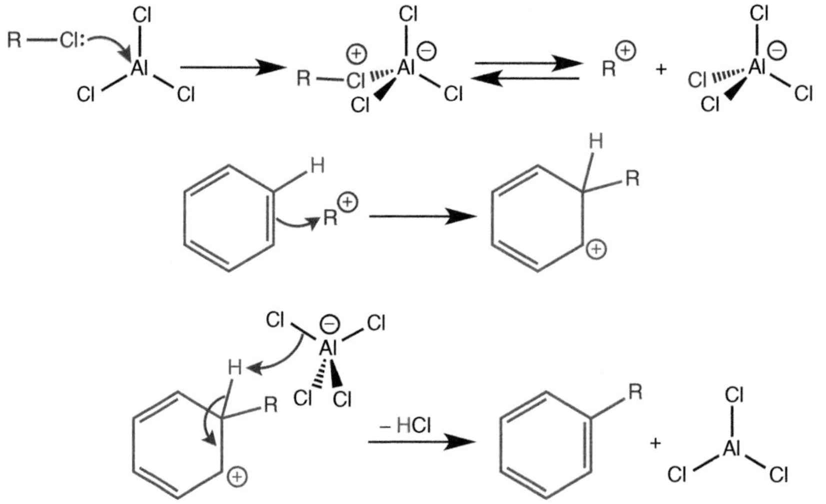

(3.2)

Observe that the Lewis acid $AlCl_3$ is regenerated in the end, as expected for a catalyst. Besides $AlCl_3$, BF_3 is another group 13-based Lewis acid that is widely employed in Friedel–Crafts reactions. A wide variety of other Lewis acids, based on p-, d-, and even f-block elements, may also be employed as catalysts in Friedel–Crafts reactions.

REVIEW PROBLEM 3.1

Aryl methyl ethers may be demethylated with boron tribromide as shown below:

$$\text{ArOMe} \xrightarrow[\text{2. } H_2O]{\text{1. } BBr_3,\ CH_2Cl_2} \text{ArOH}$$

Provide a mechanistic rationale for the transformation.

REVIEW PROBLEM 3.2

Aluminum halides are used to prepare CBr_4 or CI_4 via halide exchange from CCl_4, as shown below:

(a) $3\ CCl_4 + 4\ AlBr_3 \longrightarrow 3\ CBr_4 + 4\ AlCl_3$

(b) $CCl_4 + 4\ C_2H_5I \xrightarrow{\ AlCl_3\ } CI_4 + 4\ C_2H_5Cl$

Suggest mechanisms for the two transformations. *Note*: In (a), $AlBr_3$ is a stoichiometric reagent. In (b), $AlCl_3$ is a catalyst.

Boron Lewis acids may have played a role in the origin of life, specifically in the prebiotic formation of ribose, the sugar present in RNA. Sugars are thought to have arisen from one- and two-carbon molecules such as formaldehyde and glycolaldehyde via the so-called formose reaction, as depicted in Figure 3.2. Note that the individual chain elongation step is an aldol reaction. Thus, glycolaldehyde enolate (more accurately described as an enediolate) attacks formaldeyhde to yield glyceraldehyde. A second molecule of glycolaldehyde then condenses with glyceraldehyde to yield ribose and other pentoses (five-carbon sugars). Left to itself, however, the formose reaction quickly turns into brown tar, which is unlikely to have led to nucleosides or nucleotides in an orderly manner.

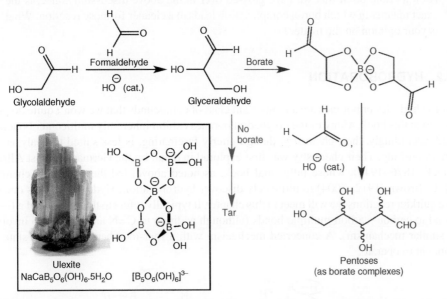

Figure 3.2 *Schematic representation of the formose reaction in the absence and presence of borate minerals. Inset: Note the three-coordinate Lewis-acidic borons in ulexite.*

Borate minerals may have come to the rescue at this point. Boric acid, $B(OH)_3$, readily forms borate esters, $B(OR)_3$, with alcohols. Indeed, this reaction was the basis of the old-fashioned flame test for borates, because trimethyl borate burns with an intense green flame. The reaction is even more facile with 1,2 diols, which form cyclic borate esters; this is shown in Figure 3.2 for glyceraldehyde. The borate-complexed glyceraldehyde is then reacts with glycolaldehyde enolate to yield a borate-complexed pentose. The borate mineral, on whose surface the reaction is thought to occur, may afford a cleaner tar-free version of the formose reaction, which is much more promising as a source of ribose. Experiments provide support for this picture. In the presence of $Ca(OH)_2$, an aqueous solution of glycolaldehyde and glyceraldehyde rapidly turns brown. In the presence of borate minerals such as ulexite (see Figure 3.2) or colemanite, the same reaction does not turn brown, even over months, while slowly producing ribose and other pentoses.

This is where the theory, advanced by the American chemist and origin-of-life researcher Steven Benner, takes a somewhat otherworldly turn. Borate minerals are scarce in the earth's crust and may not have been around in the watery environment of early Earth. Mars, however, was drier, and evidence from a meteorite suggests that it also had boron. Thus, according to Benner, "The evidence seems to be building that we are actually all Martians; that life started on Mars and came to Earth on a rock." With characteristic humor, he has added: "It's lucky that we ended up here, nevertheless—as certainly Earth has been the better of the two planets for sustaining life. If our hypothetical Martian ancestors had remained on Mars, there may not have been a story to tell."

REVIEW PROBLEM 3.3

An important point that we have glossed over in the above discussion concerns the exact manner in which boron complexation leads to a cleaner formose reaction. What is your opinion on the matter?

3.2 HYDROBORATION

Boron hydrides or boranes are an important class of compounds that we won't quite do justice to in this book. Many of the polyhedral boranes exhibit inherently multicenter bonding and, accordingly, their chemistry, though utterly fascinating, is less suited to analysis by arrow pushing. Their chemistry was first explored in detail by the German chemist Alfred Stock (1876–1946), whose influential book, as noted above, led the American chemist H. C. Brown (1912–2004) to ultimately discover hydroboration. Hydroboration is one of the quirkier reactions we will meet in this chapter. It typically refers to the addition of a B–H bond to carbon–carbon multiple bonds (although addition to C=N and C=O also involve a similar mechanism). A concerted mechanism with a four-membered transition state is thought to operate:

$$(3.3)$$

The addition is thus stereospecific, taking place across a given face of a double bond. Hydroboration is also remarkably regioselective. Where there is a choice, the boron adds preferentially to the less sterically hindered end of a double bond.

The products of hydroboration—organoboranes—can be elaborated to a variety of organic functional groups, making hydroboration one of the most powerful synthetic reactions. The strategy in all such transformations involves adding a nucleophile to the Lewis-acidic boron, which makes it into an anionic center. In hydroboration–oxidation, the nucleophile is typically alkaline hydrogen peroxide:

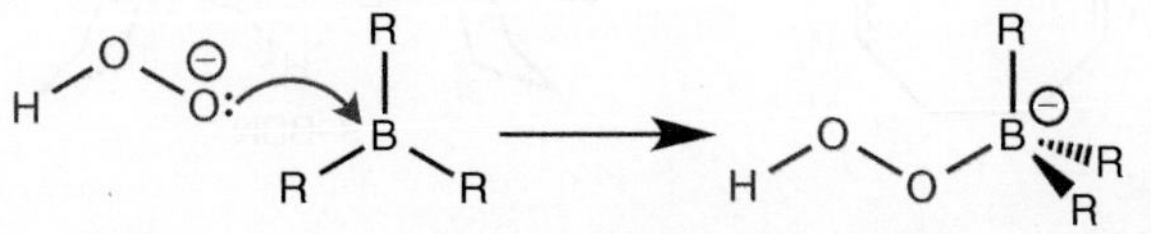

$$(3.4)$$

The anionic boron center then serves as a platform from which an alkyl anion migrates to the boron-bound peroxo oxygen, kicking out an OH^- anion. As mentioned in Section 1.18, this is an important concept that we will encounter repeatedly in various guises throughout this book: a negatively charged, relatively electropositive atom such as B, Si, Sn, Bi, and so on, can act as an effective launchpad for migrating alkyl and aryl groups.

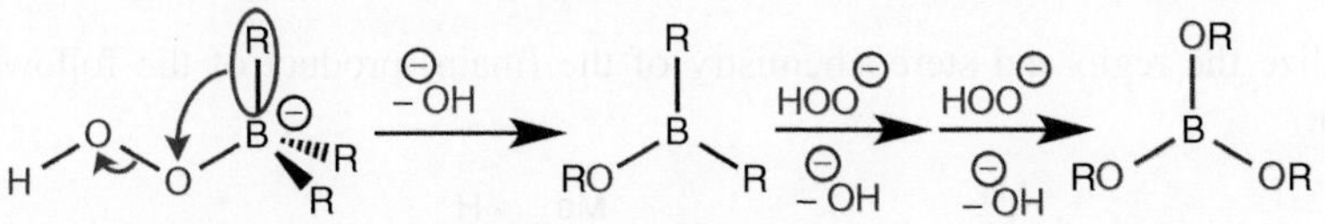

$$(3.5)$$

The product of the above reaction, the borate ester $B(OR)_3$, is hydrolyzed under the reaction conditions to yield the tetrahydroxyborate anion:

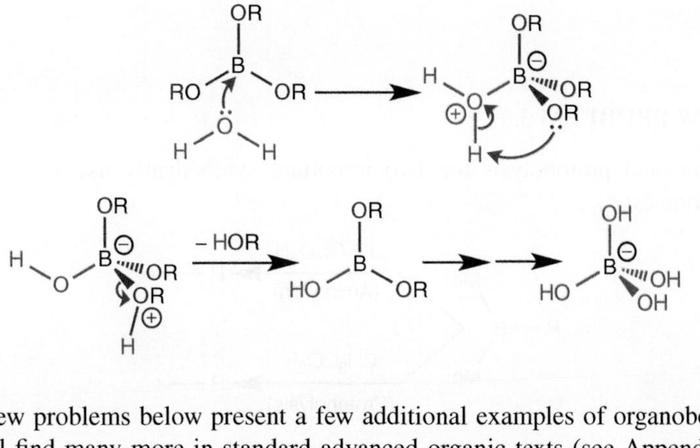

$$(3.6)$$

The review problems below present a few additional examples of organoborane reactions. You'll find many more in standard advanced organic texts (see Appendix B). The common feature of all the reactions is that a nucleophile adds to the tricoordinate boron of the organoborane to yield an anionic tetracoordinate boron center, from which the migrating alkyl group takes off. In the chapters to follow, we'll encounter a variety of other platforms, based on other relatively electropositive p-block elements, from which alkyl and aryl groups may migrate.

REVIEW PROBLEM 3.4

Organoboranes are generally air-sensitive. As a reasonably air-stable colorless solid, the sterically hindered 9-BBN (9-borabicyclo[3.3.1]nonane) is a popular reagent for hydroboration. It is prepared as follows from 1,5-cyclooctadiene:

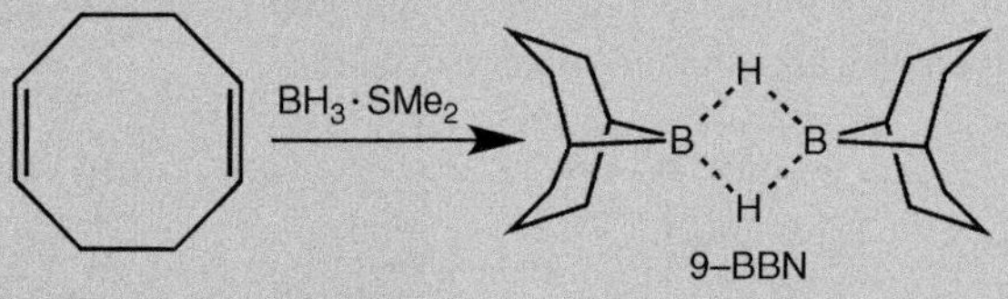

Rationalize the formation of 9-BBN with arrow pushing.

    Note: The structure shown above is a dimer. Clearly, the monomer forms first, before dimerizing.

REVIEW PROBLEM 3.5

Rationalize the regio and stereochemistry of the (main) product of the following reaction:

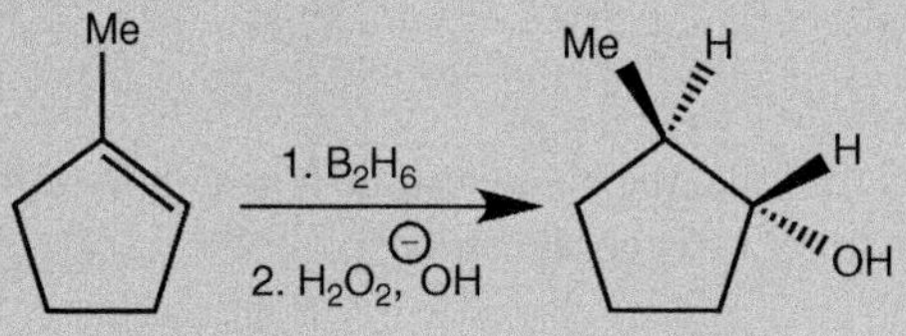

REVIEW PROBLEM 3.6

Amination and protonolysis are two important synthetically useful reactions of organoboranes:

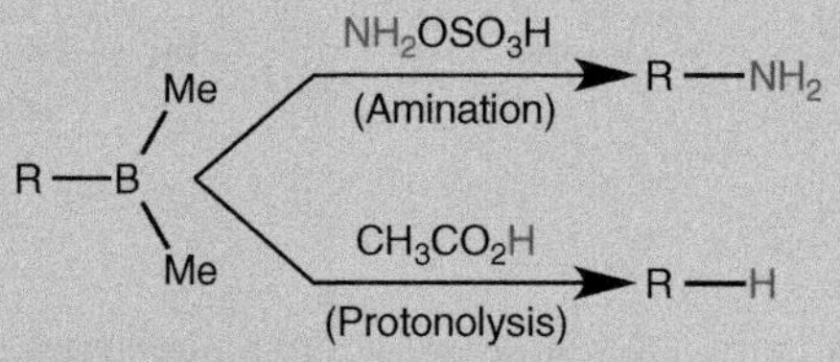

Suggest mechanisms for the two transformations. Of the two, amination, in our view, has the more straightforward mechanism. Protonolysis begins with coordination of the carbonyl oxygen of the carboxylic acid to the boron; R–H bond formation is then believed to occur via a cyclic transition state.

3.3 GROUP 13-BASED REDUCING AGENTS

Anionic hydrido complexes of the group 13 elements, MH_4^-, are widely used as reducing agents. Of these, lithium aluminum hydride ($LiAlH_4$, often abbreviated as LAH) and sodium borohydride ($NaBH_4$) are probably the best known. LAH is a powerful reducing agent (which reacts violently with water), which reduces essentially all reducible organic functional groups, including carbonyl compounds, esters, carboxylic acids, and nitriles. Shown below is the reduction of an ester, R^1COOR^2, by LAH:

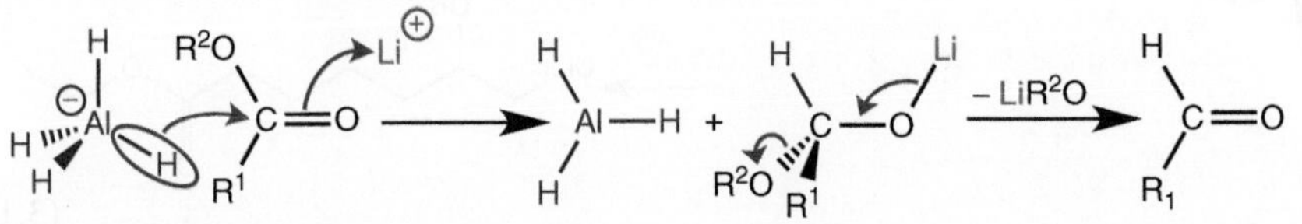

$$\tag{3.7}$$

The alkoxide anion (R^2O^-) and AlH_3 produced above now react together, forming $[Al(OR^2)H_3]^-$, which then continues to act as a hydride transfer agent:

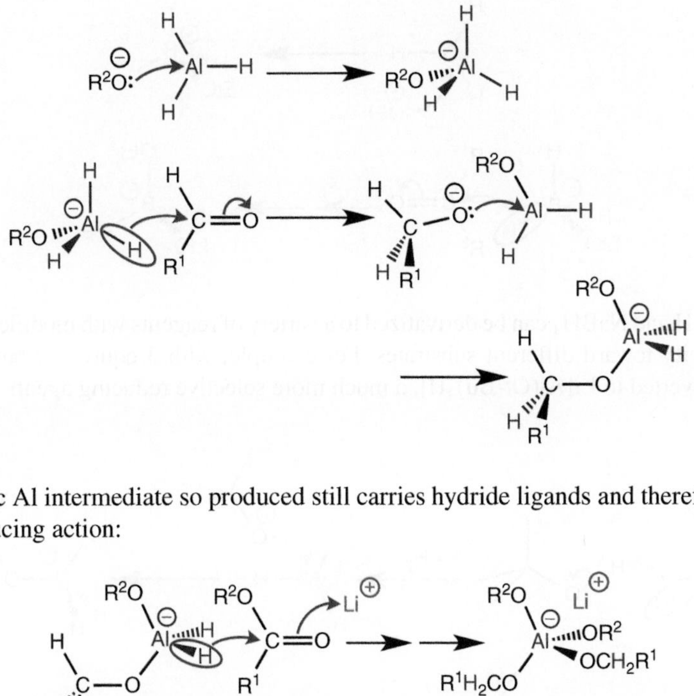

$$\tag{3.8}$$

The anionic Al intermediate so produced still carries hydride ligands and therefore continues its reducing action:

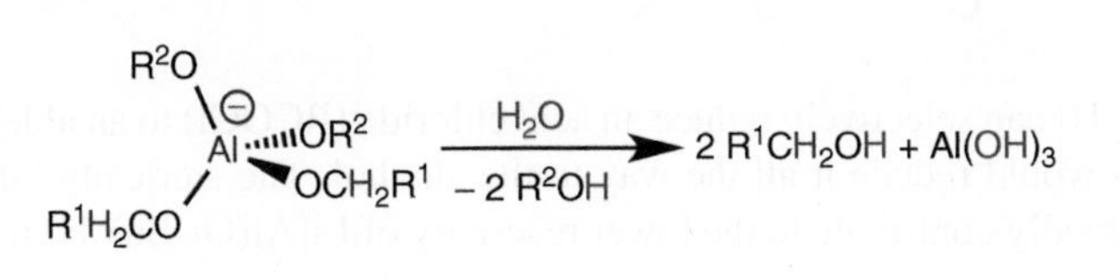

$$\tag{3.9}$$

Aqueous work-up then liberates the free alcohol:

$$\tag{3.10}$$

Sodium borohydride is a considerably milder and more selective reducing agent. Thus, in the example below, only the ketone is reduced, but the ester and alkyl halide moieties remain unaffected:

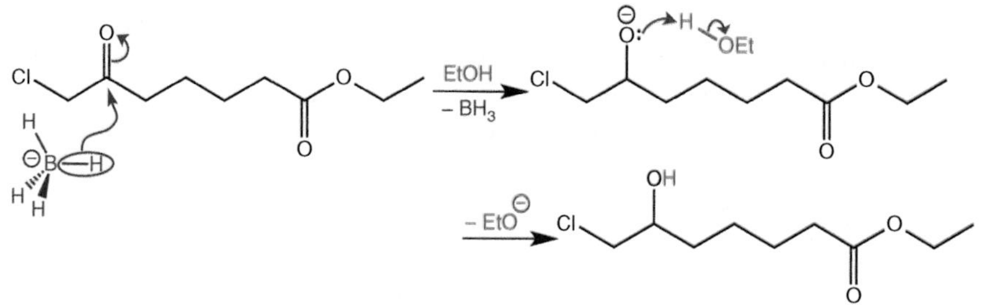

(3.11)

The BH_3 and alkoxide (in this case, EtO^-) produced above capture each other and the resulting anionic product continues to function as a hydride-transferring agent:

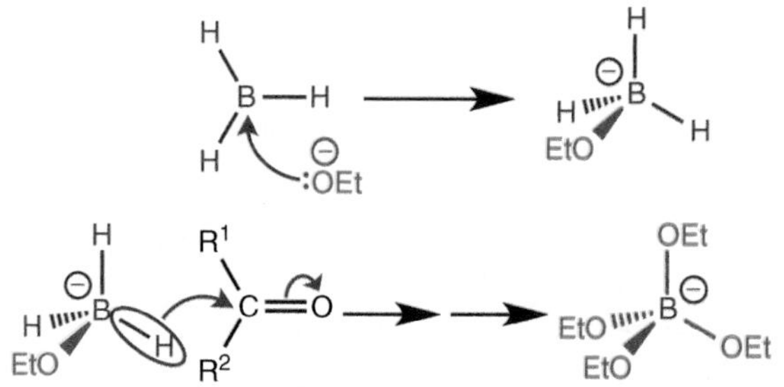

(3.12)

Both LAH and $NaBH_4$ can be derivatized to a variety of reagents with modified reactivity and selectivity toward different substrates. For example, with 3 equiv of *t*-butyl alcohol, LAH is converted to $Li[Al(Ot\text{-}Bu)_3H]$, a much more selective reducing agent:

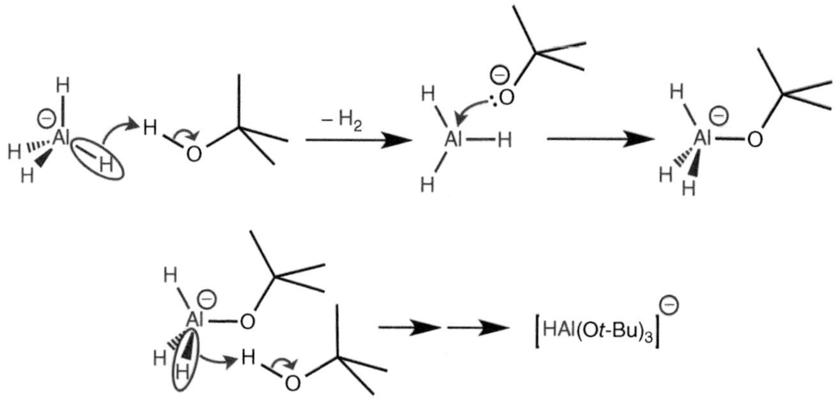

(3.13)

$Li[Al(Ot\text{-}Bu)_3H]$ can selectively reduce an acid chloride (RCOCl) to an aldehyde (RCHO), whereas LAH would reduce it all the way to the alcohol. The sterically hindering *t*-butyl groups undoubtedly contribute to the lower reactivity of $Li[Al(Ot\text{-}Bu)_3H]$ relative to LAH.

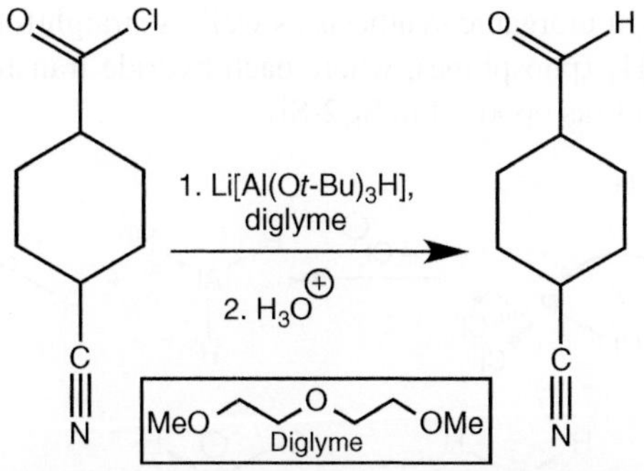

$$(3.14)$$

REVIEW PROBLEM 3.7

Superhydride, Li[Et$_3$BH], and sodium cyanoborohydride (prepared from BH$_3$ and NaCN), Na[BH$_3$CN], provide nice examples of how ligand electronic effects can tune the reactivity of hydride transfer reagents. The first is a supernucleophilic hydride transfer agent, whereas the second is mild and selective. Sodium cyanoborohydride can thus be used to reduce an aldehyde group selectively in the presence of a keto group, or a keto group selectively in the presence of an ester:

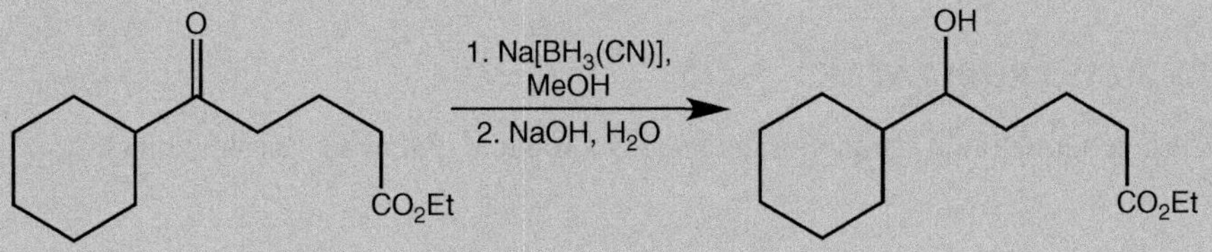

Suggest a rationale for the difference in reactivity between the two reducing agents.

Another useful reagent is the dimeric diisobutylaluminum hydride (DIBAL-H), (i-Bu$_2$AlH)$_2$, which can reduce esters and nitriles in a controlled manner. Thus, unlike for LAH, only *one* hydride is transferred to the organic substrate, so for esters the reduction stops at the aldehyde stage.

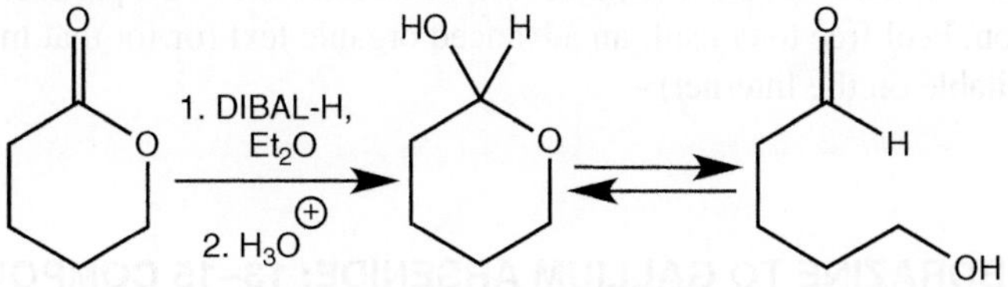

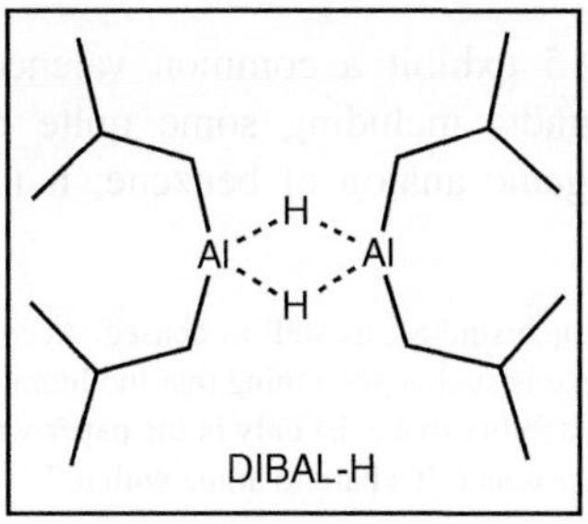

$$(3.15)$$

LAH is widely used in inorganic synthesis as well. A straightforward application is the reduction of PCl_3 to PH_3 (phosphine), where each hydride transfer is shown below as a simple S_N2 displacement (as opposed to S_N2-Si):

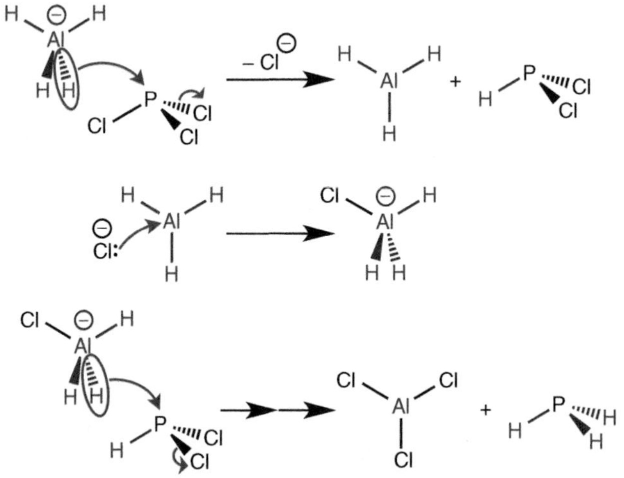

(3.16)

REVIEW PROBLEM 3.8

Carbonylation of an organoborane affords a route to aldehydes, as shown below:

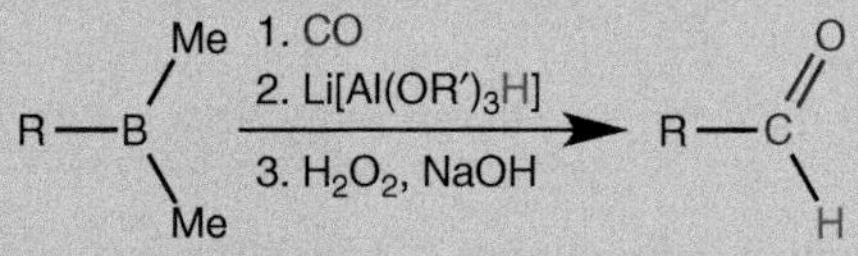

Provide a mechanistic explanation of the process, as far as you can. *Hint*: The first step, which is the actual carbonylation step, results in the formation of the R–CO bond. The additional steps serve to adjust the oxidation state of the product and cleave off the boron. Feel free to consult an advanced organic text (or for that matter information available on the Internet).

3.4 FROM BORAZINE TO GALLIUM ARSENIDE: 13–15 COMPOUNDS

Because both groups 13 and 15 exhibit a common valence of 3, they form a suite of equimolar "13–15" compounds, including some quite important ones. One such compound is borazine, an inorganic analog of benzene; it is, however, only "slightly" aromatic[1]:

---

[1] The concept of aromaticity has been much studied, as well as abused. According to Roald Hoffmann (1937-; Nobel Prize in chemistry 1981), "Benzene is such a good thing that the literature is … full of mostly imagined "aromaticities," supposed harbingers of stability that exist only in the paper-writers' hype-addled minds. But the stability of the parent C_6H_6 system is sacrosanct. It's hard to argue with it."

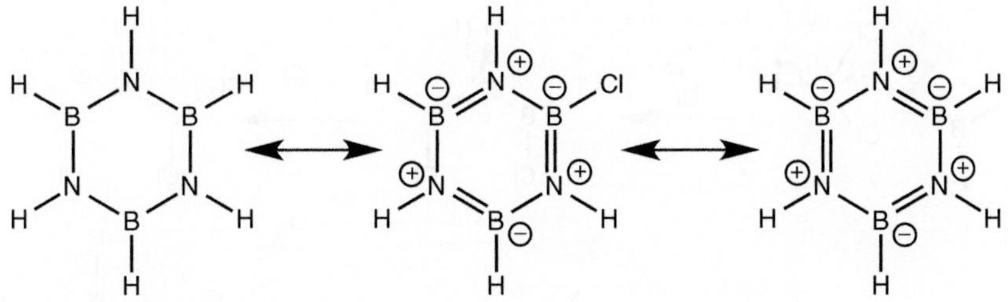

Borazine was first prepared in 1926 by Stock and Pohland by the high temperature reaction of diborane and ammonia:

$$3\ B_2H_6 + 6\ NH_3 \xrightarrow{\ 250-300\ °C\ } 2\ B_3H_6N_3 + 12\ H_2 \tag{3.17}$$

In the same vein, boron trichloride and ammonium chloride react to yield trichloro-borazine:

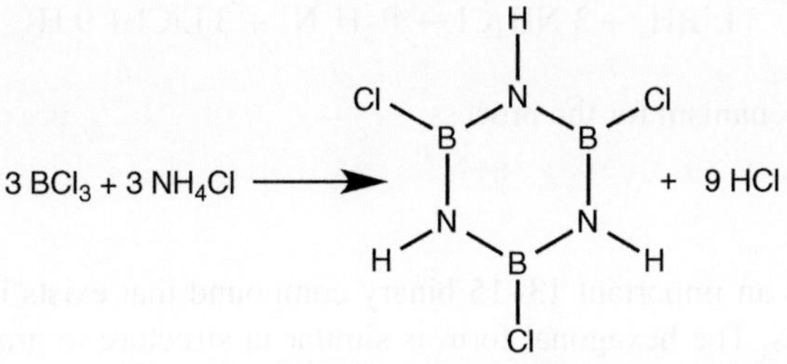

$$3\ BCl_3 + 3\ NH_4Cl \longrightarrow \qquad + \quad 9\ HCl \tag{3.18}$$

Trichloroborazine may then be reduced to borazine with $NaBH_4$, providing a two-step route to the parent compound.

The mechanism of assembly of the borazine skeleton is illustrated below for trichloroborazine. Ammonia, a nucleophile present in the system, attacks the electrophile BCl_3, producing an aminoborane:

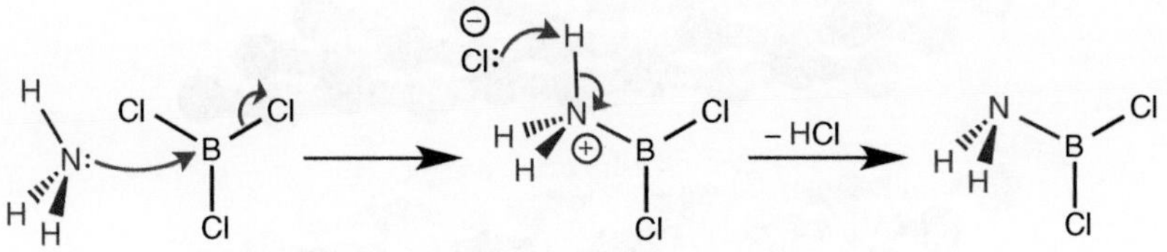

$$\tag{3.19}$$

The B–N chain then continues to grow until a cyclization leads to the borazine skeleton:

$$(3.20)$$

REVIEW PROBLEM 3.9

A modern and relatively convenient borazine synthesis uses lithium borohydride and ammonium chloride instead as the starting materials:

$$3\,LiBH_4 + 3\,NH_4Cl \rightarrow B_3H_6N_3 + 3\,LiCl + 9\,H_2$$

Sketch out a mechanism for the process.

Boron nitride is an important 13–15 binary compound that exists in a number of different crystal forms. The hexagonal form is similar in structure to graphite (as shown in Figure 3.3) and is thermodynamically the most stable. It's soft and used as a lubricant and as a sheen-producing additive in cosmetics. It's synthesized by heating boron trioxide (B_2O_3) or boric acid with ammonia at ~900 °C:

$$B_2O_3 + 2\,NH_3 \rightarrow 2\,BN + 3\,H_2O \qquad (3.21)$$

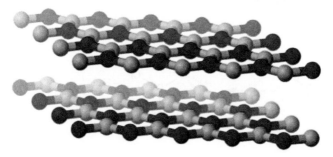

Figure 3.3 *Schematic diagram of hexagonal boron nitride sheets. (Public domain figure obtained from Wikimedia Commons.)*

$$B(OH)_3 + NH_3 \rightarrow BN + 3\,H_2O \tag{3.22}$$

An exciting recent development has been the preparation of single sheets of hexagonal BN, dubbed "white graphene" by analogy with graphene, which refers to single sheets of graphite. These are found to soak up prodigious quantities of various pollutants out of water, such as ethylene glycol (antifreeze) and engine oil. Impressively, the pollutants could then be driven out by heating in a furnace, and the boron nitride could be recovered and reused. Cubic boron nitride, on the other hand, has the same structure as diamond but is inferior to diamond only in hardness. Like diamond, it's widely used as an abrasive.

Gallium nitride and gallium arsenide are the preeminent examples of 13–15 semiconductors. The 13–15 semiconductors are overall isoelectronic with silicon and greatly expand the repertoire of available, useful properties relative to the elemental silicon and germanium. Gallium arsenide is used in the manufacture of a wide range of integrated circuits, laser diodes, laser lenses, reflectors, and solar cells. Gallium nitride, widely used as light-emitting diodes since the 1990s, is considered even more exciting. Its high thermal stability compared with GaAs makes it an ideal material for high power transistors capable of operating at high temperatures. Several preparative routes are known for both materials. For our mechanistic discussion, we will focus on GaAs synthesis via a specialized technique called *metal-organic vapor phase epitaxy* (MOVPE) or *metal-organic chemical vapor deposition* (MOCVD):

$$Ga(CH_3)_3 + AsH_3 \rightarrow GaAs + 3\,CH_4 \tag{3.23}$$

The reactants, in the form of ultrapure gases, are injected into a reactor and dosed so as to deposit a thin layer on a semiconductor wafer.

The first step of the mechanism, likely to be a simple Lewis acid–base interaction between trivalent gallium and trivalent arsenic, is as shown below:

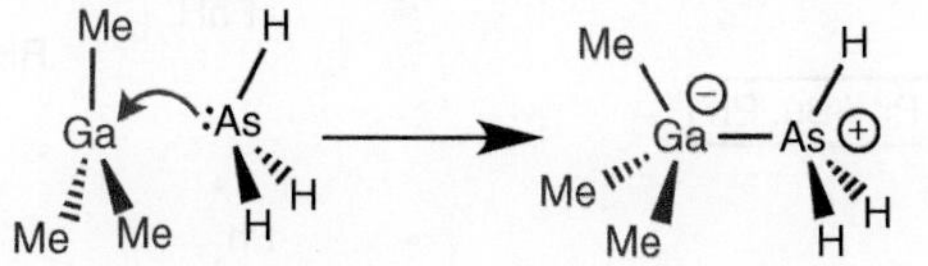

$$\tag{3.24}$$

The zwitterionic adduct is then expected to eliminate methane. The partial double bond character of the Ga–As bond, as shown below, may well play a role in this elimination:

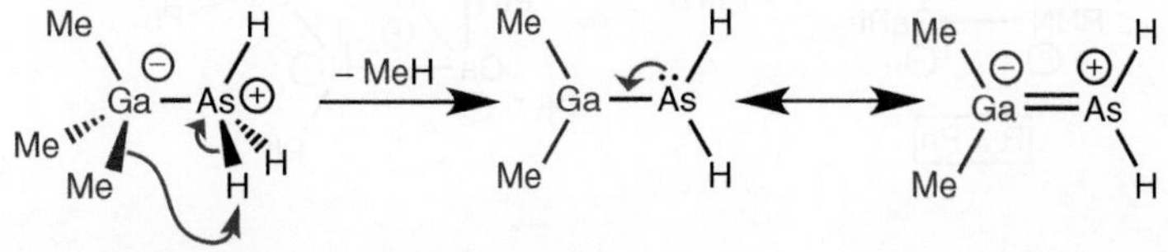

$$\tag{3.25}$$

Because of the weakness of multiple bonds between heavier p-block elements, however, we do *not* envision further eliminations that lead to full-fledged double bonds:

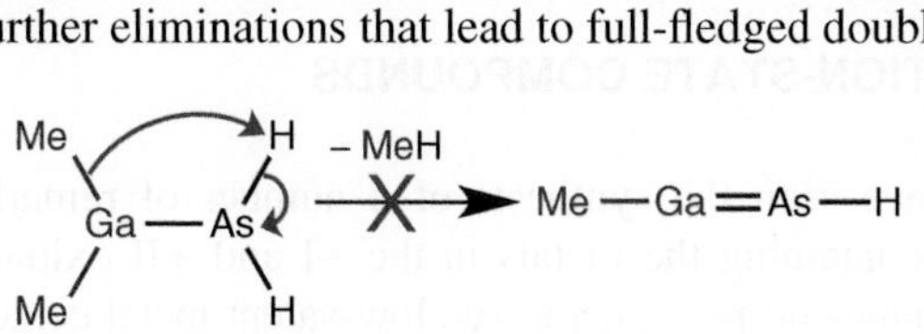

$$\tag{3.26}$$

The above process now repeats itself, leading to more extended Ga–As skeletons and ultimately GaAs:

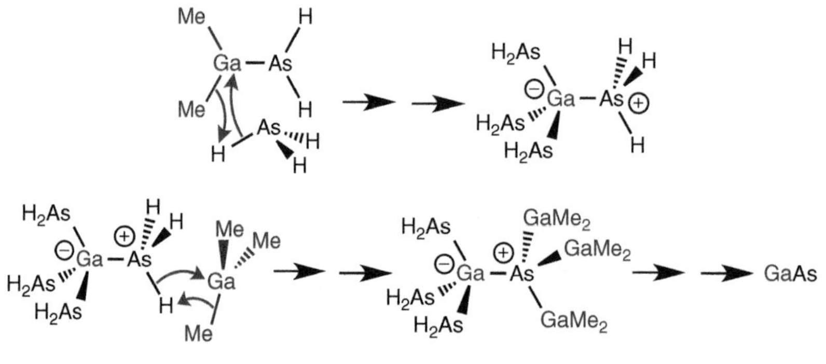

(3.27)

REVIEW PROBLEM 3.10

Gladfelter and coworkers at the University of Minnesota have synthesized and structurally characterized a variety of "imidogallane" clusters, of the form $(R^1GaNR^2)_n$, as models for intermediates on the pathway to GaN (*Inorg. Chem.* **2002**, *41*, 590–597). The synthesis of a cubane cluster is summarized below. Use arrow pushing to rationalize the formation of the various products:

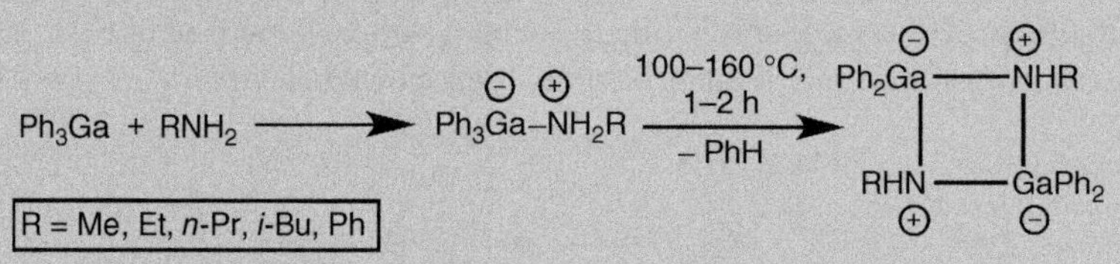

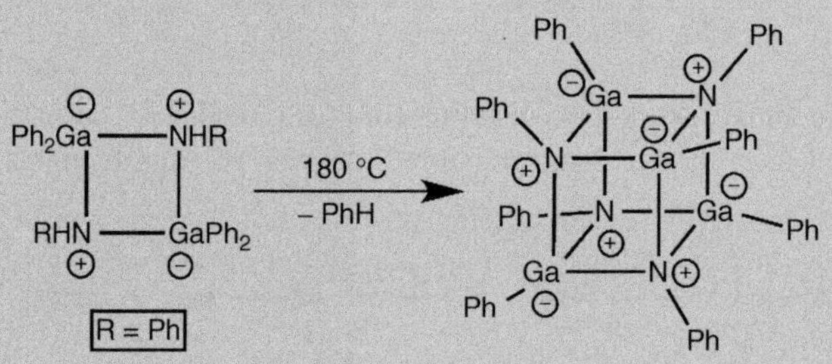

3.5 LOW-OXIDATION-STATE COMPOUNDS

The last 25 years have seen the synthesis of a number of remarkable aluminum and gallium compounds containing the metals in the +I and +II oxidation states. Although some of these compounds do not contain true low-valent metal centers, others do contain the metals in true monovalent form. (Have a look at Section 1.24 for a reminder on the distinction between valence and oxidation state.) A common route to low-oxidation-state group 13 compounds consists of the reductive dehalogenation of R_2EX and REX_2 (E = Al, Ga; X = Cl, Br) with alkali metals or Rieke magnesium.

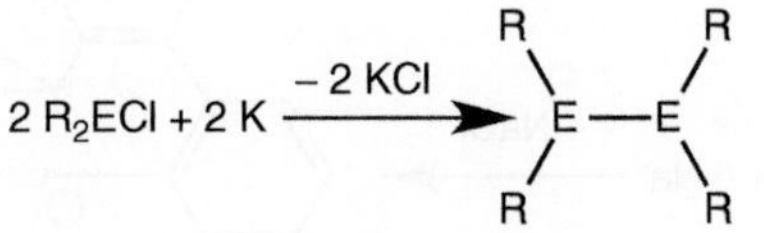

$$2\ R_2ECl + 2\ K \xrightarrow{-\ 2\ KCl} \begin{array}{c} R \\ E - E \\ R \end{array} \begin{array}{c} R \\ \\ R \end{array} \tag{3.28}$$

$$REX_2 + 4\ M \xrightarrow[n\ =\ 1-4]{-\ 4\ MX} (RE)_n \tag{3.29}$$

For stable products, the R groups should ideally be sterically hindered aryl or alkyl groups; examples follow in the discussion below. A typical example of an $(RE)_n$ oligomer is the following:

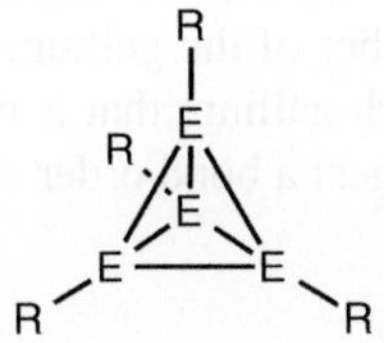

Note that the six "bonds" linking the vertices of the tetrahedron are not normal two-electron bonds but are part of a delocalized bonding network; the tetrahedron is held together by eight valence electrons, two from each group 13 element.

The tetraorganodigallanes, Ga_2R_4, can be further reduced to digallane anion radicals, a Ga_3 ring, and even a formal "digallyne" (Su, J.; Li, X.-W.; Crittendon, R. C.; Robinson, G. H. *J. Am. Chem. Soc.* **1997**, *119*, 5471–5472), as shown below:

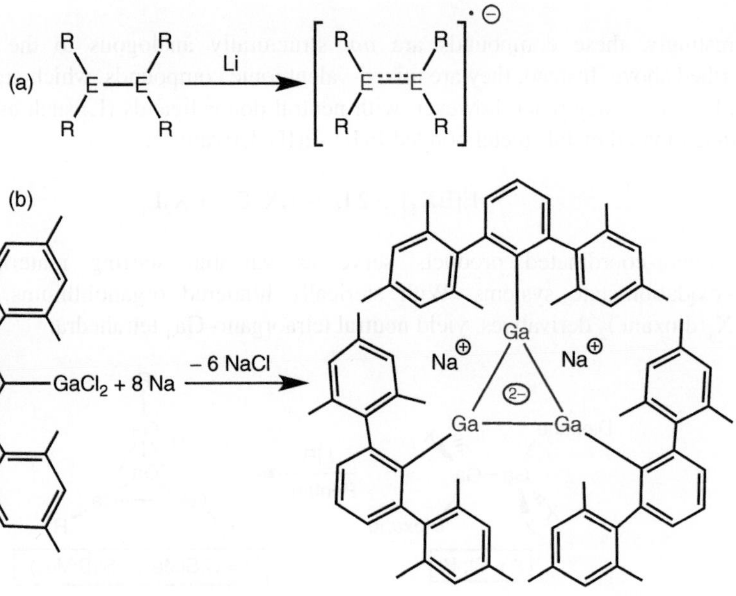

(c)

$$(3.30)$$

For the last product, we have used the triple-bonded "digallyne" formulation simply to highlight the low coordination number of the gallium centers. An alternative formulation would be to have a lone pair on each gallium that is only partially involved in π-overlap. Quantum chemical studies also suggest a bond order closer to two than to three:

Gallium and indium metals react with the metal trihalides ($X = Cl$, Br, I) to yield the so-called subhalides:

$$2\,E + 4\,EX_3 \rightarrow 3\,E_2X_4 \qquad (3.31)$$

Interestingly, these compounds are *not* structurally analogous to the E_2R_4 species described above. Instead, they are mixed-valent ionic compounds, which are better written as $E^{I}[E^{III}X_4]$. They react, however, with neutral donor ligands (L) such as pyridines and dioxane to yield metal–metal bonded E(II)–E(II) derivatives:

$$E[EX_4] + 2\,L \rightarrow LX_2E - EX_2L \qquad (3.32)$$

The donor-coordinated products serve as valuable starting materials for other low–oxidation-state systems. With sterically hindered organolithiums, for example, $Ga_2X_4(\text{dioxane})_2$ derivatives, yield neutral tetraorgano-Ga_4 tetrahedra:

$$(3.33)$$

Note that this reaction is not a simple ligand substitution but a net reduction. The yields are not particularly good, and the Ga_4 tetrahedra are obtained more efficaciously via reaction 3.29, that is, the alkali metal reduction of REX_2 derivatives.

A true monovalent, mononuclear aluminum compound has been obtained through potassium metal reduction of a β-diketiminatoaluminum(III) diiodide, as shown below (Cui, C., et al. *Angew. Chem. Int. Ed.* **2000**, *39*, 4274–4276):

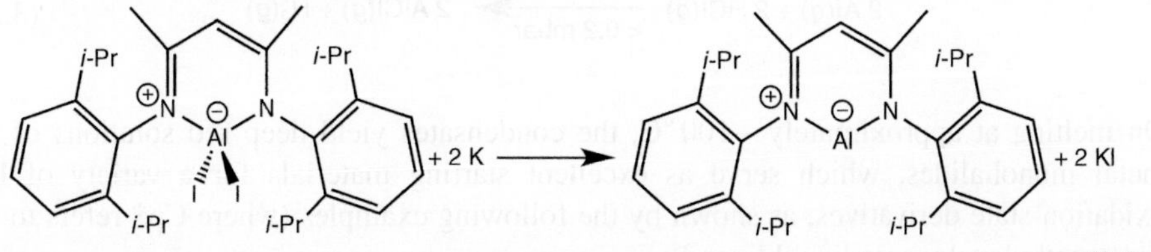

(3.34)

For the aluminum center in the product, we encourage you to satisfy yourself that: valence = no. of bonds + formal charge = 2 + (−1) = 1 (see Section 1.24 for a reminder). As you might imagine, the monovalent Al center is quite reactive and activates a variety of small molecules, as shown below. For brevity, we have depicted the mononegative β-diketiminato as an arc in all the products, which are all uncharged:

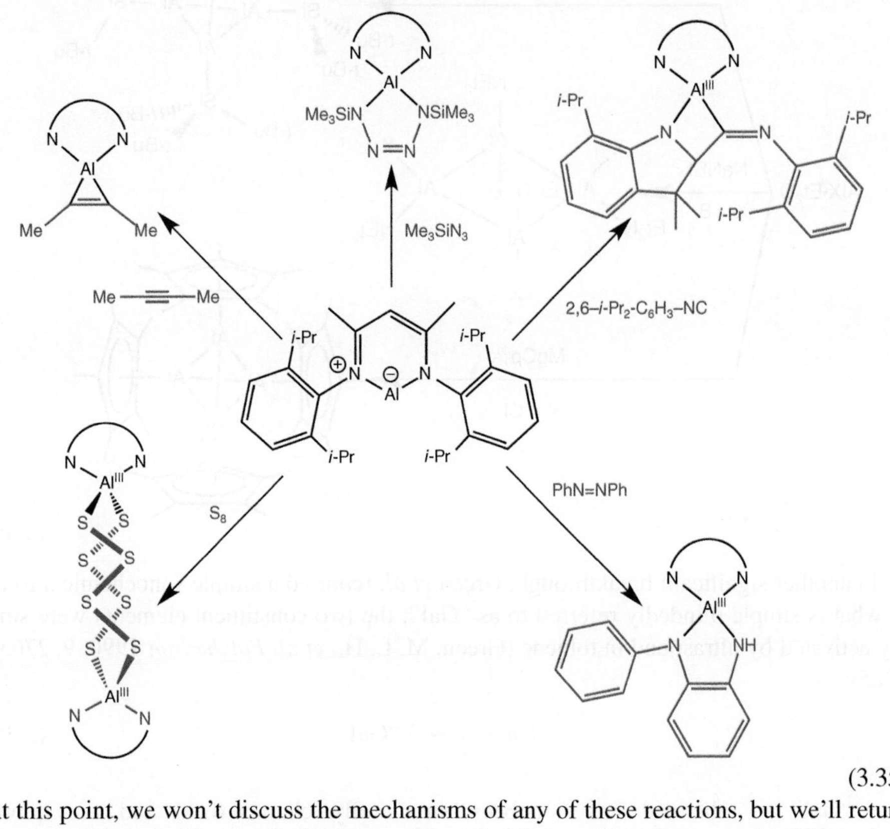

(3.35)

At this point, we won't discuss the mechanisms of any of these reactions, but we'll return to some of them in Chapter 5a (review problem 5a.14), by which time you should have had a bit more experience with p-block chemistry.

In 1989, in a major development, Tacke and Schnöckel showed that metastable solutions of Al and Ga monohalides could be prepared in organic solvents (Tacke, M.; Schnöckel, H. *Inorg. Chem.* **1989**, *28*, 2895–2896). The monohalides were prepared from the molten metal and a hydrogen halide in a high temperature reactor and subsequently condensed at −196 °C with toluene, with various Lewis-base additives.

$$2\,Al(g) + 2\,HCl(g) \xrightarrow[< 0.2\ \text{mbar}]{1000\ °C} 2\,AlCl(g) + H_2(g) \tag{3.36}$$

On melting at approximately −100 °C, the condensates yield deep red solutions of the metal monohalides, which serve as excellent starting materials for a variety of low oxidation-state derivatives, as shown by the following examples (where Cp* refers to the pentamethylcyclopentadienyl ligand):

$$(3.37)$$

In another significant breakthrough, Green *et al.* reported a simple sonochemical route to what is simple-mindedly referred to as "GaI"; the two constituent elements were simply activated by ultrasound in toluene (Green, M. L. H., *et al. Polyhedron* **1990**, *9*, 2763–2765):

$$2\,Ga + I_2 \rightarrow 2\ \text{"GaI"} \tag{3.38}$$

The "GaI" so obtained has not been structurally characterized and the quotation marks indicate the likelihood of oligomeric structures. With the monoanionic hydridotris[3,5-bis(*t*-butyl)pyrazolyl]borate (Tp*t*-Bu$_2$) ligand, "GaI" yields Ga[Tp*t*-Bu$_2$], a monovalent gallium compound:

$$\text{``GaI'' + NaT}pt\text{-Bu}_2 \longrightarrow \quad\quad\quad\quad\quad \tag{3.39}$$

In case you find the 2– formal charge on the gallium unsettling, Ga[Tpt-Bu$_2$] is more conventionally depicted as follows:

Ga^{I}[Tpt-Bu$_2$]

The bonds marked in red are intended to help with electron counting; they are N → Ga dative bonds. Regardless of the representation you favor, we encourage you to verify that the gallium valence is 1: the key point to realize is that the closed-shell Tpt-Bu$_2$ ligand does not contribute to the valence. With a lone pair on the gallium, Ga[Tpt-Bu$_2$] acts as a Lewis base toward GaI$_3$, forming a Ga–Ga-bonded Lewis acid–base adduct:

$$\xrightarrow{\text{GaI}_3} \tag{3.40}$$

Once again, you may be more comfortable with a more conventional picture in terms of oxidation states, rather than formal charges:

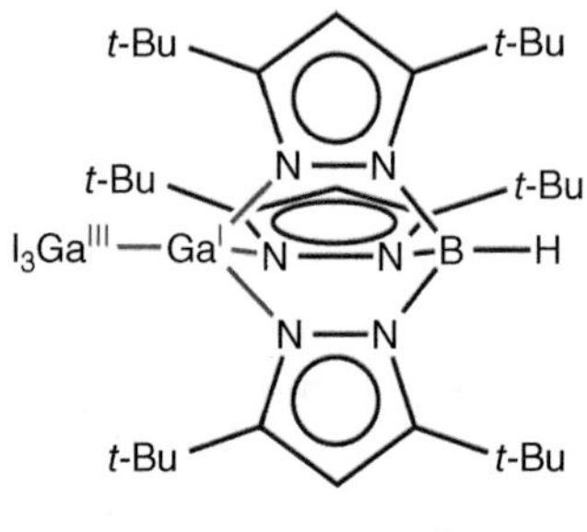

$[\text{Tp}t\text{-Bu}_2]\text{Ga}^\text{I}\text{Ga}^\text{III}\text{I}_3$

REVIEW PROBLEM 3.11

What are the valences of the two Ga centers in the above Ga–Ga bonded adduct? This is a bit of a tricky question; for a good discussion, see, Parkin, G. *J. Chem. Educ.* **2006**, *83*, 791–799.

REVIEW PROBLEM 3.12*

Robinson and coworkers have reported a remarkably general route to low–oxidation-state p-block element homodimers, exploiting *N*-heterocyclic carbenes (NHCs) as capping ligands (see Lintl, G.; Schnöckel, H. *Coord. Chem. Rev.* **2000**, *206*, 285–319):

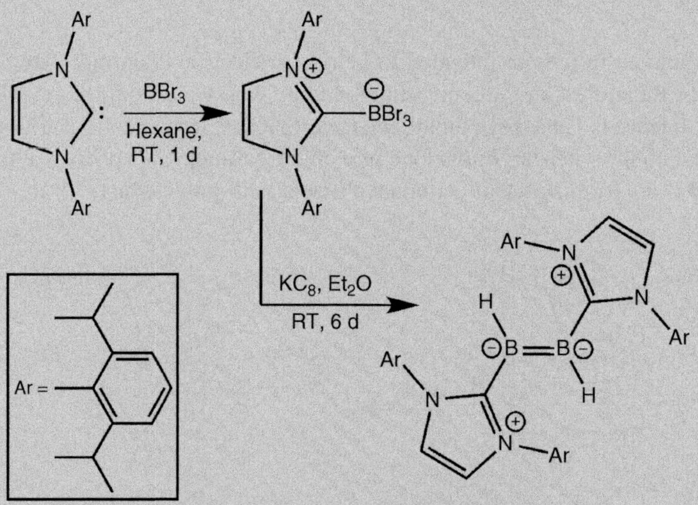

What are the valence, oxidation state, and coordination number of each boron in the product? What might account for the efficacy of the NHC ligands in this approach?

3.6 THE BORYL ANION

Unlike carbanions and silyl anions, boryl anions (BR_2^-), in the form of boryllithiums, have been unknown until recently. In 2006, Japanese researchers reported the synthesis of a boryllithium via the reduction of a cyclic (diamino)bromoborane with lithium/naphthalene in THF, that is, $Li^+Np^{\bullet-}$, where $Np^{\bullet-}$ is the anion radical of naphthalene (Segawa, Y.; Yamashita, M.; Nozaki, K. *Science* **2006**, *314*, 113–115):

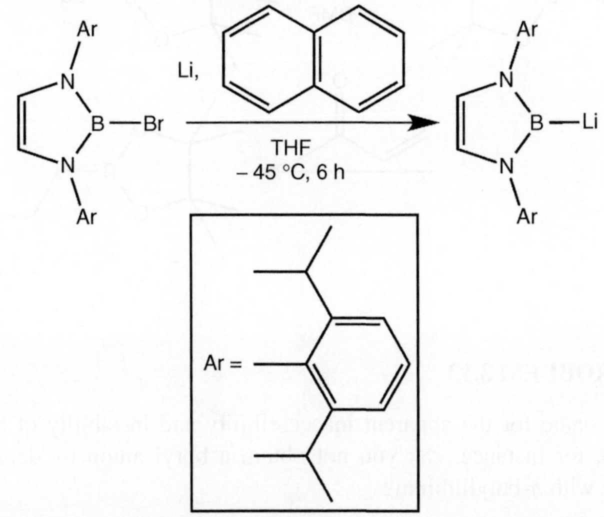

(3.41)

Note the close structural similarity of the boryllithium with an NHC. This is not a coincidence: the two species are in fact isoelectronic.

In spite of its uniqueness, the above boryllithium was found to behave more or less like a regular organolithium. Thus, it undergoes transmetallation with $MgBr_2 \cdot OEt_2$ to yield a boryl-Grignard reagent. Similarly, with electrophiles, it reacts like a conventional carbanion:

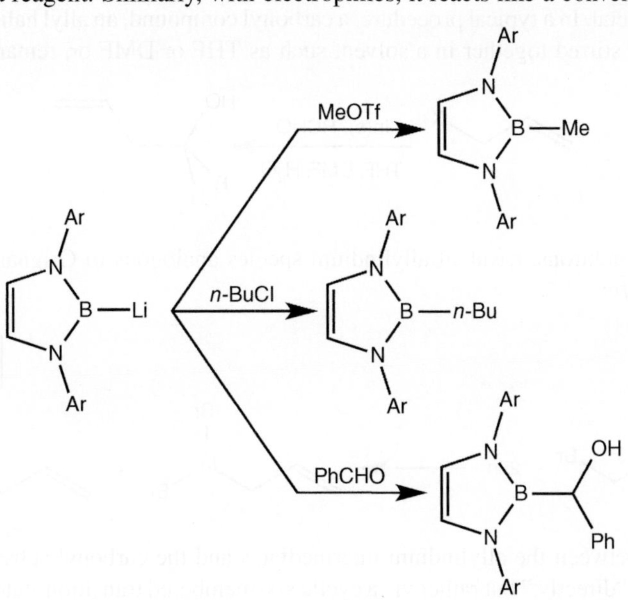

(3.42)

Subsequently, borylcopper reagents were also synthesized, as shown below. As softer nucleophiles than boryllithiums, borylcopper reagents exhibit significantly different reactivity, undergoing, for instance, conjugate additions with α,β-unsaturated carbonyl compounds:

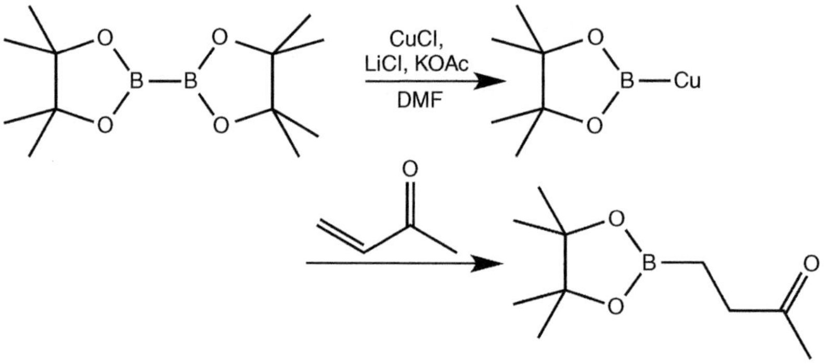

(3.43)

REVIEW PROBLEM 3.13

Suggest a rationale for the apparent inaccessibility and instability of boryl anions (BR_2^-). Why, for instance, can you not obtain a boryl anion by deprotonating a dialkylborane with *n*-butyllithium?

3.7 INDIUM-MEDIATED ALLYLATIONS

Indium-mediated allylations of carbonyl compounds have proved to be an important application of the metal. In a typical procedure, a carbonyl compound, an allyl halide, and indium powder are all stirred together in a solvent such as THF or DMF or, remarkably enough, water:

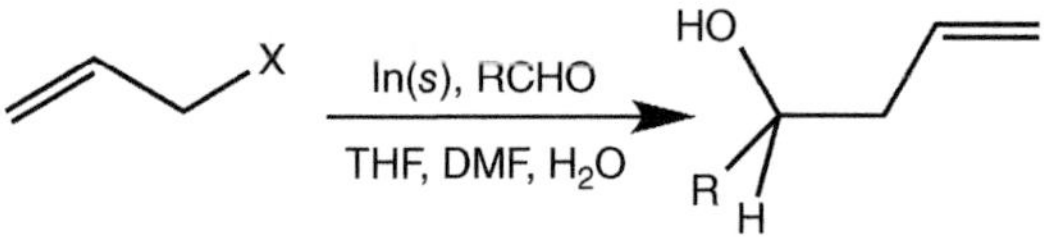

(3.44)

Under these conditions, trivalent allylindium species analogous to Grignard reagents are generated *in situ*:

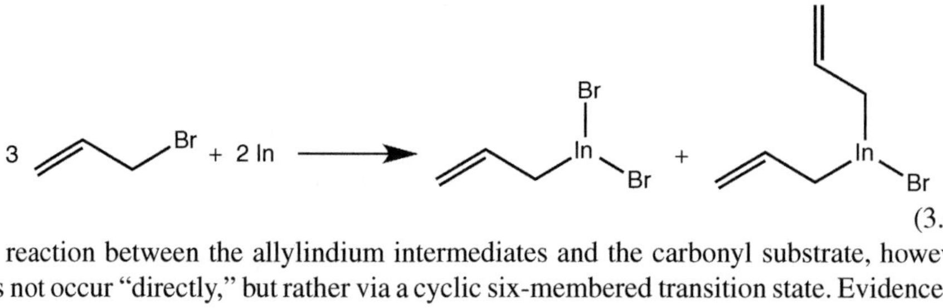

(3.45)

The reaction between the allylindium intermediates and the carbonyl substrate, however, does not occur "directly," but rather via a cyclic six-membered transition state. Evidence for a cyclic pathway comes from the substitution pattern of the product, a homoallylic alcohol (an unsaturated alcohol with two saturated carbons between the C=C double bond and the OH group), when the starting material is a substituted allyl halide, as in the example below:

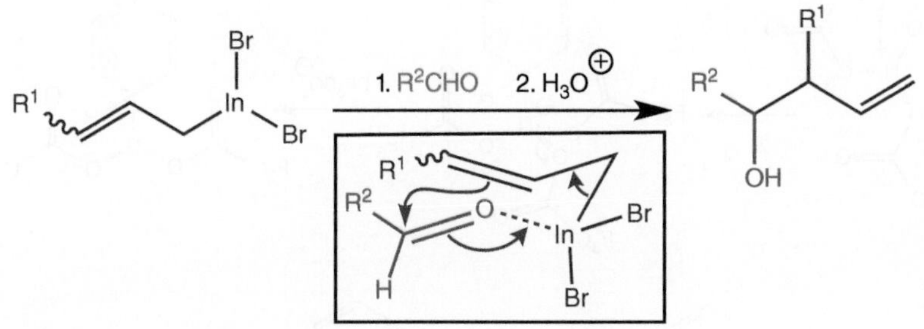

$$(3.46)$$

A direct addition via an acyclic pathway would have led to a homoallylic alcohol with the substitution pattern shown below, which is not observed:

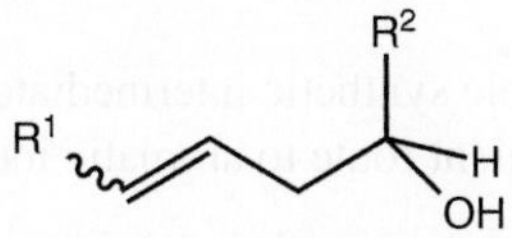

A key distinction from Grignard reagents is that the organoindium reagents are water-tolerant, to the extent that indium-mediated allylations are often carried out in water. The low toxicity of indium also makes this reaction an excellent example of a "green" synthetic reaction.

3.8 THALLIUM REAGENTS

Compared with the exotic low oxidation-state aluminum and gallium compounds, thallium(I) compounds are both stable and common. Because of their toxicity, however, they should be handled with great care. With thallium, the inert pair effect is in full play and the monovalent state is generally the stable state, whereas the trivalent state is oxidizing. For the thallium compounds that are of interest here, the oxidation state and valence are generally the same, so we may, for a brief while, use the two terms interchangeably. In spite of their toxicity, trivalent thallium compounds have proved remarkably useful as reagents in organic chemistry.

Consider thallation, an electrophilic substitution of an aromatic compound by a trivalent thallium compound, depicted below for benzene and $Tl(O_2CCF_3)_3$:

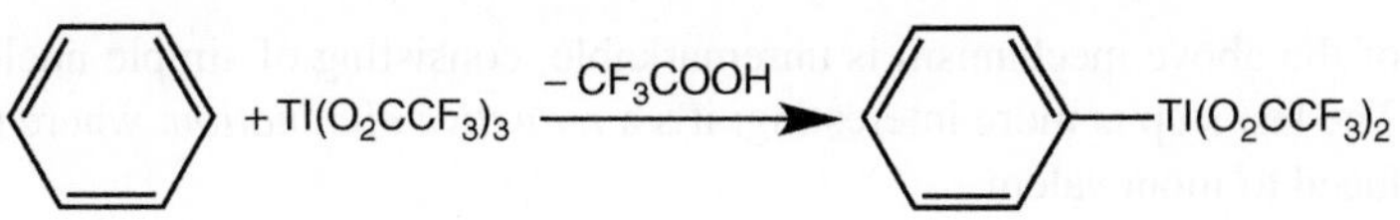

$$(3.47)$$

The first steps of the mechanism are standard for electrophilic aromatic substitution:

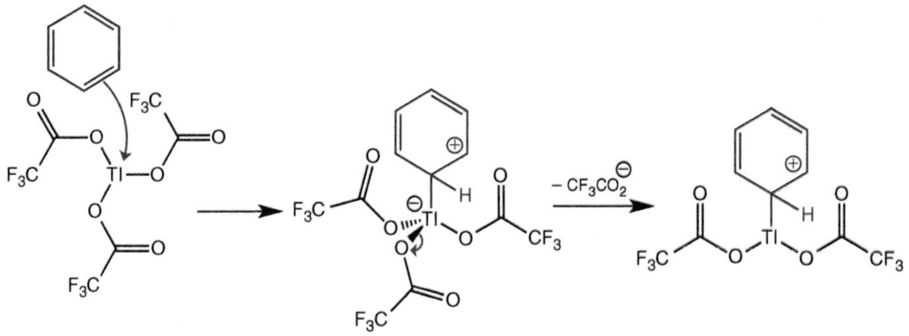

(3.48)

The thallated benzene is a valuable synthetic intermediate. With two equivalents of potassium iodide, it provides a convenient route to aromatic iodides, as shown below:

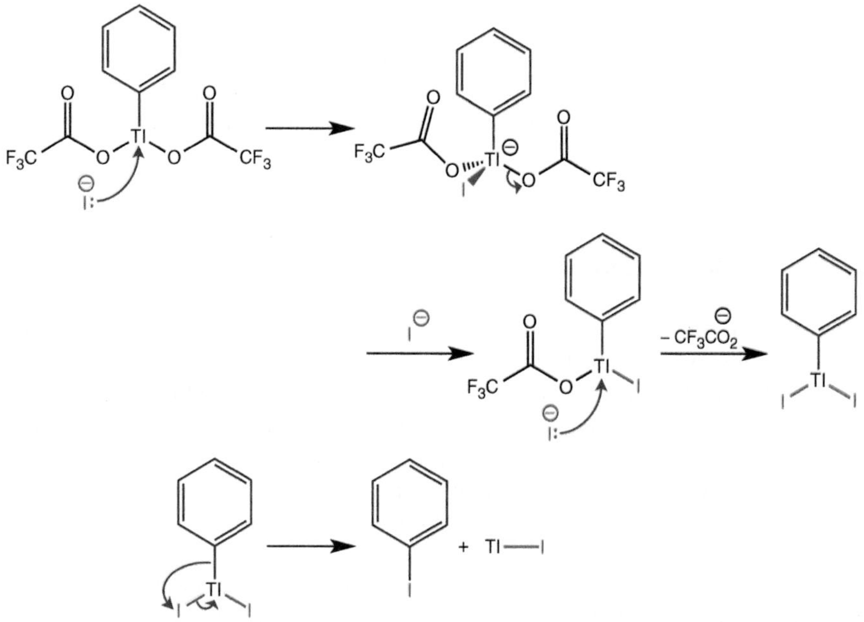

(3.49)

The first part of the above mechanism is unremarkable, consisting of simple nucleophilic substitutions. The last step is more interesting: it's a *reductive elimination*, where trivalent thallium is reduced to monovalent.

REVIEW PROBLEM 3.14*

The compound TlI_3 exists but it does not contain trivalent thallium. Can you suggest an explanation? (*Hint*: What might be another reasonable valence state for Tl?)

Aside from thallation, trivalent thallium is best known as a versatile oxidant in organic chemistry; many of the reactions proceed with unique rearrangements. Let us first consider one that does not involve rearrangement. Thus, thallium(III) triflate (see Section 1.14 for the structure of the triflate anion), which may be obtained from the more common thallium(III) nitrate by treatment with trifluoromethanesulfonic (triflic) acid in DMF, oxidizes acetophenone to its α-trifluoromethanesulfonyloxy derivative:

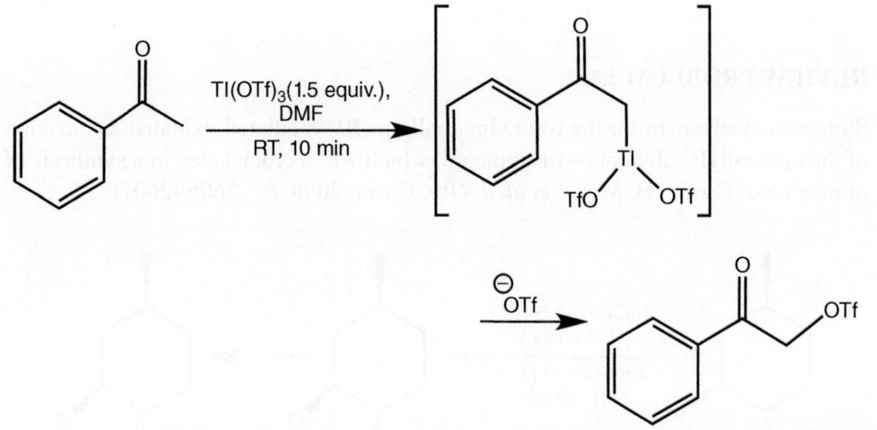

(3.50)

We won't write a mechanism for the formation of the organothallium intermediate: it forms via thallation of the enol form of acetophenone, and we urge you to write out the details. A triflate anion, which got kicked out in the first step, then displaces the thallium, which simultaneously undergoes reductive elimination to produce Tl^IOTf:

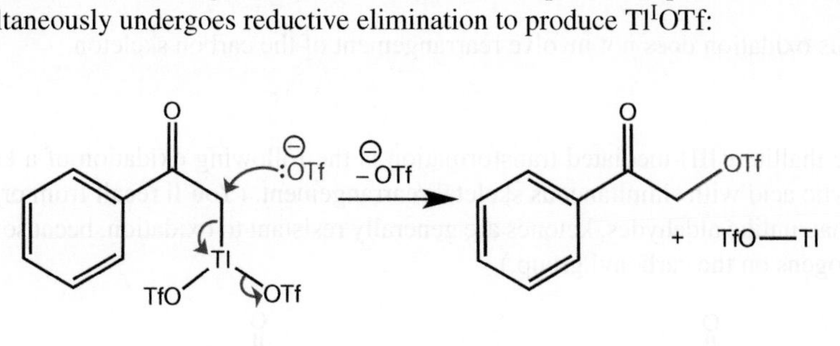

(3.51)

Triflate, being an excellent leaving group, is then typically displaced by DMF under the reaction conditions; the resulting iminium salt may then be hydrolyzed under work-up, as shown below:

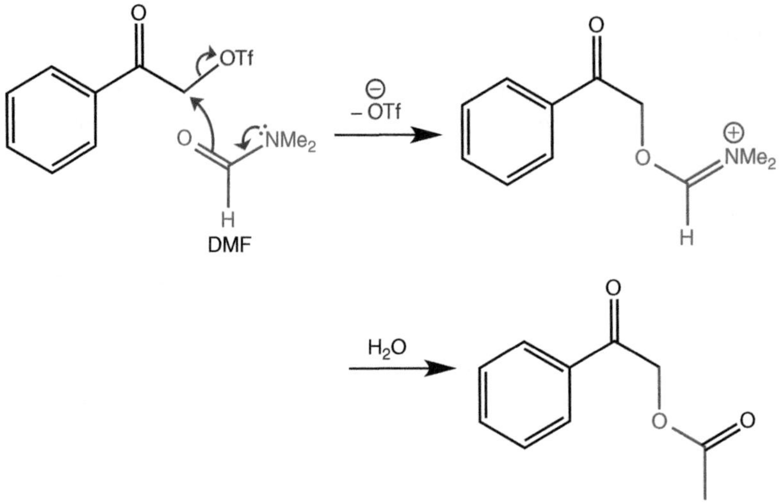

(3.52)

REVIEW PROBLEM 3.15*

Suggest a mechanism for the following thallium(III)-mediated oxidative cyclization of the homoallylic alcohol (−)-isopulegol, which was a crucial step in a synthesis of mintlactone (Ferraz, H. M. C., *et al. J. Org. Chem.* **2000**, *65*, 2606–2607):

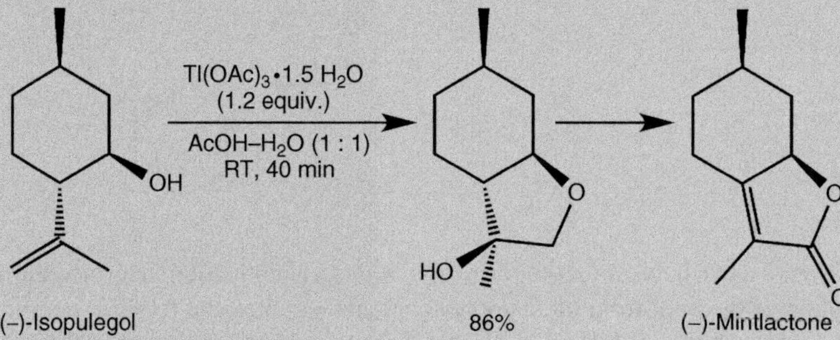

Note: This oxidation does not involve rearrangement of the carbon skeleton.

A classic thallium(III)-mediated transformation is the following oxidation of a ketone to a carboxylic acid with simultaneous skeletal rearrangement. (You'll recall from organic chemistry that, unlike aldehydes, ketones are generally resistant to oxidation, because there are no hydrogens on the carbonyl group.)

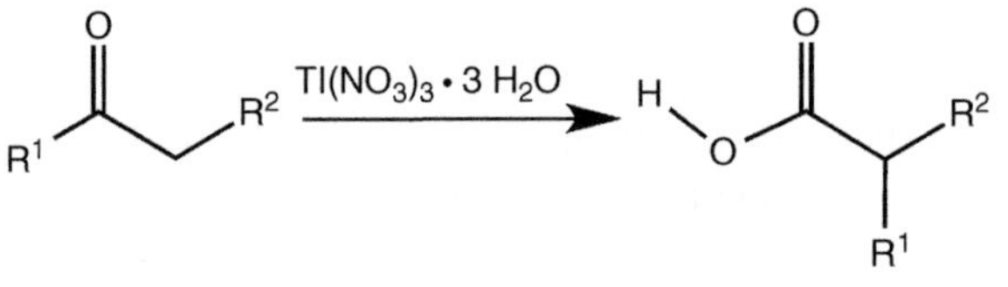

(3.53)

The mechanism of this reaction is a bit more complex than the one above, so it's all right if you want to skip it on first reading! The first few steps are expected to involve thallation of the enol form of the ketone:

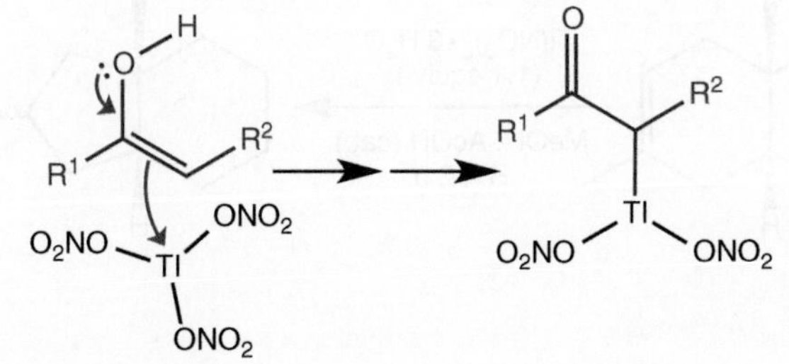

(3.54)

Water then attacks the protonated carbonyl group, and the *gem*-diol intermediate falls apart to the rearranged carboxylic acid product, while reductively eliminating monovalent thallium, as shown below:

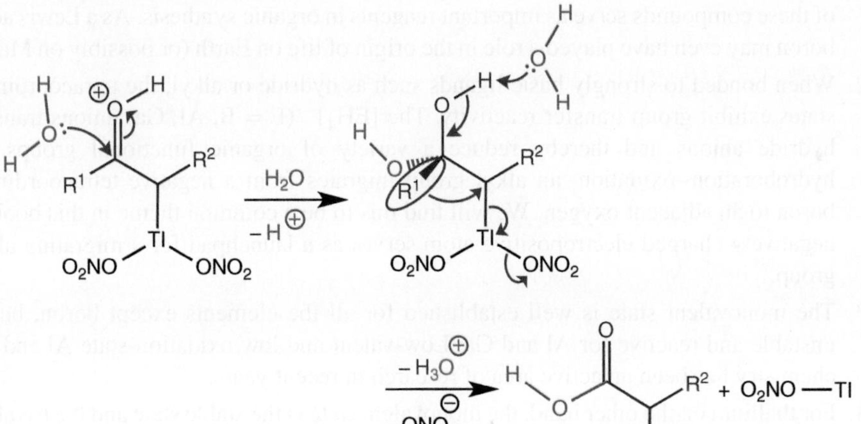

(3.55)

From a practical point of view, this is a mild and efficient way of accomplishing the overall transformation. The reaction is particularly useful for aryl alkyl ketones, as in the example below:

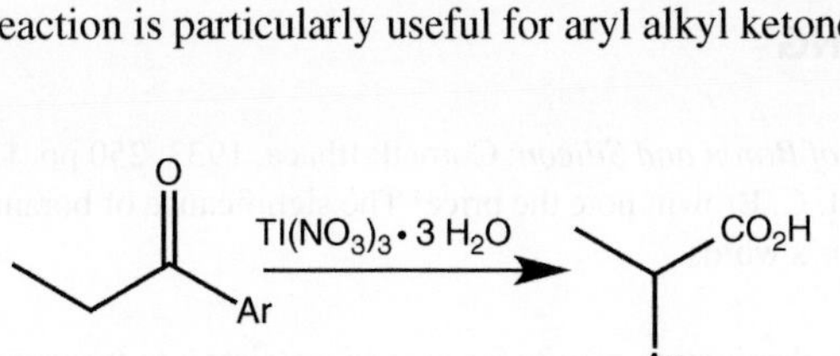

(3.56)

REVIEW PROBLEM 3.16*

Suggest a mechanism for the following oxidation–rearrangement:

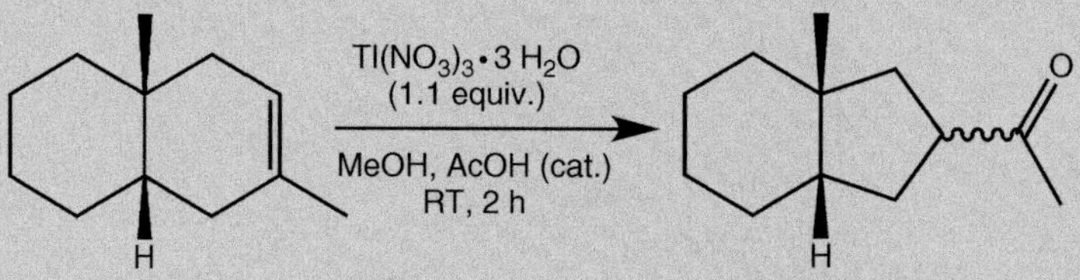

3.9 SUMMARY

The following are some highlights of what we discussed:

1. The trivalent state is the normal state for all the elements, except thallium. When the coordination number is also 3, the compounds act as powerful Lewis acids. A number of these compounds serve as important reagents in organic synthesis. As a Lewis acid, boron may even have played a role in the origin of life on Earth (or possibly on Mars).

2. When bonded to strongly basic ligands such as hydride or alkyl, the tetracoordinate states exhibit group transfer reactivity. The $[EH_4]^-$ (E = B, Al, Ga) anions transfer hydride anions and thereby reduce a variety of organic functional groups. In hydroboration–oxidation, an alkyl group migrates from a negative tetracoordinate boron to an adjacent oxygen. We will find this to be a common theme in this book: a negatively charged electropositive atom serves as a launchpad for a migrating alkyl group.

3. The monovalent state is well established for all the elements except boron, but is unstable and reactive for Al and Ga. Low-valent and low oxidation-state Al and Ga chemistry has been an active area of research in recent years.

4. For thallium on the other hand, the monovalent state is the stable state and the trivalent state is oxidizing one and prone to reductive elimination, thanks to the inert pair effect. Nevertheless, in spite of their toxicity, trivalent thallium reagents enjoy some unique applications as oxidants in organic chemistry.

FURTHER READING

1. Stock, A. *Hydrides of Boron and Silicon*; Cornell: Ithaca, 1933; 250 pp. $2.00. This is the historic book that inspired H. C. Brown; note the price! The significance of boranes (and silanes) is worth quoting in the author's words:

   In nature boron's dominating affinity for oxygen restricts it to the monotonous role of boric acid and the borates, and prevents it from competing with carbon, its neighbor in the periodic system.

It is very stimulating to observe the change from the rich chemical possibilities of carbon, the fountain of terrestrial life, as one passes to its three neighboring elements boron, silicon and nitrogen.

To be sure, the three refractory non-metals carbon, boron and silicon are so strikingly similar in their elemental form that even the older chemistry included them in a limited group. Their chemical behavior, however, showed hardly any recognizable similarities.

2. Benner, S. A.; Kim, H.-J-; Kim, M.-J-; Ricardo, A. "Planetary Organic Chemistry and the Origins of Biomolecules," *Cold Spring Harb. Perspect. Biol.* **2010**, *2*, a003467.
3. Brown, H. C. *Hydroboration*; 2nd ed. Benjamin: San Francisco, 1980; 336 pp. *A fun, early historical account.*
4. Aldrige, S.; Downs, A., eds. *The Group 13 Metals: Aluminium, Gallium, Indium and Thallium: Chemical Patterns and Peculiarities*; John Wiley & Sons, Inc.: Hoboken, NJ, 2011; 726 pp. *Chapters 1–6 provide detailed treatments of many of the topics discussed in this chapter.*
5. Lintl, G.; Schnöckel, H. "Low-Valent Aluminum and Gallium Compounds: Structural Variety and Coordination Modes to Transition Metal Fragments," *Coord. Chem. Rev.* **2000**, *206*, 285–319. *An excellent review by major contributors to the field.*
6. Wang, Y.; Robinson, G. "Carbene Stabilization of Highly Reactive Main-Group Molecules," *Inorg. Chem.* **2011**, *50*, 12326–12337. *A summary of an ingenious route to low–oxidation-state compounds pioneered by Robinson and his coworkers.*
7. Yamashita, M.; Nozaki, K. "Recent Developments of Boryl Anions: Boron Analogues of Carbanion," *Bull. Chem. Soc. Jpn.* **2008**, *81*, 1377–1392. *A history of these elusive anions.*
8. Silva, Jr., L. F.; Carneiro, V. M. T. "Thallium(III) in Organic Synthesis," *Synthesis* **2010**, 1059–1074. *A relatively short but excellent review article.*

4

Group 14 Elements

… sticky messes with no particular use …

Frederic Stanley Kipping, referring to his
newly discovered silicones, *Proc. Chem.
Soc.* **1905**, *21*, 65–66.

Since the few [organosilicon compounds]
which are known are very limited in their
reactions, the prospect of any immediate and
important advances in this section of organic
chemistry does not seem very hopeful.

Frederic Stanley Kipping, Bakerian lecture of the Royal Society, 1936

Our focus in this chapter is on group 14 elements other than carbon, namely, silicon, germanium, tin, and lead. As in the case of group 13, the properties change considerably as we go down the group, with the greatest discontinuity between carbon and silicon. Thus, carbon is a quintessential nonmetal, whereas silicon and germanium are semiconductors and tin and lead are metals. Compared with the s-block elements, both Sn and Pb are far less electropositive; unsurprisingly, therefore, all group 14 elements have an extensive covalent chemistry. The following are a few important group trends:

- The standard group valence is 4. For Si, Ge, and Sn, however, coordination numbers exceeding 4 occur regularly. The central atoms in such species are thus hypervalent, that is, they have more than an octet of valence electrons. For example, 2,2-bipyridyl reacts with triphenylchlorosilane to yield a relatively stable pentacoordinate silicon complex. Observe that, although the Si carries a −1 formal charge, the complex as a whole is cationic:

Arrow Pushing in Inorganic Chemistry: A Logical Approach to the Chemistry of the Main-Group Elements,
First Edition. Abhik Ghosh and Steffen Berg.
© 2014 John Wiley & Sons, Inc. Published 2014 by John Wiley & Sons, Inc.

$$(4.1)$$

By contrast, recall that coordination numbers greater than 4 are essentially unknown for carbon. Among the familiar species, the closest approach to pentacoordinate carbon is probably the S_N2 *transition state*, where the incoming nucleophile (Nu) and leaving group (LG) are *trans* with respect to each other and partially bonded to the central carbon:

- As mentioned in Section 1.10, the classic S_N2 pathway is not a major reaction pathway for atoms below the second period of the periodic table. Thus, for nucleophilic displacement at Si and at its heavier congeners, the two-step S_N2-Si mechanism is generally expected to operate. This is a fundamental issue that's worth thinking about: Why does the S_N2-Si mechanism operate for silicon but not for carbon? Stated differently, why is the S_N2 transition state of organic chemistry not an actual intermediate with five full bonds? The reason is believed to be primarily steric. Carbon and other second-row p-block elements are simply too small to accommodate five full bonds—unlike silicon and other heavier p-block elements.

- The group 14 elements are all rather oxophilic. Silicon is particularly so. Much of silicon chemistry and nearly all of silicon's numerous applications in organic synthesis hinge around the thermodynamic imperative to form highly stable Si–O bonds (bond dissociation energy (BDE) ~368 kJ/mol). That said, an Si–F (BDE ~540 kJ/mol) bond is even stronger by a considerable margin. Feel free to have a quick browse of Table 1.6 to appreciate the magnitudes of these BDEs relative to those for other bonds involving p-block elements.

- As might be expected for the carbon group, catenation is an important property for the elements, especially Si and Sn. However, the parent E_nH_{2n+2} (E = Si, Sn) compounds tend to be reactive and unstable; the compounds are much more stable when most or all of the hydrogens are replaced by alkyl or aryl groups. Thus, a variety of organopolysilanes and organopolystannanes are known.

- Because of the increased separation of the valence s and p orbital energies, the stability of the divalent state increases down the group. Thus, whereas divalent carbon (carbenes) and silicon (silylenes) compounds are generally unstable, there are many relatively stable divalent compounds of Ge, Sn, and Pb compounds. Divalent Ge and Sn compounds, however, are strong reducing agents. By contrast, divalent Pb compounds are stable and nonreducing. On the other hand, many tetravalent lead compounds are strongly oxidizing; lead tetraacetate (LTA), for example, is an important oxidizing agent in organic chemistry.

- Silyl anions are comparatively stable and may be generated under surprisingly mild conditions, relative to unstabilized carbanions. This may seem surprising, given that silicon is more electropositive than carbon. Once again, the reason appears to be the greater valence s–p energy gap for Si, relative to carbon; thus the Si lone pair of a silyl anion is thought to have greater s character and less p character than the lone pair of an analogous carbanion.

- Unlike in organic chemistry, multiple bonds are rare (indeed essentially unknown, except for a few designer molecules) for silicon as well as for other p-block elements below the second period. Taking Si as an example, a Si–Si bond is simply too long for effective sideways overlap of p orbitals that would result in a stable π-bond. This also explains why CO_2 is a gas under ordinary conditions, whereas SiO_2 is a high-melting polymeric solid, essentially pure sand!

With these general considerations in mind, we should now be in a position to explore group 14 chemistry.

REVIEW PROBLEM 4.1

Dissolved carbon dioxide consists largely of hydrogen-bonded CO_2 molecules, existing in equilibrium with a small quantity of bicarbonate and carbonate anions. Indeed, until recently, true carbonic acid, H_2CO_3, was thought not to exist. It does exist, however, and has been recently isolated as the pure, unhydrated solid at low temperature (Loerting, T., *et al. Angew. Chem. Int. Ed.* **2000**, *39*, 892–894). Despite this remarkable proof of kinetic stability, carbonic acid is thermodynamically unstable with respect to dehydration, and a water-catalyzed pathway involving a six-membered transition state is thought to provide one of the lower-energy paths for the process.

$$H_2CO_3 + H_2O \rightarrow CO_2 + 2\,H_2O$$

Use arrow pushing to suggest a mechanism for the process.

REVIEW PROBLEM 4.2

If you are a science fiction fan, you may have heard about the prospect of silicon-based life. What is your opinion on this issue, especially for an aerobic planetary home?

4.1 SILYL PROTECTING GROUPS

Substituted silyl groups (R_3Si-, where R is alkyl or aryl) are widely used as protecting groups for alcohols. A variety of reagents are available for this purpose, from simple and inexpensive to highly sophisticated. A simple approach consists of treating the alcohol with a substituted silyl chloride, in the presence of a base (e.g., pyridine or triethylamine), which neutralizes the HCl produced:

$$ROH + Me_3SiCl \xrightarrow[\text{– HCl}]{\text{Pyridine}} ROSiMe_3 \qquad (4.2)$$

As noted before (Section 1.10), the mechanism is S_N2-Si, leading first to a five-coordinate intermediate:

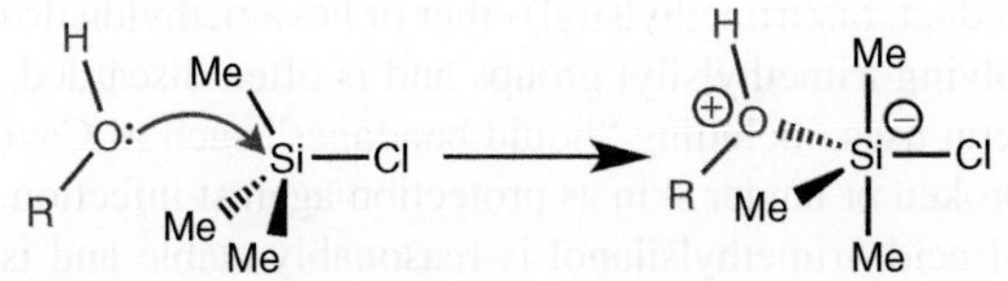

$$(4.3)$$

The base then plucks off the proton from the coordinated alcohol; chloride leaves, possibly at the same time or in the next step, producing a silyl ether of the alcohol as the product.

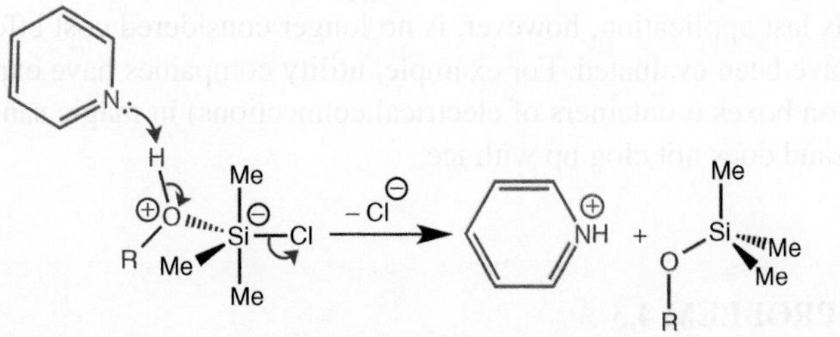

$$(4.4)$$

The thermodynamic driving force for the reaction lies in silicon's strong oxophilicity, a property that is also shared by phosphorus. Besides trimethylsilyl chloride (the reagent used above), a variety of other silylating agents are also used. Depending on the steric bulk of the silyl group, the reagent can be used to selectively protect different kinds (1°, 2°, or 3°) of alcohols.

Once the protected alcohol has been manipulated in any of a myriad ways, one typically needs to take off the silyl group, or "deprotect" the alcohol, to use the technical term. This is often done simply with aqueous acid. Water attacks the protonated silyl ether, as shown below, freeing the alcohol and generating trimethylsilanol:

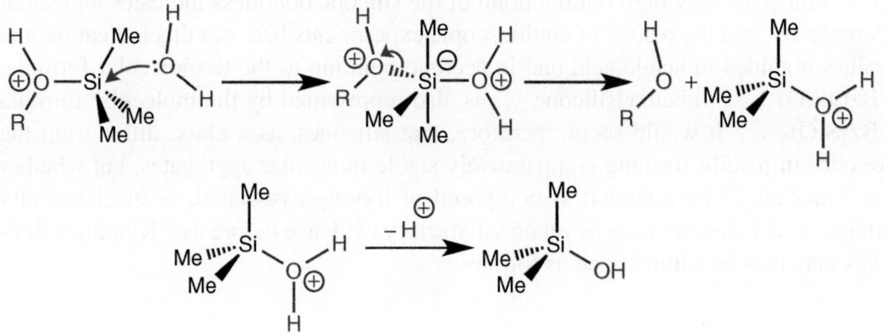

$$(4.5)$$

Silanols are unstable under acidic conditions and readily self-condense as shown below:

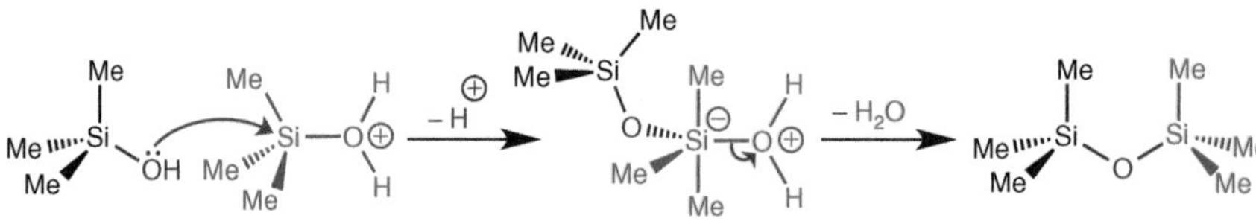

$$\tag{4.6}$$

The highly stable product, bis(trimethylsilyl) ether or hexamethyldisiloxane, is produced in many reactions involving trimethylsilyl groups and is often discarded. Its inertness, however, makes for several uses, including "liquid bandages" such as Cavilon™ spray, which can be sprayed on broken or tender skin as protection against infection.

In the absence of acid, trimethylsilanol is reasonably stable and is used for applying hydrophobic coatings on silicate surfaces (including sand), which it covers with a layer of hydrophobic groups. A fun example is magic sand, which can be used to build sand castles under water but is dry when taken out of the water. Magic sand was originally developed for trapping oil spills. Sprinkled on floating petroleum, it adsorbs oil, clumps together, and sinks. This last application, however, is no longer considered cost effective, but other applications have been evaluated. For example, utility companies have experimented with burying junction boxes (containers of electrical connections) in magic sand. Being free of water, magic sand does not clog up with ice.

REVIEW PROBLEM 4.3

The British chemist Frederic Stanley Kipping studied the hydrolysis of organosilicon dichlorides (R_2SiCl_2) and named the products *silicones*, by analogy with ketones. As shown by the depressing quotes at the beginning of this chapter, Kipping continued to be disappointed with his discoveries as late as 1936. About the hydrolysis product of $PhCH_2(C_2H_5)SiCl_2$ ($= BzEtSiCl_2$; $Bz = PhCH_2$) he wrote (Robinson, R.; Kipping, F. S. *J. Chem. Soc.* **1908**, *93*, 439): " … as benzylethylsilicon dichloride is decomposed by water, giving benzylethylsilicone, we have studied the behavior of this silicone in order to ascertain whether it shows any similarities to the corresponding ketone. We may say at once that it does not; benzyl ethyl ketone boils at 226° under atmospheric pressure; benzylethylsilicone at 305–315° under a pressure of 22 mm. This very high boiling point of the silicone doubtless indicates molecular complexity, and the results of ebullioscopic experiments bear out this indication, the values obtained in acetic acid and in acetone pointing to the termolecular formula, $(BzEtSiO)_3$. … dibenzylsilicone … is also represented by the molecular formula $(Bz_2SiO)_3$ … It would seem, therefore, that silicones, as a class, differ from the ketones in readily forming comparatively stable molecular aggregates, but whether the latter are to be regarded as composed of loosely associated, or of chemically united, molecules, we have as yet no satisfactory evidence before us." Kipping's findings may thus be summarized as follows:

$$Bz = PhCH_2$$

Based on your mechanistic knowledge, provide an explanation of this chemistry that Kipping found so baffling.

As mentioned, silicon-based protecting group chemistry hinges on silicon's well-known oxophilicity. That said, although Si–O bonds are strong, Si–F bonds are much stronger. Thermodynamic considerations thus explain why sand and glass dissolve in HF and why you shouldn't use glass vessels to handle HF. What this means for protection/deprotection chemistry is that fluoride may be used to deprotect silyl ethers. Typically, a milder surrogate of HF such as pyridinium fluoride (py·HF) or tetrabutylammonium fluoride (TBAF) in THF/H_2O is used for this purpose:

$$Me_3SiOR \xrightarrow[\text{TBAF, THF/H}_2\text{O}]{\overset{\text{py·HF}}{\text{or}}} ROH + Me_3SiF \qquad (4.7)$$

Note that there must be a source of protons for generating the alcohol, so TBAF alone would *not* do the job:

$$(4.8)$$

REVIEW PROBLEM 4.4

Carbon tetrafluoride and tetrachloride do not react with water under ordinary conditions. Silicon tetrachloride, by contrast, is instantly hydrolyzed by water.

$$SiCl_4 + 4\,H_2O \rightarrow Si(OH)_4 + 4\,H^+ + 4\,Cl^-$$

Silicon tetrafluoride also reacts vigorously with water, but the reaction may be described as an incomplete hydrolysis:

$$2\,SiF_4 + 4\,H_2O \rightarrow Si(OH)_4 + SiF_6{}^{2-} + 2\,H^+ + 2\,HF$$

Explain the different behaviors exhibited by the different group 14 halides.

Note: $Si(OH)_4$ is a simplified formula; the actual product is a mixture of polymeric silanols, homologs of $(HO)_3Si-O-Si(OH)_3$.

Reagents of the form R_3SiY, where Y^- is a nucleophile, are an important class of organosilicon reagents. They are primarily used for addition of Y^- to carbonyl groups. Space will not permit a proper discussion here, but a good example of such a reagent is trimethylsilyl cyanide (Me_3SiCN), which we discussed in Section 1.14. Another excellent example is the Ruppert–Prakash reagent (Me_3SiCF_3), discussed below in Review problem 4.5.

Enolizable carbonyl compounds (i.e., those with a C–H bond next to the carbonyl group) undergo a somewhat special reaction with trimethylsilyl chloride to yield silyl enol ethers.

$$(4.9)$$

We urge you to try to write a mechanism for the above process. You may find it useful to think of silyl groups as "fat protons." Start off with "silylating" the carbonyl oxygen and proceed from there. Silyl enol ethers function as less reactive analogs of enolate anions. Thus, with titanium tetrachloride as a catalyst, silyl enol ethers undergo aldol condensations with a variety of carbonyl compounds. This is the Mukaiyama aldol condensation:

$$(4.10)$$

A likely cyclic transition state makes the Mukaiyama aldol condensation diastereoselective, that is, strongly favoring one diastereomer of the product. Addition of a chiral ligand or chelator (such as a chiral diamine) can make the process enantioselective as well, that is, favoring one enantiomer.

REVIEW PROBLEM 4.5

Trimethyl(trifluoromethyl)silane (Me_3SiCF_3), commonly referred to as the *Ruppert–Prakash reagent*, serves as a trifluoromethyl anion equivalent, which is remarkable because of the instability of the anion with respect to decomposition to difluorocarbene and fluoride:

$$CF_3^- \rightarrow CF_2 + F^-$$

The reagent works well for a variety of carbonyl compounds, both aldehydes and ketones:

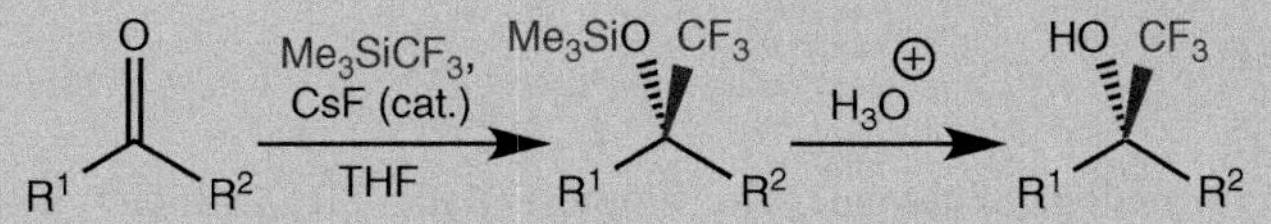

The reaction depends on a fluoride ion source such as cesium fluoride or TBAF as catalyst. Provide a mechanistic rationale for the transformation, as well as for the catalytic role of fluoride.

4.2 A CASE STUDY: PETERSON OLEFINATION

In Peterson olefination, an α-silylcarbanion is reacted with a carbonyl compound to give a β-hydroxysilane, after mild acidic work-up (e.g., with aqueous NH_4Cl):

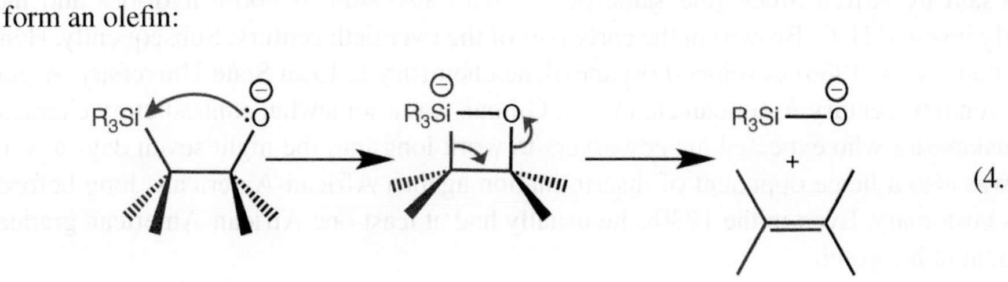

$$(4.11)$$

Under basic conditions, the alkoxide derived from the β-hydroxysilane eliminates silanolate to form an olefin:

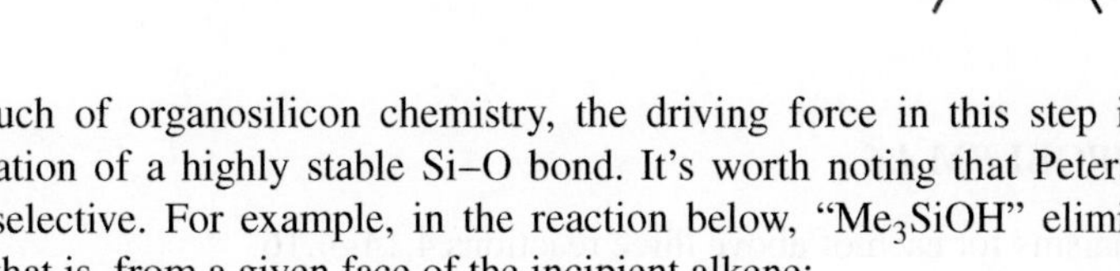

$$(4.12)$$

As in much of organosilicon chemistry, the driving force in this step is provided by the formation of a highly stable Si–O bond. It's worth noting that Peterson olefination is stereoselective. For example, in the reaction below, "Me_3SiOH" eliminates in a *syn* manner, that is, from a given face of the incipient alkene:

$$(4.13)$$

Yield: 96%
$E : Z = 95 : 5$

Unfortunately, limitations of space do not permit a more detailed discussion of the origin of this stereoselectivity.

4.3 SILANES

Silanes are silicon analogs of alkanes. The simplest silane, SiH_4, which is a pyrophoric gas, can be prepared in the laboratory by hydrolysis of magnesium silicide:

$$Mg_2Si + 4\, HCl \rightarrow SiH_4 + 2\, MgCl_2 \qquad (4.14)$$

Organosilanes, where the Si atoms also carry alkyl groups, may be synthesized via standard organic reactions. Thus, phenylsilane ($PhSiH_3$) may be prepared by lithium aluminum hydride reduction of $PhSiCl_3$.

$$4\, PhSiCl_3 + 3\, LiAlH_4 \rightarrow 4\, PhSiH_3 + 3\, LiCl + 3\, AlCl_3 \qquad (4.15)$$

Trimethylsilyl chloride and silicon tetrachloride may be combined via an "organolithium chemistry" approach to yield tetrakis(trimethylsilyl)silane:

$$(4.16)$$

   A quick historical digression might be of interest. The groundwork of silane chemistry was laid by Alfred Stock (the same person who also studied boron hydrides and indirectly inspired H. C. Brown) in the early part of the twentieth century. Subsequently, Henry Gilman (1893–1986) developed organosilane chemistry at Iowa State University. A giant of twentieth-century American chemistry, Gilman was a somewhat contradictory character. A taskmaster who expected his coworkers to work long into the night seven days a week, he was also a fierce opponent of discrimination against African-Americans long before it was customary. Even in the 1930s, he usually had at least one African-American graduate student in his group.

REVIEW PROBLEM 4.6

Draw mechanisms for each of above three reactions 4.14–4.16.

Silicon being more electropositive than hydrogen (Pauling electronegativities: Si 1.90, H 2.20), silanes with Si–H bonds can reduce a variety of functional groups such as carbonyl compounds, imines, and even alcohols by transferring a hydride anion. Triethylsilane is a common reagent, and typical reaction conditions also include either a Brønsted or a Lewis acid, as shown below for the reduction of a ketone:

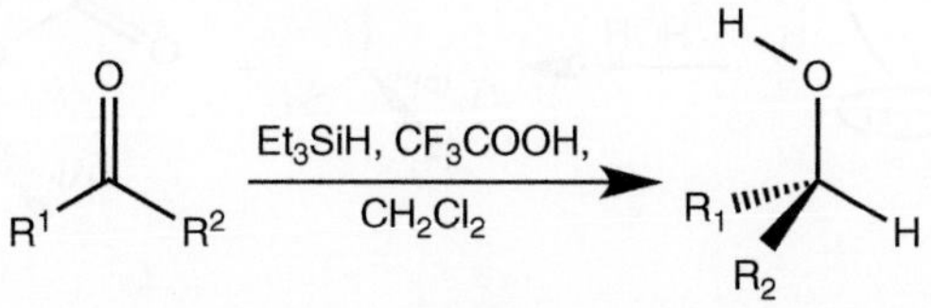

$$
\begin{array}{c}
\text{(4.17)}
\end{array}
$$

The first step of the mechanism is expected to be a protonation:

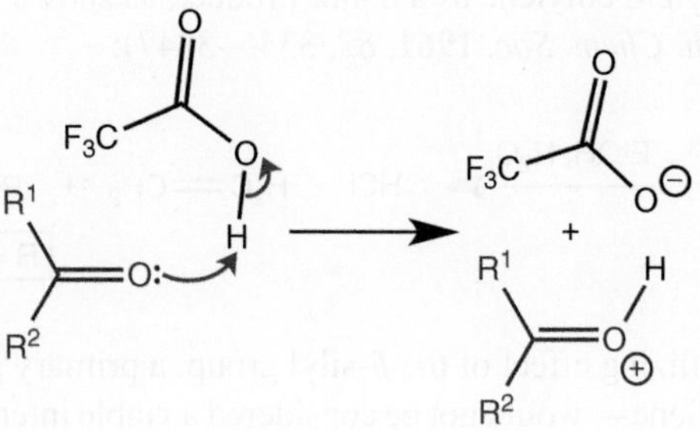

$$
\begin{array}{c}
\text{(4.18)}
\end{array}
$$

The trifluoroacetate anion now coordinates the Si, producing an anionic Si center, which then transfers a hydride anion:

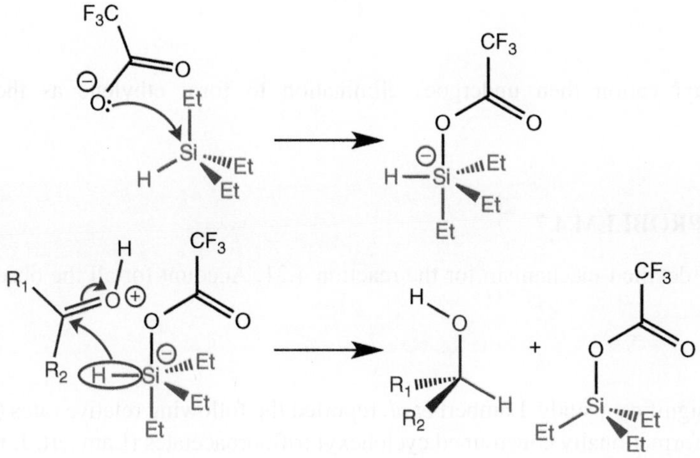

$$
\begin{array}{c}
\text{(4.19)}
\end{array}
$$

Triethylsilane can even reduce alcohols to alkanes. This is illustrated below for a protonated secondary alcohol, R^1R^2CHOH:

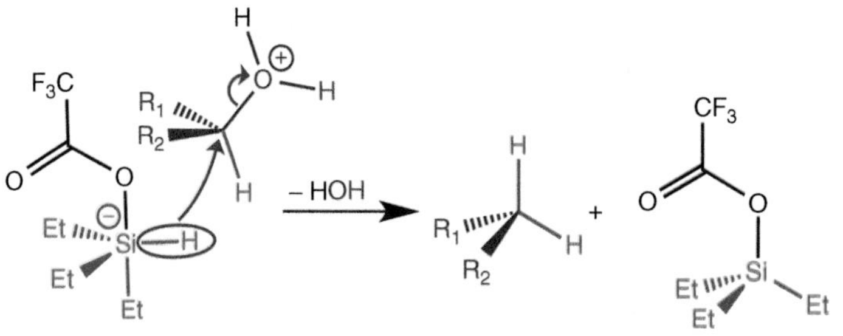

(4.20)

4.4   THE β-SILICON EFFECT: ALLYLSILANES

In Section 1.11, we referred to the stabilizing effect of β-silyl substituents. In an important early experiment, β-trimethylsilyl-substituted chloroethane was found to undergo solvolysis in ethanol/water to yield ethylene as a major product, as shown below (Sommer, L. H.; Baughman, G. A. *J. Am. Chem. Soc.* **1961**, *83*, 3346–3347):

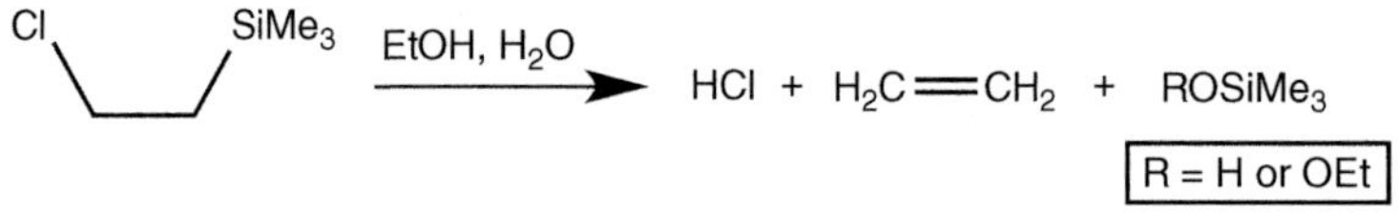

$$HCl + H_2C{=}CH_2 + ROSiMe_3$$

$$\boxed{R = H \text{ or OEt}}$$

(4.21)

Were it not for the stabilizing effect of the β-silyl group, a primary alkyl carbocation—the expected source of ethylene—would not be considered a viable intermediate. As it happens, the β-silyl group strongly stabilizes the primary carbocation via hyperconjugative electron donation:

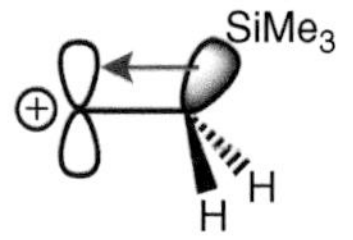

The β-silylethyl cation then undergoes elimination to form ethylene as the major product.

REVIEW PROBLEM 4.7

Write out a detailed mechanism for the reaction 4.21. Account for all the observed products.

In another significant study, Lambert *et al.* reported the following relative rates (k_{rel}) for a series of conformationally constrained cyclohexyl trifluoroacetates (Lambert, J. B., *et al.*

J. Am. Chem. Soc. **1987**, *109*, 7838–7845)[1]:

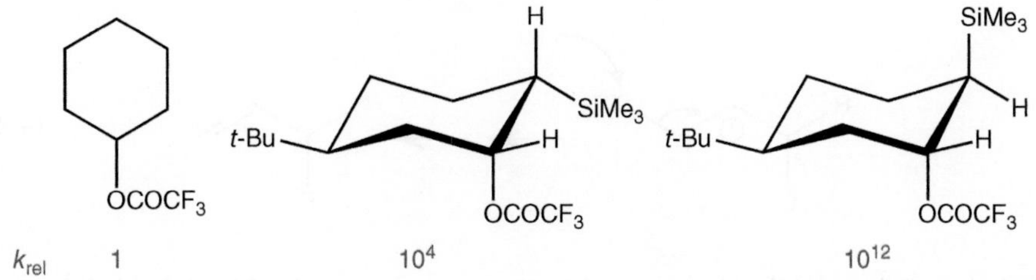

(4.22)

The dramatic rate enhancement for the third compound may be attributed to the almost perfectly parallel alignment of the carbocation's empty p orbital and the β-C–Si bond that allows for maximum hyperconjugation, as shown below:

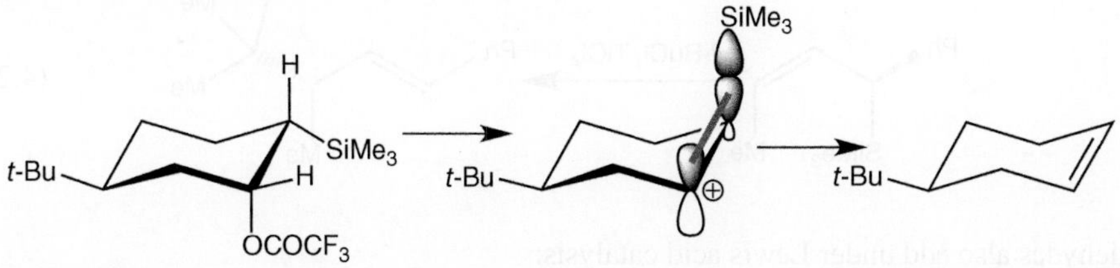

(4.23)

As expected for carbocation intermediates, these reactions produced 4-*t*-butylcyclohexene as a major product.

The β-silicon effect on carbocation stability has been widely exploited in organic synthesis, perhaps most notably in the reactions of allylsilanes toward electrophiles. In this, allylsilanes behave remarkably like silyl enol ethers, as shown by the following generic transformations:

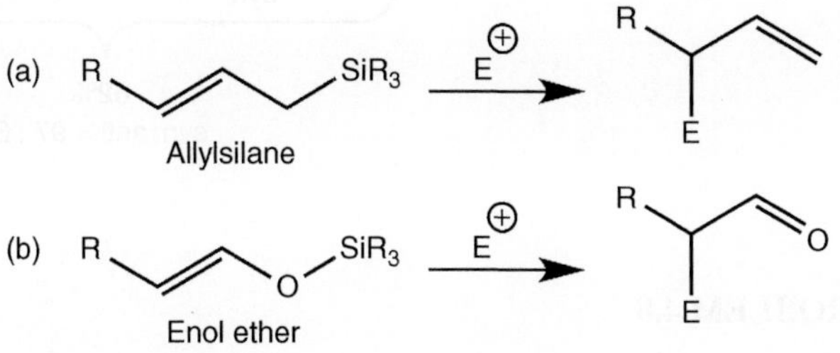

(4.24)

Addition of an electrophile (E^+) to an allylsilane typically generates a stable β-silylcarbocation, as shown below:

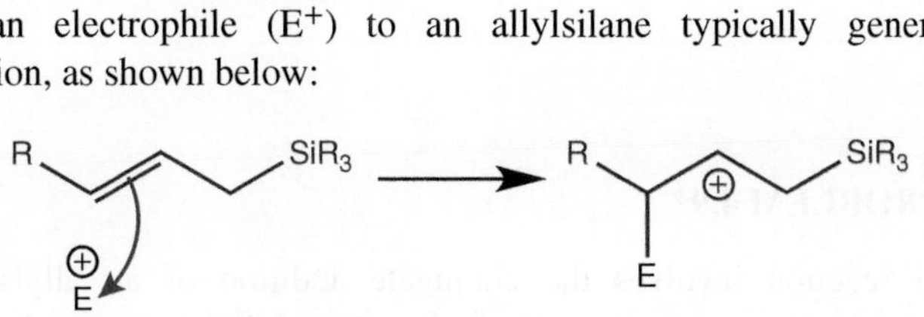

(4.25)

[1] Bulky substituents such as *t*-butyl have a strong preference for an equatorial position on a cyclohexane ring. This prevents the ring from flipping, which would result in an axial position for the bulky group. Feel free to revisit your organic text, if you need to refresh your knowledge of cyclohexane conformations.

Subsequent elimination of the silyl group leads to the alkene product:

$$(4.26)$$

The choice of electrophiles that react with allylsilanes is considerable. Alkyl halides typically need to be tertiary, allylic, or benzylic and they also require a Lewis acid catalyst, as in the example below:

$$(4.27)$$

Aldehydes also add under Lewis acid catalysis:

62%
syn : anti = 97 : 3

$$(4.28)$$

REVIEW PROBLEM 4.8

Write out detailed mechanisms for reactions 4.27 and 4.28.

REVIEW PROBLEM 4.9*

The Sakurai reaction involves the conjugate addition of an allylsilane to an α,β-unsaturated ketone in the presence of a Lewis acid. Suggest a mechanism for the following intramolecular Sakurai reaction:

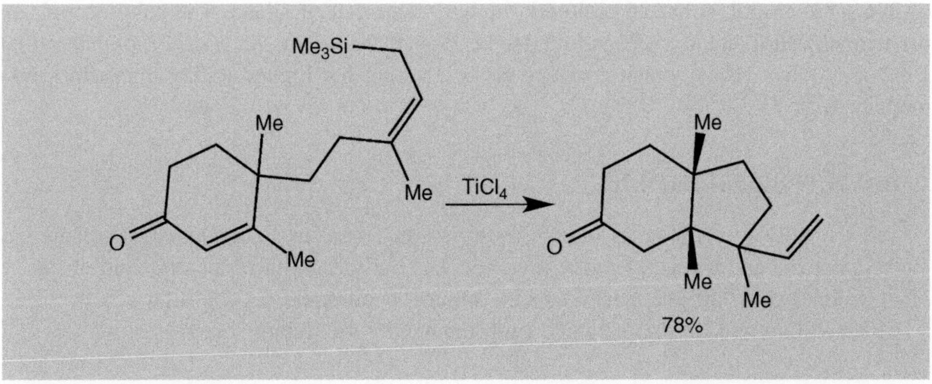

78%

4.5 SILYL ANIONS

Silyl anions are not commonly mentioned in introductory inorganic courses, but they should be! Many silyl anions are considerably more stable than their carbanion analogs and many can be generated with surprising ease. Thus a tertiary amine suffices to deprotonate trichlorosilane:

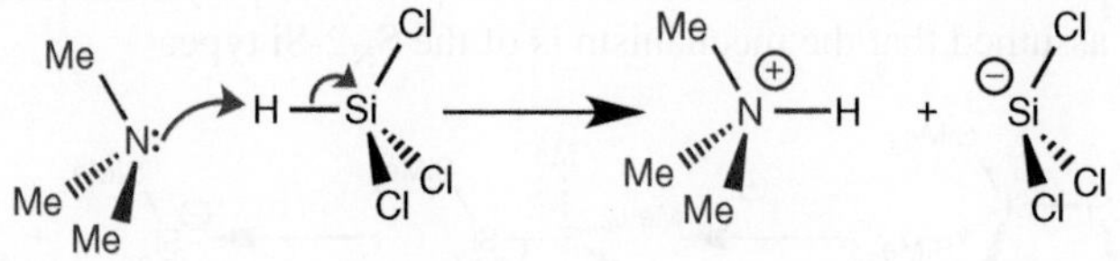

$$(4.29)$$

The trichlorosilyl anion so produced can be put to a variety of uses. With a simple alkyl halide, it carries out an S_N2 displacement, producing an alkyltrichlorosilane:

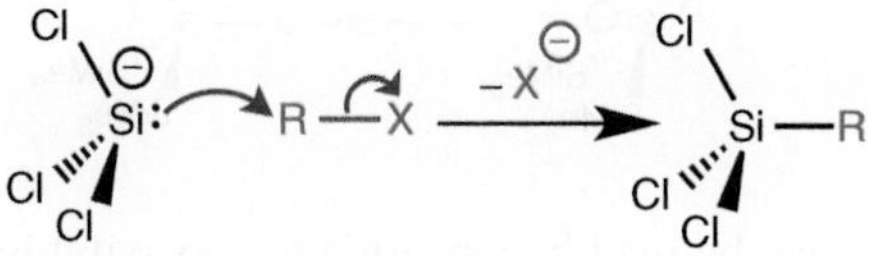

$$(4.30)$$

With hexachloroethane, the Cl_3Si^- anion does an E2 elimination to produce tetra-chloroethylene, as shown below:

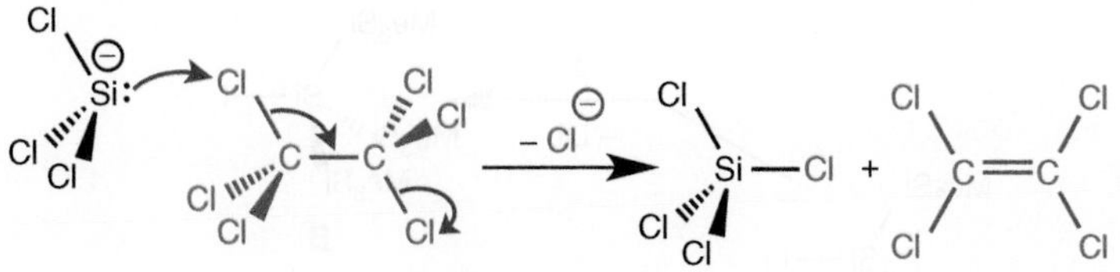

$$(4.31)$$

Let's put the matter in an energetics perspective. The gas-phase acidity of silane (i.e., the $\Delta H°$ for the reaction $SiH_4 \rightarrow SiH_3^- + H^+$) is about 372 ± 2 kcal/mol, which is much "higher" than that of methane (416 kcal/mol). Note that a lower number implies a higher acidity. Methyl groups decrease the acidity of silanes, but trimethylsilyl groups have a dramatic acidity-enhancing effect. Tris(trimethylsilyl)silane thus has a gas-phase acidity of 346 ± 3 kcal/mol, some 26 kcal/mol "higher" than that of silane. The solution pK_a of tris(trimethylsilyl)silane has been found to be 29.4 (Korogodsky, G., *et al. Organometallics* **2002**, *21*, 3157–3161), which corresponds to a somewhat higher acidity than triphenylmethane (pK_a 31.3).

REVIEW PROBLEM 4.10

Tris(trimethylsilyl)silane is much more acidic than tris(trimethylsilyl)methane ($pK_a = 36.8$) and triphenylsilane ($pK_a = 35.1$). Can you think of an explanation for these relative acidities? Also, can you suggest a rationale for the high acidity of trichlorosilane (recall that it can be deprotonated by triethylamine).

In 1968, Gilman and Smith showed that tetrakis(trimethylsilyl)silane can be cleaved by methyllithium to generate tris(trimethylsilyl)silyllithium (**A** in the reaction below), sometimes also called *supersilyllithium* (*J. Organomet. Chem.* **1968**, *14*, 91):

$$Si(SiMe_3)_4 + MeLi \longrightarrow (Me_3Si)_3SiLi + Me_4Si$$
$$\mathbf{A}$$
(4.32)

This discovery proved crucial to the further development of polysilane chemistry. Below, we have tacitly assumed that the mechanism is of the S_N2-Si type:

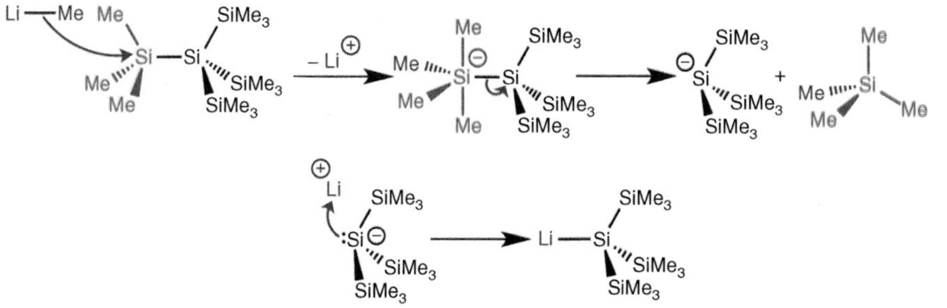

(4.33)

Supersilyllithium (**A**) may be readily converted to supersilyl halides (**B** and **C** below), which are useful for introducing supersilyl groups into organic molecules. Supersilyllithium is also a key building block in the synthesis of branched organosilanes (e.g., **D**) and of even bulkier silyllithiums (e.g., **E**). All these processes are depicted below.

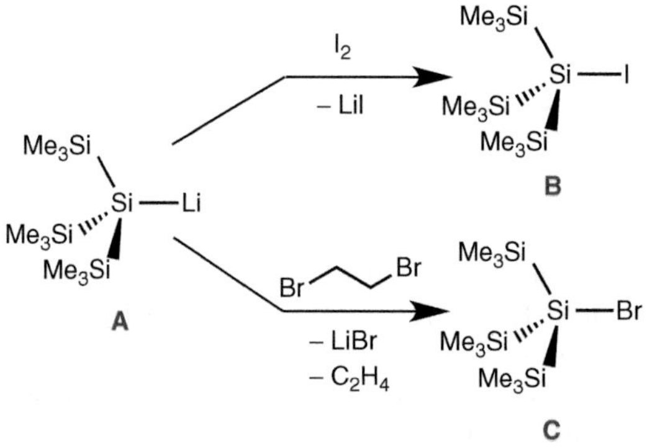

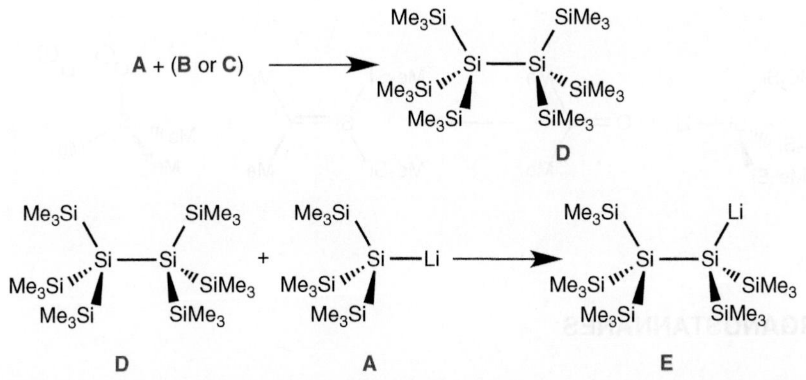

$$A + (B \text{ or } C) \longrightarrow \overset{\displaystyle Me_3Si \quad\quad SiMe_3}{\underset{\displaystyle Me_3Si \quad\quad SiMe_3}{Me_3Si^{\backslash\backslash\backslash}\overset{\displaystyle |}{Si}-\overset{\displaystyle |}{Si}^{\cdots\backslash\backslash}SiMe_3}}$$

D

(4.34)

$$\underset{D}{\overset{\displaystyle Me_3Si \quad\quad SiMe_3}{\underset{\displaystyle Me_3Si \quad\quad SiMe_3}{Me_3Si^{\backslash\backslash\backslash}\overset{\displaystyle |}{Si}-\overset{\displaystyle |}{Si}^{\cdots\backslash\backslash}SiMe_3}}} + \underset{A}{\overset{\displaystyle Me_3Si}{\underset{\displaystyle Me_3Si}{Me_3Si^{\backslash\backslash\backslash}\overset{\displaystyle |}{Si}-Li}}} \longrightarrow \underset{E}{\overset{\displaystyle Me_3Si \quad\quad Li}{\underset{\displaystyle Me_3Si \quad\quad SiMe_3}{Me_3Si^{\backslash\backslash\backslash}\overset{\displaystyle |}{Si}-\overset{\displaystyle |}{Si}^{\cdots\backslash\backslash}SiMe_3}}}$$

REVIEW PROBLEM 4.11

Suggest reasonable mechanisms for the set of reactions 4.34.

Recently, it has been found that one doesn't need to resort to organolithium reagents to generate stoichiometric quantities of silyl anions. Potassium *t*-butoxide in THF, which is easier to handle than methyllithium, rapidly generates supersilylpotassium from tetrakis(trimethylsilyl)silane, as shown below:

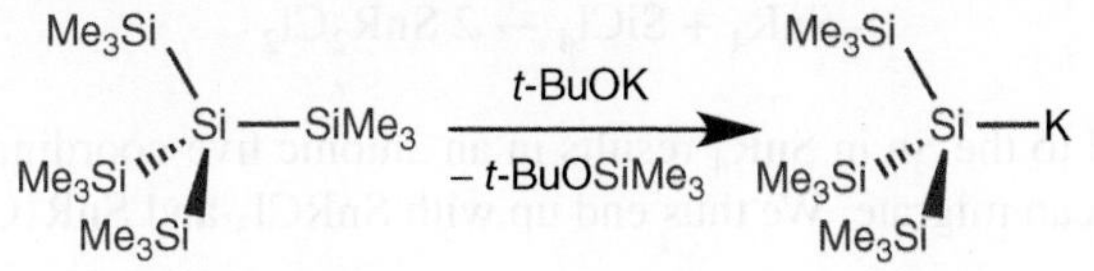

$$\underset{\displaystyle Me_3Si}{\overset{\displaystyle Me_3Si}{Me_3Si^{\backslash\backslash\backslash}\overset{\displaystyle |}{Si}-SiMe_3}} \xrightarrow[\displaystyle -\,t\text{-BuOSiMe}_3]{\displaystyle t\text{-BuOK}} \underset{\displaystyle Me_3Si}{\overset{\displaystyle Me_3Si}{Me_3Si^{\backslash\backslash\backslash}\overset{\displaystyle |}{Si}-K}}$$

(4.35)

Like its lithium analog, supersilylpotassium has proved to be an excellent building block for organopolysilanes, as shown below:

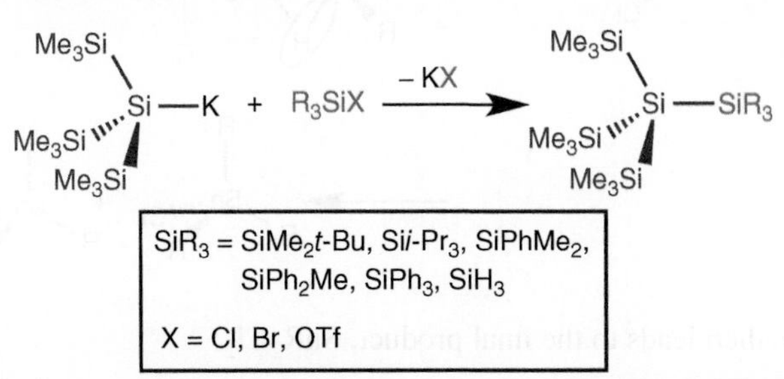

$$\underset{\displaystyle Me_3Si}{\overset{\displaystyle Me_3Si}{Me_3Si^{\backslash\backslash\backslash}\overset{\displaystyle |}{Si}-K}} + R_3SiX \xrightarrow{-KX} \underset{\displaystyle Me_3Si}{\overset{\displaystyle Me_3Si}{Me_3Si^{\backslash\backslash\backslash}\overset{\displaystyle |}{Si}-SiR_3}}$$

(4.36)

> SiR$_3$ = SiMe$_2$*t*-Bu, Si*i*-Pr$_3$, SiPhMe$_2$,
> SiPh$_2$Me, SiPh$_3$, SiH$_3$
>
> X = Cl, Br, OTf

REVIEW PROBLEM 4.12

In a creative application of Peterson olefination, supersilyllithium was reacted with acetone to give a silaalkene, as shown below. Suggest a mechanism for the reaction:

4.6 ORGANOSTANNANES

Organostannanes are readily synthesized from the corresponding Grignard reagents and tributyltin chloride:

$$(4.37)$$

Ligand exchange reactions are an important reaction pathway for organostannanes, for example:

$$SnR_4 + SiCl_4 \rightarrow 2\, SnR_2Cl_2 \tag{4.38}$$

Coordination of Cl to the Sn in SnR_4 results in an anionic five-coordinate Sn center, from which an R group can migrate. We thus end up with $SnRCl_3$ and SnR_3Cl:

$$(4.39)$$

A similar process then leads to the final product, SnR_2Cl_2:

$$(4.40)$$

On account of the electronegativity difference between carbon and tin (Pauling electronegativities: C 2.55, Sn 1.96), organostannanes add to carbonyl compounds, particularly aldehydes (the reaction with ketones is typically slow). With allylstannanes, the reaction creates two stereogenic centers in one stroke, generally with high diastereoselectivity. The selectivity can be explained with a chair transition state (as for many aldol reactions), much as we saw for indium-mediated allylations:

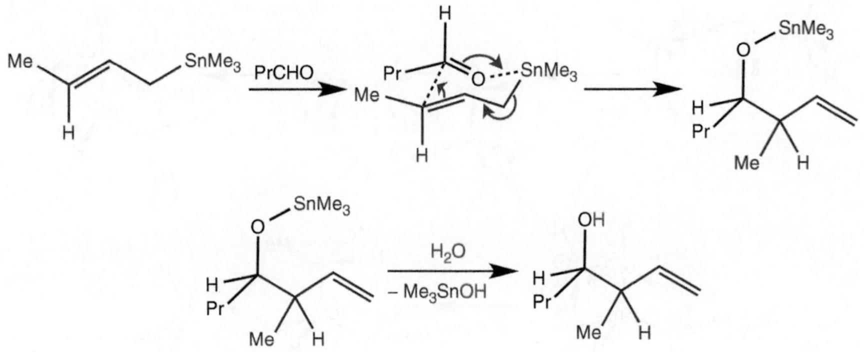

$$(4.41)$$

4.7 POLYSTANNANES

There are numerous ways of forming bonds between two group 14 atoms. An important one, discussed above, involves the nucleophilic attack of a silyl anion on a silyl halide. Another method, called *hydrostannolysis*, involves condensation of a Sn–H bond and a Sn–NMe$_2$ linkage, with the elimination of HNMe$_2$. The following reaction leads to a branched Sn$_4$ framework:

$$RSnH_3 + 3\ Me_3SnNMe_2 \rightarrow (Me_3Sn)_3SnR + 3\ HNMe_2 \qquad (4.42)$$

By analogy with silyl anion chemistry, it's not unreasonable to postulate stannyl anion intermediates, the first of which could be produced as depicted below:

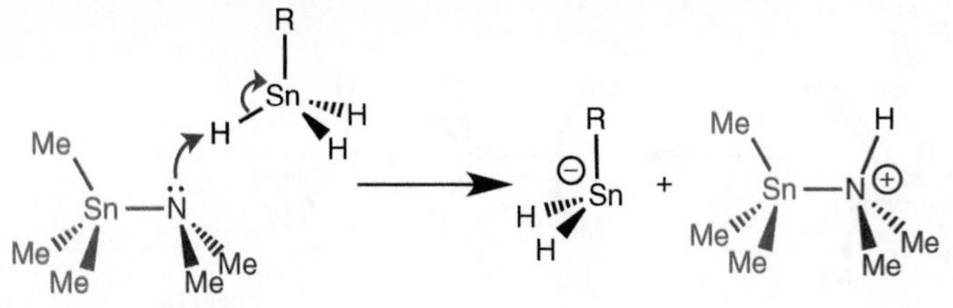

$$(4.43)$$

The stannyl anion is an effective nucleophile that may kick out the first HNMe$_2$ molecule. A two-step S$_N$2-Si mechanism is as follows:

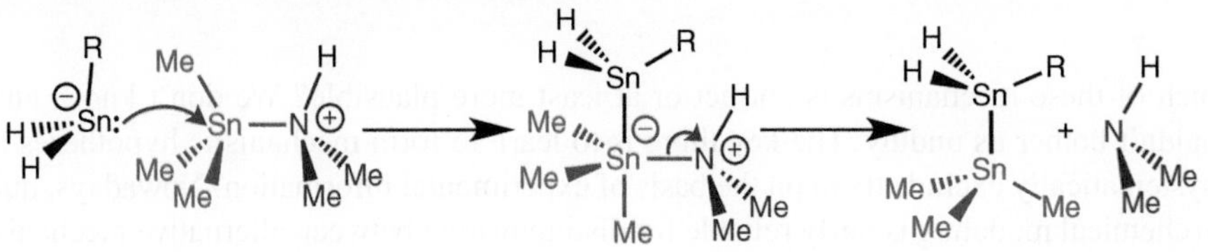

$$(4.44)$$

The distannane so produced may be deprotonated again, and the resulting anion may carry out a second nucleophilic displacement to produce a tristannane. The process continues until all the Sn–H bonds have been consumed and replaced by Sn–Sn bonds.

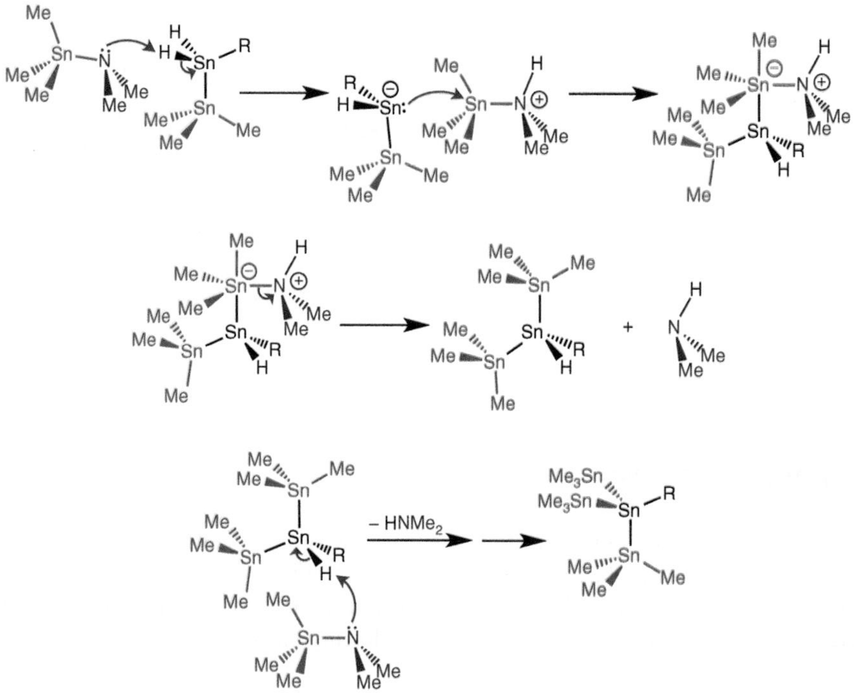

(4.45)

An alternative to the above mechanism (involving stannyl anions) is to consider a concerted cyclic mechanism, as shown below:

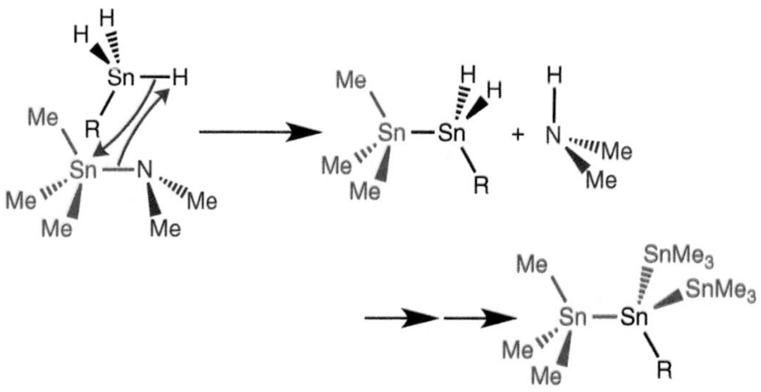

(4.46)

Which of these mechanisms is correct or at least more plausible? We don't know, and it shouldn't bother us unduly. The key thing is to learn to form mechanistic hypotheses and to systematically exclude them on the basis of experimental information. Nowadays, quantum chemical modeling is fairly reliable for distinguishing between alternative mechanistic pathways.

REVIEW PROBLEM 4.13

Dialkyltin dichlorides couple via the Wurtz reaction (see Section 2.3) to yield long-chain polystannanes, as shown below, along with cyclic byproducts.

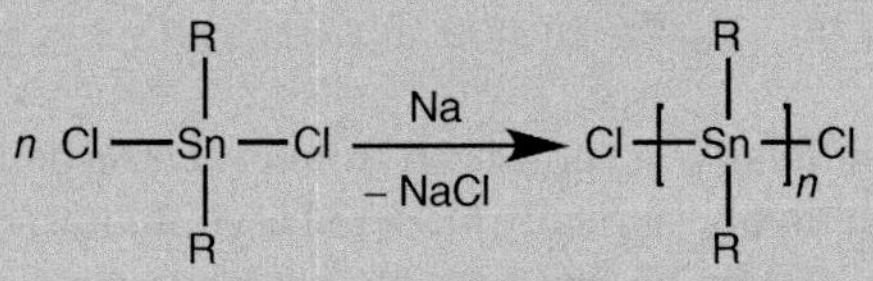

The polystannanes thus obtained are of interest for their semiconducting and liquid-crystalline properties. Provide a mechanistic rationale for the polymerization.

4.8* CARBENE AND ALKENE ANALOGS

Germanium, tin, and lead form stable dihalides with the general formula EX_2. In the gas phase or in noble gas matrixes, they exhibit bent structures, reflecting their "stereo-chemically active" lone pairs (or based on VSEPR considerations). Like trivalent boron compounds, they are Lewis acids and readily accept a lone pair to complete their octets, as shown below for the formation of the $GeCl_2 \cdot$dioxane complex:

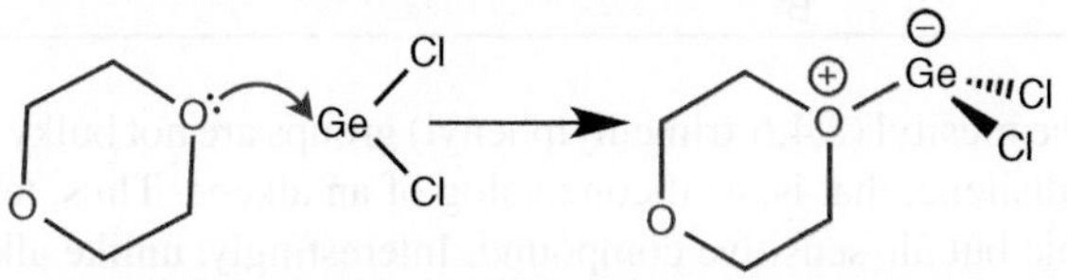

(4.47)

In the solid state as well, the dihalides associate so as to complete their octets as shown below for crystalline $SnCl_2$, $SnCl_2 \cdot 2\,H_2O$, and $SnCl_2 \cdot CsCl$:

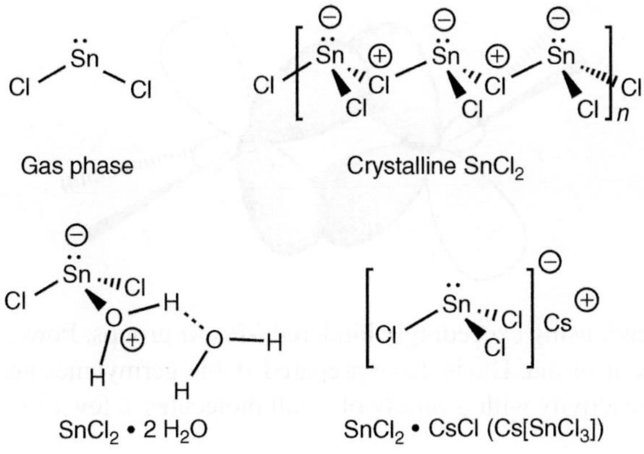

Gas phase

Crystalline $SnCl_2$

$SnCl_2 \cdot 2\,H_2O$

$SnCl_2 \cdot CsCl\ (Cs[SnCl_3])$

The divalent Ge and Sn halides are moderately reactive and insert into or add to various reactive bonds, as in the following cycloaddition:

$$\tag{4.48}$$

Divalent organocompounds can be synthesized from the corresponding group 14 dihalides and organolithiums. With highly sterically hindered aryllithiums, one can obtain relatively stable monomeric group 14 diaryl derivatives. By analogy with carbene, these compounds are known as *silylenes* (Si), *germylenes* (Ge), *stannylenes* (Sn), and *plumbylenes* (Pb):

$$EX_2 + 2\,ArLi \longrightarrow \begin{array}{c} Ar \\ \backslash \\ E + 2\,LiX \\ / \\ Ar \end{array}$$

X = Cl, Br, I, N(SiMe$_3$)$_2$

$$\tag{4.49}$$

Ar =

A = Me, Et, cyclohexyl, *i*-Pr, Ph, Mes
B = H, Me, *i*-Pr, *t*-Bu

For silylenes, the mesityl (2,4,6-trimethylphenyl) groups are not bulky enough to prevent dimerization to a disilene, that is, a silicon analog of an alkene. Thus, tetramesityldisilene is a thermally stable but air-sensitive compound. Interestingly, unlike alkenes, the disilene does not have a perfectly planar core, but a "beach chair" conformation, as shown below. A Si–Si single bond is too long to permit effective sideways overlap of Si 3p orbitals. Instead, each silylene fragment might be thought of as feeding its lone pair into the empty p orbital of the other silylene:

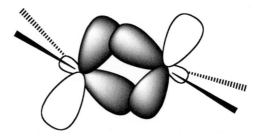

On the other hand, using exceedingly hindered *i*-Pr$_4$Ar groups, Power and coworkers at the University of California, Davis, have prepared stable germylenes and stannylenes and also studied their reactivity with a variety of small molecules, a few examples of which are as follows:

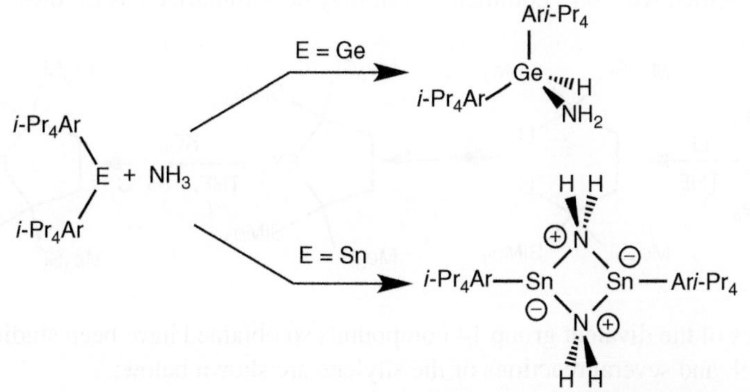

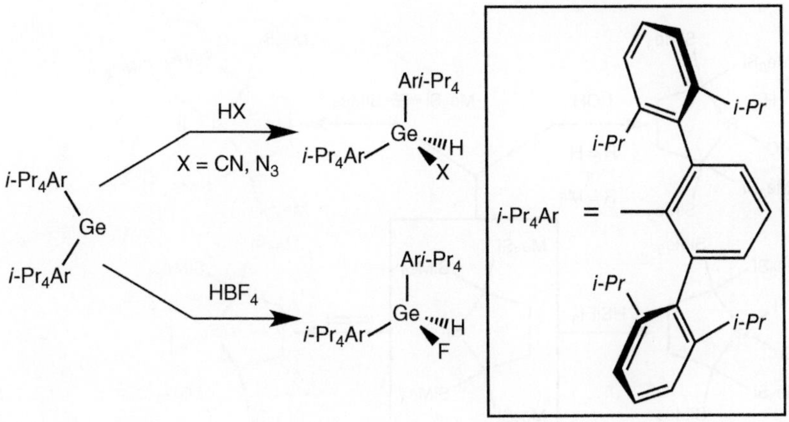

(4.50)

The diarylgermylene also reacts with H_2, giving $(i\text{-Pr}_4\text{Ar})_2\text{GeH}_2$.

REVIEW PROBLEM 4.14*

Suggest simple polar mechanisms for each of the above reactions of $(i\text{-Pr}_4\text{Ar})_2\text{E}$ (E = Ge, Sn).

Another structural framework that is sterically hindered enough to afford genuine divalent group 14 (Si, Ge, Sn) compounds is the following (Kira, N., *et al. J. Am. Chem. Soc.* **1999**, *121*, 9722):

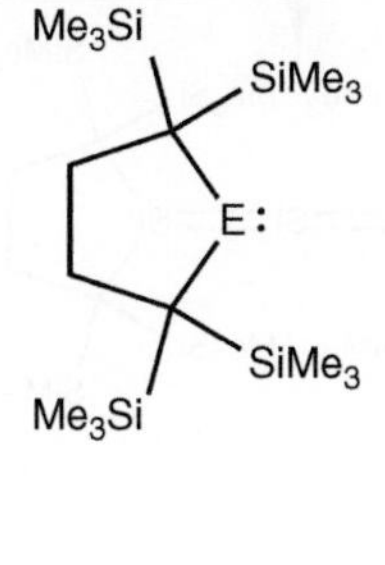

The synthesis, which we won't comment upon, may be summarized as follows:

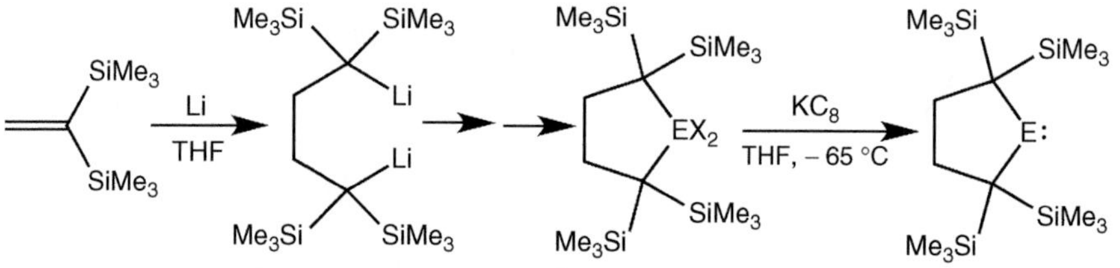

$$(4.51)$$

The reactivities of the divalent group 14 compounds so obtained have been studied in considerable detail, and several reactions of the silylene are shown below:

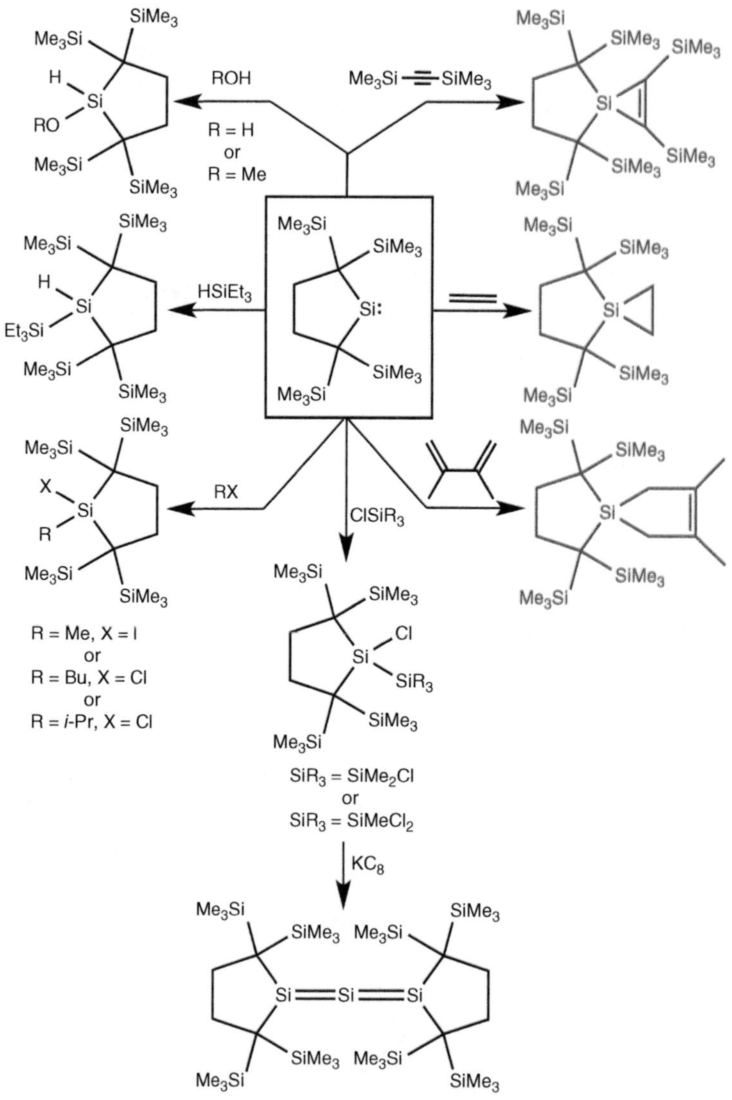

$$(4.52)$$

The reactions can be broadly classified into two categories. The products shown in green arise via a cycloaddition reaction; the others form by oxidative addition. Note that the steric protection afforded by the trimethylsilyl group even allows the formation of a trisilaallene (the product at the bottom).

REVIEW PROBLEM 4.15*

Assume that the oxidative additions (shown in black) and the cycloadditions (shown in green) depicted above (reaction set 4.52) are concerted one-step processes. Use arrow pushing to rationalize one reaction from each of the two categories.

REVIEW PROBLEM 4.16*

Protchenko *et al.* have reacted a sterically hindered boryllithium (see Section 3.6 for a reminder) with a sterically hindered (amino)tribromosilane to produce a thermally stable (amino)borylsilylene, as shown below (*Angew. Chem. Int. Ed.* **2013**, *52*, 568–571):

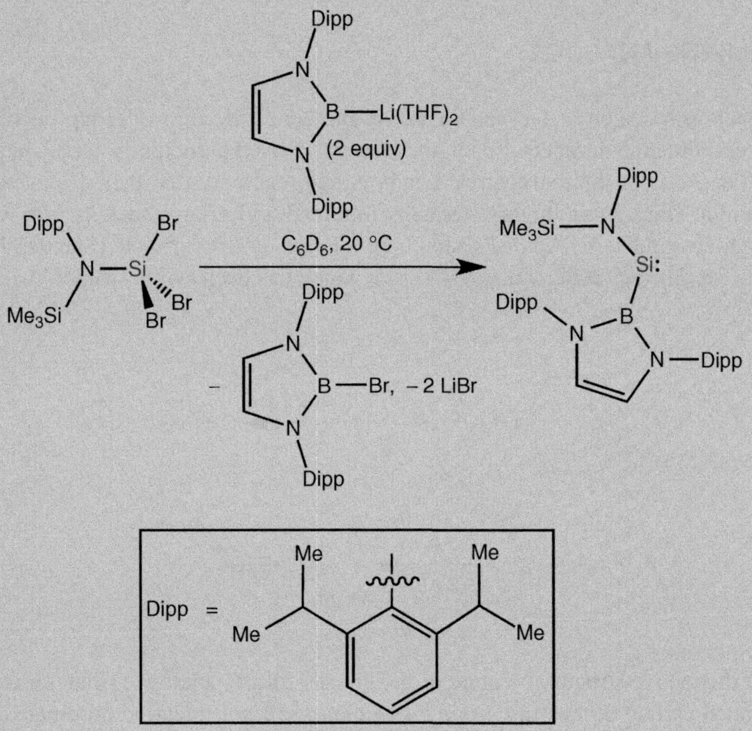

Suggest a mechanism for the reaction. Can you think of a rationale behind the researchers' choice of amino and boryl substituents for the silylene?

REVIEW PROBLEM 4.17*

Using Jones's Mg(I)–Mg(I) reagent (see Section 2.7 for the structure), Power and coworkers have reduced a dibromodimercaptosilane $SiBr_2(SAr^{Me6})$ to an acyclic silicon dithiolate, as shown below:

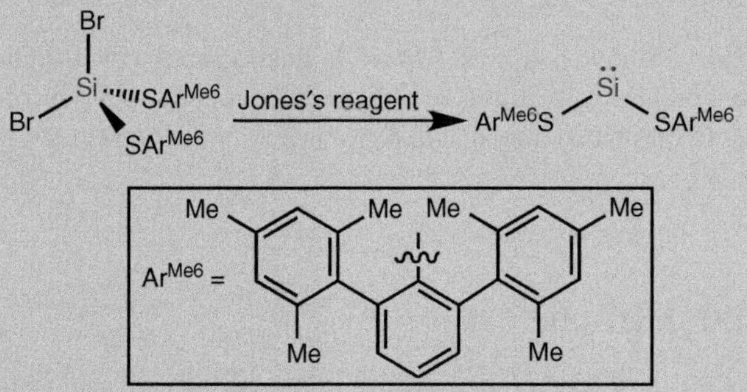

Carefully write out a mechanism for the reaction.

4.9*   ALKYNE ANALOGS

Alkyne analogs based on Si, Ge, and Sn are yet another exotic class of group 14 compounds with low coordination numbers. So far, they haven't proved particularly useful, but they are so "cool" in terms of their structures, bonding, and reactivity that they deserve at least a brief mention. Once again, highly sterically hindered aryl groups such as $i\text{-}Pr_4Ar$ proved critical to their synthesis. The discussion here focuses on a digermyne (Stender, M., *et al.* *Angew. Chem. Int. Ed.* **2002**, *41*, 1785–1787), which has been synthesized in two steps, as shown below:

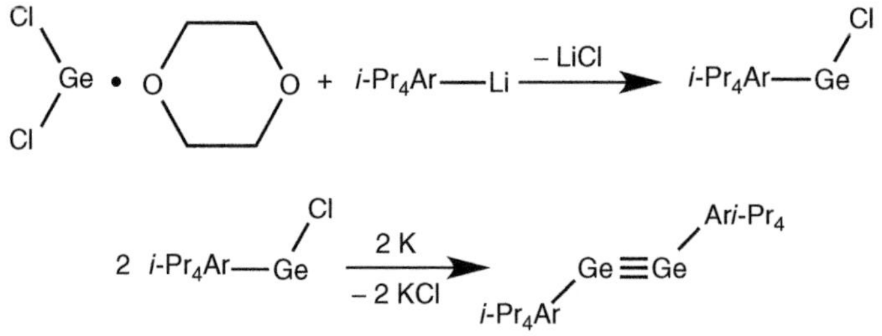

$$(4.53)$$

The most distinctive structural feature of the heavier alkyne analogs is that, unlike typical triple-bonded carbon atoms, the group 14 elements are not linearly coordinated. For the digermyne, the bent geometry suggests that the Ge–Ge interaction is not a true triple bond, but that the molecule might be better viewed as a bis(germylene):

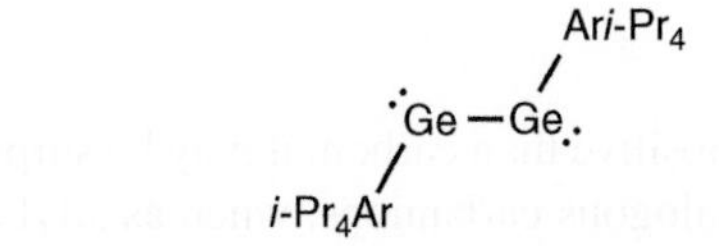

Like a diarylgermylene, the digermyne reacts readily with dihydrogen, yielding multiple products:

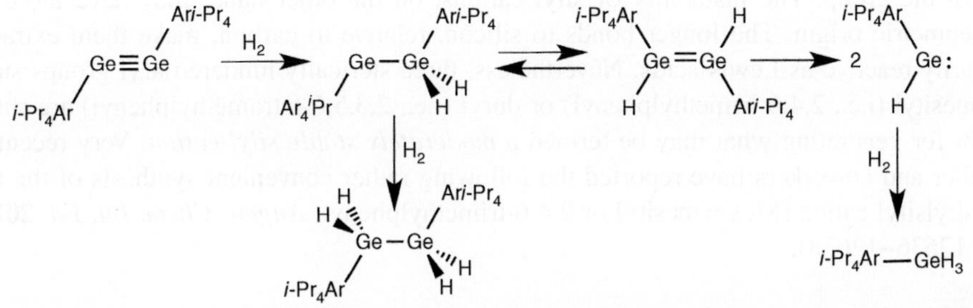

$$(4.54)$$

The Ge and Sn alkyne analogs also undergo cycloaddition reactions with carbon–carbon double bonds. The following reactions are common to both the digermyne and distannyne derivatives:

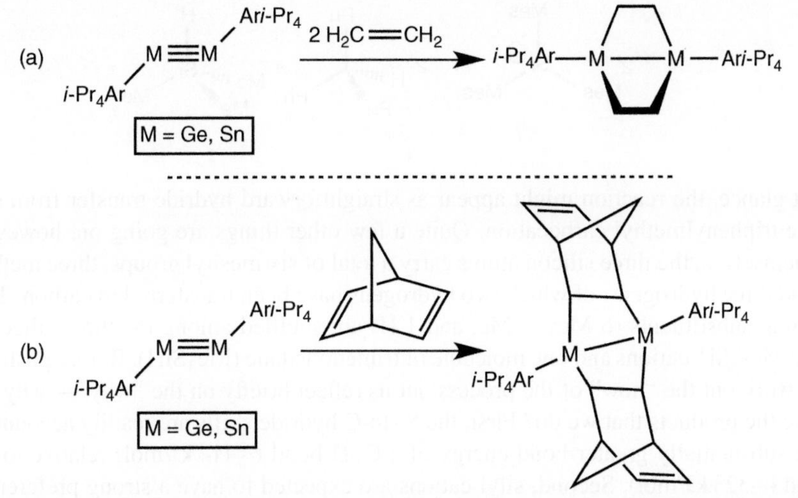

(a)

$$(4.55)$$

(b)

REVIEW PROBLEM 4.18

Write simple concerted mechanisms for reaction 4.55a and 4.55b.

4.10   SILYL CATIONS

Given that silicon is more electropositive than carbon, it may be surprising that silyl anions are much more stable than the analogous carbanions, whereas silyl cations are less stable than the analogous carbocations. Multiple factors account for these differences. As mentioned, the stability of silyl anions reflects the relative stability of the 3s electron pair, relative to the 3p electrons; the 3s–3p energy gap is significantly higher than the 2s–2p gap. The lone pair of a silyl anion thus has substantial 3s character. This is essentially the same reason why the divalent state becomes progressively more stable as one goes down the group. The instability of silyl cations, on the other hand, may have more of a geometric origin. The longer bonds to silicon, relative to carbon, make them extraordinarily reactive as Lewis acids. Nevertheless, three sterically hindered aryl groups such as mesityl (i.e., 2,4,6-trimethylphenyl) or duryl (i.e., 2,3,5,6-tetramethylphenyl) are sufficient for generating what may be termed a *moderately stable silyl cation*. Very recently, Müller and coworkers have reported the following rather convenient synthesis of the trimesitylsilyl cation (Mes = mesityl or 2,4,6-trimethylphenyl; *Angew. Chem. Int. Ed.* **2011**, *50*, 12636–12638):

$$(4.56)$$

At first glance, the reaction might appear as straightforward hydride transfer from silicon to the triphenylmethyl carbocation. Quite a few other things are going on, however. Among themselves, the three silicon atoms carry a total of six mesityl groups, three methyl groups, and three hydrogens, of which two hydrogens have been transferred to carbon. The 10 remaining substituents (6 Mes, 3 Me, and 1 H) are shuffled among the three silicons, giving two Mes_3Si^+ cations and one molecule of trimethylsilane (Me_3SiH). Before pushing arrows to work out the "how" of the process, let us reflect briefly on the "why"—why do we observe the products that we do? First, the Si-to-C hydride shifts are readily accounted for by the substantially greater bond energy of a C–H bond (~414 kJ/mol) relative to an Si–H bond (~323 kJ/mol). Second, silyl cations are expected to have a strong preference for mesityl substituents relative to methyl. Mesityl substituents provide for significant resonance stabilization, whereas any hyperconjugative stabilization afforded by methyl groups is expected to be weak on account of the long Si–C bonds, which explains the formation of the Mes_3Si^+ cations.

In writing out the mechanism, the two Si-to-C hydride transfers mentioned above are a rather obvious first step:

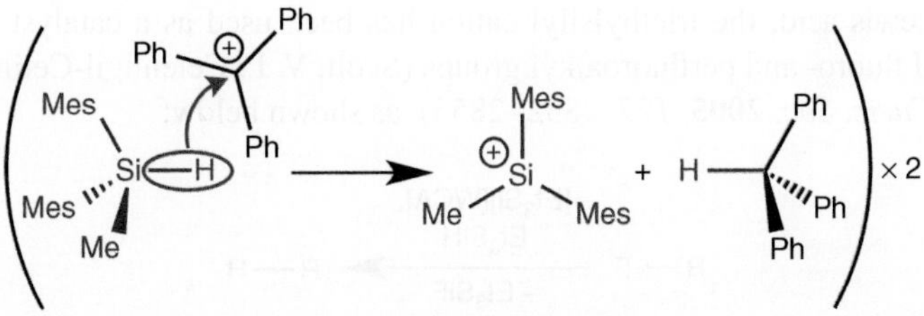

$$(4.57)$$

Now, to produce an Mes_3Si^+ cation, we need to carry out an Mes/Me ligand exchange between the $MeMes_2Si^+$ cation produced above and the third molecule of $MeMes_2SiH$ that has not reacted so far. A mesityl-bridged intermediate appears plausible for the process:

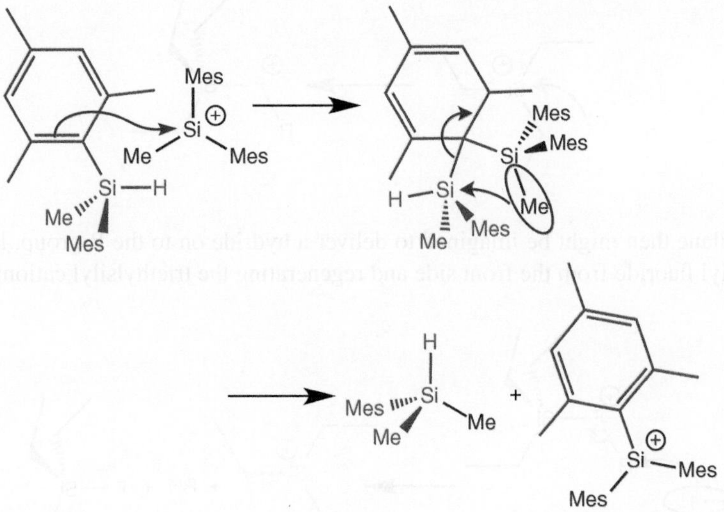

$$(4.58)$$

We have thus produced the first (of two) Mes_3Si^+ cation and a new silane $Me_2MesSiH$. The latter can now engage in a second Mes/Me ligand exchange reaction with an $MeMes_2Si^+$ cation to produce the second Mes_3Si^+ cation as well as Me_3SiH as the final products. The mechanism of this step is analogous to that shown above and is left as an exercise.

Use of extremely weakly coordinating anions (including several based on "carboranes," whose structures we will not discuss for reasons of space), hereafter simply referred to as $[WCA]^-$, allows the synthesis of even relatively unhindered silyl cations, such as the triethylsilyl cation:

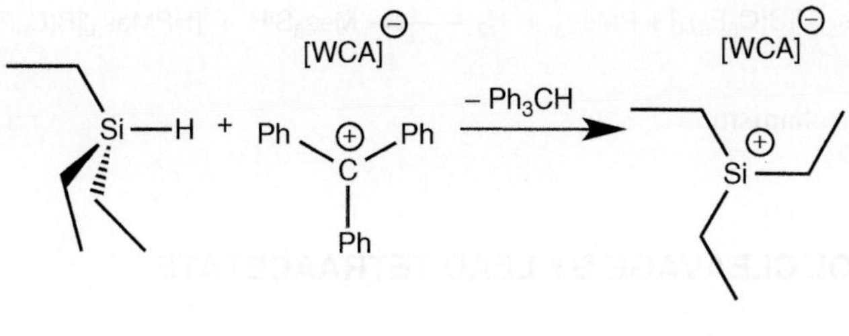

$$(4.59)$$

A powerful Lewis acid, the triethylsilyl cation has been used as a catalyst for hydrode-fluorination of fluoro- and perfluoroalkyl groups (Scott, V. J.; Celenligil-Cetin, R.; Ozerov, O. V. *J. Am. Chem. Soc.* **2005**, *127*, 2852–2853), as shown below:

$$R\!-\!F \xrightarrow[- Et_3SiF]{\substack{[Et_3Si][WCA],\\ Et_3SiH}} R\!-\!H \tag{4.60}$$

Given that triethylsilyl fluoride is a product, coordination of RF to the silyl cation appears to be a plausible first step:

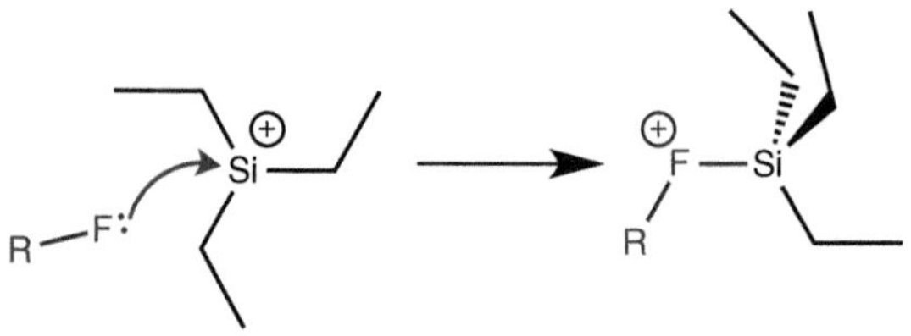

$$\tag{4.61}$$

The triethylsilane then might be imagined to deliver a hydride on to the R group, kicking out triethylsilyl fluoride from the front side and regenerating the triethylsilyl cation:

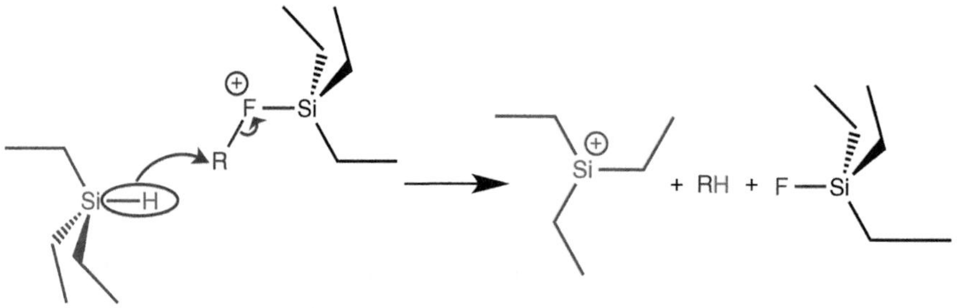

$$\tag{4.62}$$

REVIEW PROBLEM 4.19

The trimesitylsilyl cation and trimesitylphosphine constitute a frustrated Lewis pair capable of dihydrogen activation, as shown below:

$$[Mes_3Si][B(C_6F_5)_4] + PMes_3 + H_2 \longrightarrow Mes_3SiH + [HPMes_3][B(C_6F_5)_4]$$

Suggest a mechanism.

4.11   GLYCOL CLEAVAGE BY LEAD TETRAACETATE

Whereas $GeCl_2$ and $SnCl_2$ are strong reducing agents, many divalent lead compounds, such as the halides and the diacetate, are perfectly stable, thanks to the inert pair effect. Like trivalent thallium, many tetravalent lead compounds are oxidizing. The functioning of the now-discontinued antiknock agent tetraethyllead depends on its facile dissociation to lead atoms and ethyl radicals, which results in smooth, controlled combustion of gasoline. In organic chemistry, LTA in acetic acid is widely used as an oxidizing agent, particularly for cleaving glycols (1,2-diols):

$$
\underset{\substack{R^2 \\ R^1}}{\overset{\text{HO} \quad \text{OH}}{C-C}} \overset{R^4}{\underset{R^3}{}} + \underset{\substack{\text{AcO} \\ \text{AcO}}}{\overset{\text{AcO}}{Pb}}-\text{OAc} \xrightarrow[-\text{HOAc}]{-\text{Pb(OAc)}_2} \underset{R^2}{\overset{R^1}{C}}=O + O=\underset{R^3}{\overset{R^4}{C}} \qquad (4.63)
$$

The mechanism involves the formation of a cyclic tetravalent "plumbate ester" intermediate:

$$(4.64)$$

The five-membered ring then disintegrates via a pericyclic process to generate two carbonyl fragments and lead diacetate.

$$(4.65)$$

Interestingly, a cyclic mechanism is not an absolute requirement for explaining the formation of the observed products. An E2-like elimination from a plumbate monoester can also account for the observed products.

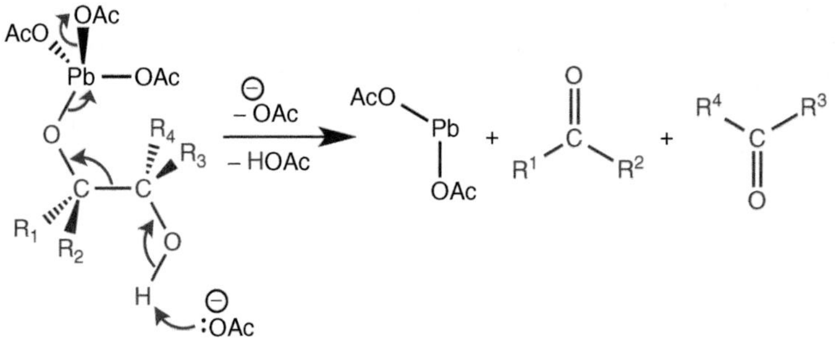

$$(4.66)$$

Support for this alternative mechanism comes from LTA-mediated cleavage of a conformationally restricted cyclohexane-1,2-diol (see below), where the OH groups are locked in a *trans* orientation[2] that does not permit the formation of a cyclic lead intermediate:

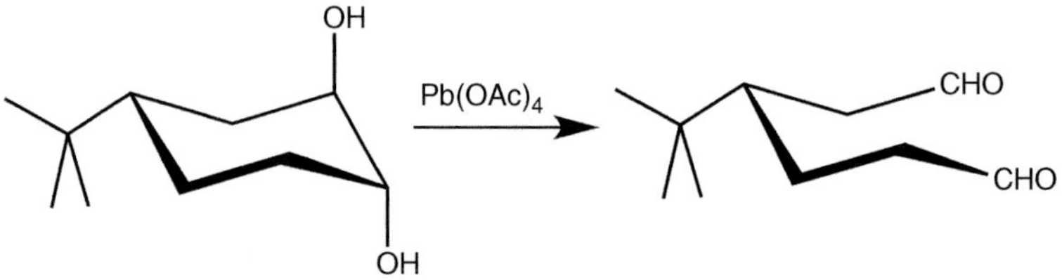

$$(4.67)$$

This reaction is considerably slower than a normal LTA cleavage, indicating that the acyclic mechanism is slower than the cyclic one.

Although diol cleavage is its best known application, LTA mediates a wide range of oxidations and rearrangements in organic chemistry. In spite of its less than green credentials, LTA is therefore likely to continue as a significant reagent in organic chemistry. The following review problems provide a small sampler of its many applications.

REVIEW PROBLEM 4.20

LTA oxidizes certain hydrazones under basic conditions to the corresponding diazoalkanes. The hydrazone of hexafluoroacetone is a good example:

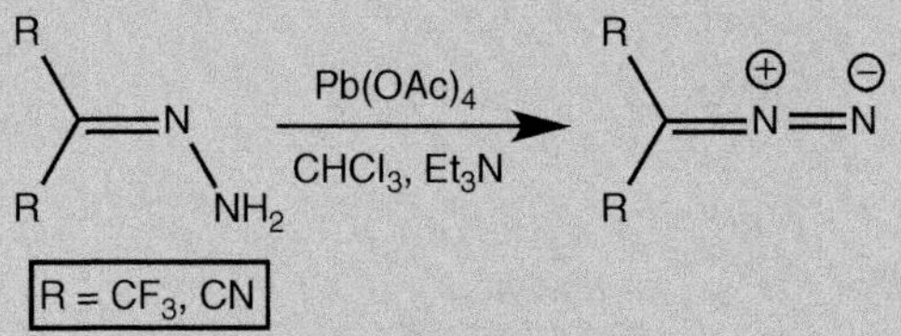

Suggest a mechanism for this conversion.

---

[2] As in a related example in Section 4.4, the *t*-butyl group acts as a conformational "lock". Its strong preference for an equatorial position prevents flipping of the cyclohexane ring, thereby freezing the OH groups into the positions shown above.

REVIEW PROBLEM 4.21*

A major application of LTA is in the decarboxylation of carboxylic acids. The mechanisms vary from radical mechanisms for simple monocarboxylic acids to likely polar ones for α-hydroxyacids and vicinal dicarboxylic acids. The last two cases are depicted below:

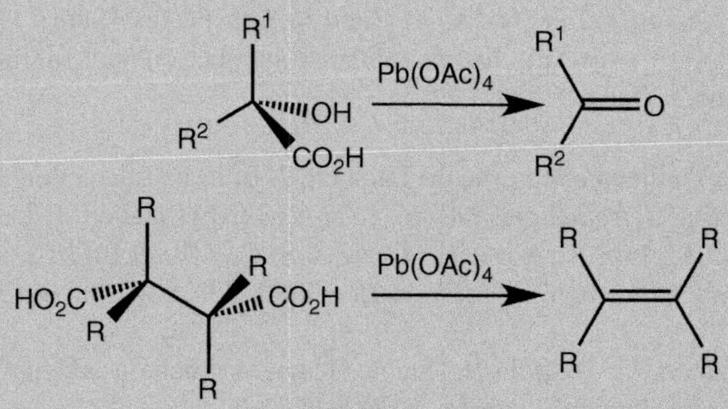

In each case, the byproducts are $Pb(OAc)_2$, CO_2, and HOAc. Suggest polar mechanisms for the above two reactions.

4.12 SUMMARY

The following are some of the take-home lessons from the vignettes presented above.

1. Silicon finds a plethora of applications in organic chemistry, as silyl-based protecting groups and as other types of reagents. The great majority of these applications hinge around silicon's oxophilicity.

2. It's worth emphasizing again that for Si and other elements below period 2, the S_N2 displacement is not a major reactivity paradigm. Instead, nucleophilic displacement typically occurs via the two-step S_N2-Si pathway. The first step of the S_N2-Si path is an **A** reaction, which typically leads to a hypervalent intermediate. Hypervalent molecules and ions are very common for Si and all its heavier congeners.

3. The stability of the divalent state increases down the group. Thus, divalent Ge, Sn, and Pb halides are all rather stable, especially when coordinated to a Lewis base. This trend reflects the increasing stability of the valence s orbital relative to the valence p orbital, that is, an increased s–p energy gap, as one goes down the group. Divalent organocompounds, however, are much more reactive. With highly sterically hindered aryl substituents, however, EAr_2 derivatives, where $E = Si$, Ge, Sn, and Pb, have all been successfully synthesized and even structurally characterized.

4. The heavier carbene analogs provide an entry into other low—coordination-number heavy element derivatives such as alkene and alkyne analogs.

5. The increased stability of the valence s orbital also accounts for the relative stability of silyl anions relative to analogous carbanions. Silyl anions provide powerful tools for the assembly of complex polysilane frameworks.

6. Presumably because of the longer bonds to silicon, silyl cations are much more reactive than analogous carbocations. They are extraordinarily powerful Lewis acids and have recently been applied as such to novel synthetic problems. One such application that we have discussed is fluoride ion abstraction.

FURTHER READING

For most topics, we recommend the textbooks listed in Appendices 1 and 2. The following focus on more specialized topics, especially low coordination-number compounds, which have yet to be discussed in current inorganic texts:

1. For reviews on the β-silicon effect, see (a) Lambert, J. B. "The Interaction of Silicon with Positively Charged Carbon" *Tetrahedron* **1990**, *46*, 2677–2689. (b) Lambert, J. B.; Zhao, Y.; Emblidge, R. W.; Salvador, L. A.; Liu, X.; So, J. H.; Chelius, E. C. "The β Effect of Silicon and Related Manifestations of σ Conjugation" *Acc. Chem. Res.* **1999**, *32*, 183–190. *And references contained therein.*
2. Mizuhata, Y.; Sasamori, T.; Tokitoh, N. "Stable Heavier Carbene Analogues," *Chem. Rev.* **2009**, *109*, 3479–3511. *A comprehensive survey of the field.*
3. Power, P. "Interaction of Multiple Bonded and Unsaturated Heavier Main Group Compounds with Hydrogen, Ammonia, Olefins, and Related Molecules," *Acc. Chem. Res.* **2011**, *44*, 627–637.
4. Reed, C. A. "The Silylium Ion Problem, R_3Si^+. Bridging Organic and Inorganic Chemistry," *Acc. Chem. Res.* **1998**, *31*, 325–332. *A review by a major contributor to the field.*
5. Klare, H. F. T.; Oestreich, M. "Silylium Ions in Catalysis," *Dalton Trans.* **2010**, *39*, 9176–9184. *A short and excellent introduction to a topic of current interest.*

5A

Nitrogen

Brot aus Luft (Bread out of air)

Fritz Haber, Referring to his discovery of the industrial synthesis of ammonia.

Nitrogen, in many ways, is a paradigm of a lighter p-block element. In other words, it embodies a wide range of properties that we typically associate with the upper part of the p block: multiple valence states, a range of oxidation states, both electrophilic and nucleophilic behavior in its different compounds, an ability to form strong bonds with many different elements, and an extensive capacity for multiple bonding. In addition, the most familiar nitrogen compounds, such as oxides, oxoacids, hydrides, and halides, engage in some of the most mechanistically interesting reactions that we will encounter in this book. Nitrogen is also one of the key elements of organic chemistry. These considerations led us to dedicate a separate chapter to nitrogen, a distinction that we have not accorded to any other element in this book. A good grounding in nitrogen chemistry is not only worthwhile on its own merits, but it also affords a better appreciation of the distinctive features of the heavier p-block elements. A few broad generalizations on nitrogen chemistry are as follows.

- With a $2s^2 2p^3$ valence configuration, nitrogen exhibits two common valence states—tri- and penta-valent. Examples of trivalent nitrogen include NH_3, N_2, and NO_2^-; examples of the pentavalent state include NH_4^+ and HNO_3. As a first-row (period 2) element, nitrogen generally cannot accommodate more than four bonds, so the equation

$$\text{valence} = \text{number of bonds} + \text{formal charge}$$

implies that pentavalent nitrogen atoms must carry a formal charge of at least +1; indeed, a +1 formal charge is common for pentavalent nitrogen.

- The Pauling electronegativity of nitrogen (3.04) is one of the highest, after fluorine (3.98) and oxygen (3.44), and about the same as chlorine (3.16). Nitrogen is much more electronegative than the other group 15 elements (the Pauling electronegativities

Arrow Pushing in Inorganic Chemistry: A Logical Approach to the Chemistry of the Main-Group Elements,
First Edition. Abhik Ghosh and Steffen Berg.
© 2014 John Wiley & Sons, Inc. Published 2014 by John Wiley & Sons, Inc.

of P and As are 2.19 and 2.18, respectively) and significantly more than both carbon (2.55) and sulfur (2.58). Against this backdrop, we can appreciate that ammonia and organic amines are considered hard bases in the HSAB sense as well as fairly good nucleophiles. The amide ion NH_2^- and its organic analogs R_2N^- are very strong bases (far too strong to exist in aqueous solution) and are widely used as such in organic chemistry.

- As suggested above, part of the richness of p-block chemistry derives from the fact that a given element can act as either a nucleophile or an electrophile depending on its valence and coordination state. This is particularly true of nitrogen. Many of the reactions discussed in this chapter involve a nitrogen nucleophile attacking a nitrogen electrophile, thereby forming a N–N bond. In some cases, subsequent elimination leads to N–N multiple bonds.

- Like carbon and oxygen, nitrogen forms highly stable multiple bonds, most notably with carbon, oxygen, and other nitrogen atoms. Not surprisingly, the great thermodynamic stability of dinitrogen (N_2) dictates the course of a large number of reactions involving nitrogen compounds.

5A.1   AMMONIA AND SOME OTHER COMMON NITROGEN NUCLEOPHILES

Ammonia is a base and a nucleophile. It reacts with alkyl halides to give alkylamines, as shown below for methyl iodide:

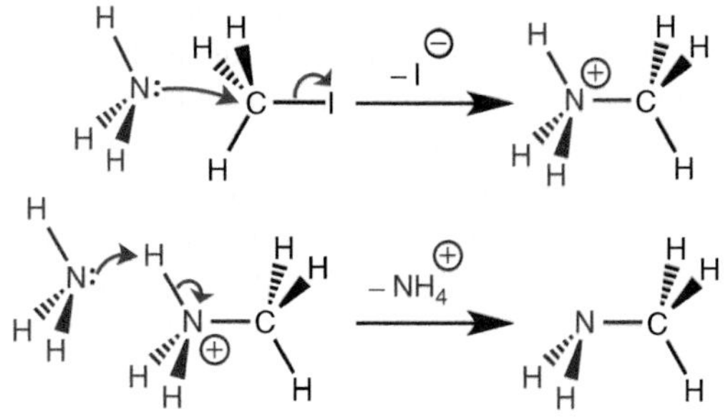

(5A.1)

Polyalkylation is common and difficult to prevent, and with an excess of methyl iodide, the final product is the tetramethylammonium cation.

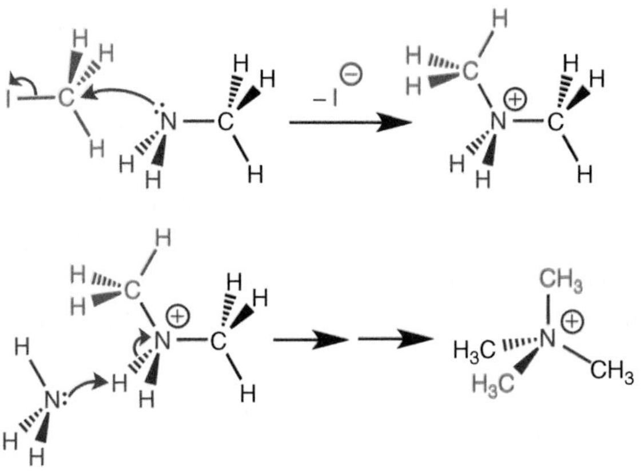

(5A.2)

Under anhydrous conditions, ammonia acts as a nucleophile toward Cl_2, giving successively NH_2Cl, $NHCl_2$, and NCl_3, as shown below:

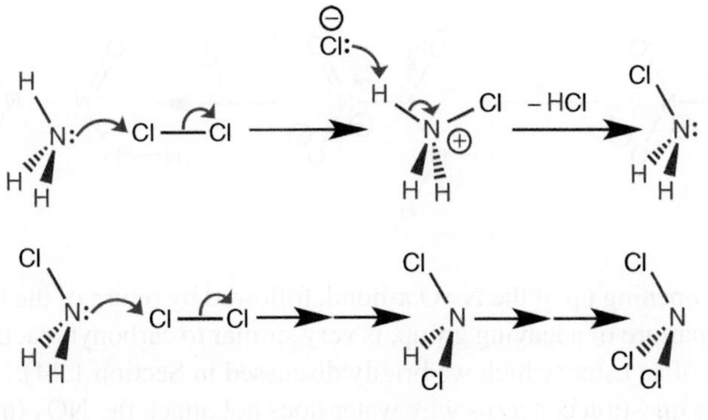

$$(5A.3)$$

Chloramine (NH_2Cl) is increasingly used as a disinfectant in swimming pools. It persists longer than molecular chlorine and has a much lower tendency to generate carcinogenic chlorocarbons such as $CHCl_3$ and CCl_4. Chloramine does, however, produce traces of nitrogen trichloride (NCl_3, also known as trichloramine), which has been linked to childhood asthma.

REVIEW PROBLEM 5A.1

NBr_3 exists but is very reactive (indeed explosive) and has been prepared at very low temperatures ($\sim -100\,°C$) via the following reaction:

$$(Me_3Si)_2NBr + 2\,BrCl \rightarrow NBr_3 + 2\,Me_3SiCl$$

Suggest a mechanism.

5A.2 SOME COMMON NITROGEN ELECTROPHILES: OXIDES, OXOACIDS, AND OXOANIONS

A fair number of nitrogen electrophiles are known. Many of these are nitrogen-oxo species, for example, N_2O_3, N_2O_5, NO^+, NO_2^+, and so on, where the higher electronegativity of oxygen implies that the nitrogen atoms are the positive ends of dipoles and therefore also the sites of attack by nucleophiles.

The two oxides N_2O_3 and N_2O_5 are classic acidic oxides; on exposure to water, they hydrolyze immediately to nitrous acid (HNO_2) and nitric acid (HNO_3), respectively:

$$N_2O_3 + H_2O = 2\,HNO_2 \qquad (5A.4)$$

$$N_2O_5 + H_2O = 2\,HNO_3 \qquad (5A.5)$$

For N_2O_3, the NO (nitroso) nitrogen is the presumptive electrophilic site, as shown below.

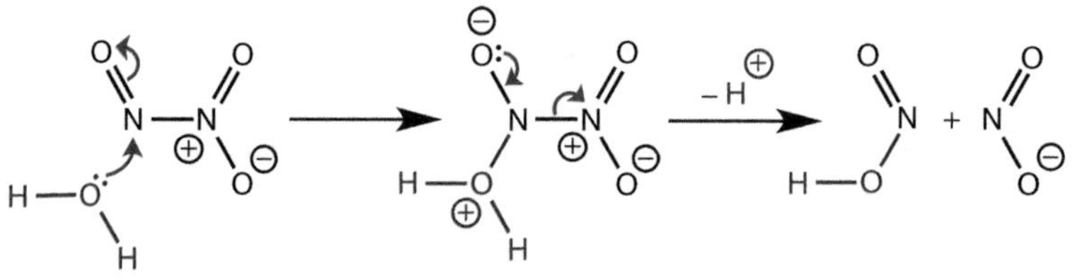

$$\text{(5A.6)}$$

Observe that the opening up of the $N=O$ π-bond, followed by return of the O lone pair, with simultaneous departure of a leaving group, is very similar to carbonyl reactivity, such as the basic hydrolysis of an ester (which we briefly discussed in Section 1.14).

An interesting question concerns why water does not attack the NO_2 (nitro) nitrogen of N_2O_3. A possible answer is that, even if water is added to the nitro group, the tetrahedral intermediate would only revert back to N_2O_3, as shown below, because $(O=N)^-$ is not expected to be a particularly good leaving group.

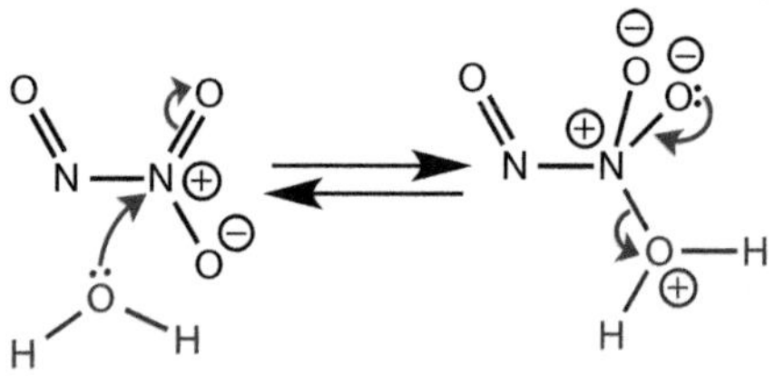

$$\text{(5A.7)}$$

Two common cationic nitrogen-oxo electrophiles are the nitrosonium (NO^+) and nitronium (NO_2^+) ions:

The NO^+ ion always exists in equilibrium with HNO_2 in an aqueous solution of the latter:

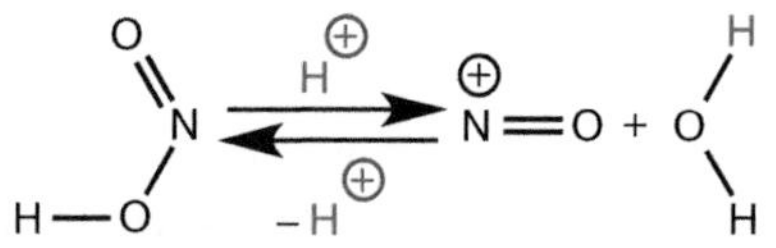

$$\text{(5A.8)}$$

The NO_2^+ cation is obtained via protonation of nitric acid under strongly acidic conditions, such as in a mixture of concentrated nitric and sulfuric acids:

$$HNO_3 + 2\,H_2SO_4 \rightarrow NO_2^+ + H_3O^+ + 2\,HSO_4^- \qquad \text{(5A.9)}$$

The NO_2^+ ion is believed to be the actual reagent involved in classic aromatic nitrations (see Section 1.13).

REVIEW PROBLEM 5A.2

Nitronium tetrafluoroborate can be isolated as a solid salt and is even commercially available. It is prepared as follows:

$$N_2O_5 \xrightarrow[\text{CH}_3\text{NO}_2]{\text{BF}_3,\ \text{HF}} NO_2^+\ BF_4^-$$

Suggest a mechanism for this reaction.

REVIEW PROBLEM 5A.3

Solid nitrosonium and nitronium salts are moisture sensitive and need to be protected even from moist air. Can you explain why?

REVIEW PROBLEM 5A.4

Anhydrous nitric acid (*dangerous!*) serves as a starting material for a number of reactive pentavalent nitrogen compounds, including nitryl chloride (NO_2Cl) and chlorine nitrate ($ClONO_2$), as shown below:

1. $HNO_3 + ClSO_3H \rightarrow ClNO_2 + H_2SO_4$
2. $HNO_3 + ClF \rightarrow ClONO_2 + HF$

Suggest mechanisms for the two reactions.

5A.3 N–N BONDED MOLECULES: SYNTHESIS OF HYDRAZINE

Given the variety of nitrogen-based nucleophiles and electrophiles, it's no surprise that reactions involving nitrogen-on-nitrogen attack, leading to N–N bond formation, are fairly common. We will discuss several such reactions, beginning with the Olin–Raschig process for the synthesis of hydrazine (NH_2NH_2). The synthesis involves two operational steps. In the first step, ammonia interacts with chilled (5 °C) aqueous sodium hypochlorite (NaOCl) to form chloramine (NH_2Cl):

$$NH_3 + NaOCl \rightarrow NH_2Cl + NaOH \tag{5A.10}$$

The resulting solution is then added to anhydrous ammonia under pressure and heated to 130 °C, to yield hydrazine:

$$NH_2Cl + NH_3 + NaOH \rightarrow N_2H_4 + NaCl + H_2O \tag{5A.11}$$

The overall synthesis may be represented as follows:

$$2\,NH_3 + NaOCl \rightarrow N_2H_4 + NaCl + H_2O \tag{5A.12}$$

The logic underlying the synthesis is that, since the formation of hydrazine involves the formation of a N–N bond, we should first transform ammonia to a nitrogen electrophile of the form NH_2X, which can then be attacked by ammonia. With a low N–Cl bond dissociation energy (BDE) of about 200 kJ/mol, chloramine (NH_2Cl) fits the bill for such an intermediate perfectly. To form NH_2Cl, however, we must first create an electrophile that can chlorinate ammonia. Hypochlorous acid (HOCl) appears well suited for that role:

$$\tag{5A.13}$$

Once NH_2Cl has been formed, hydrazine formation should occur via a straightforward nitrogen-on-nitrogen S_N2 attack, followed by a final deprotonation:

$$\tag{5A.14}$$

A final point: Is it reasonable to invoke un-ionized hypochlorous acid (HOCl) as an active intermediate in a solution of bleach, which is an alkaline medium? With a pK_a of 7.53, hypochlorous acid is a weak acid, about three orders of magnitude weaker than acetic acid, so it's not an unreasonable proposal. On the other hand, it is possible to envision concerted pathways, such as the following, where the hypochlorite anion acts as the actual chlorinating agent:

$$\tag{5A.15}$$

REVIEW PROBLEM 5A.5

Another synthesis of hydrazine involves the oxidation of urea, $(NH_2)_2C=O$, with sodium hypochlorite:

$$(H_2N)_2C=O + NaOCl + 2\,NaOH \rightarrow N_2H_4 + H_2O + NaCl + Na_2CO_3$$

Suggest a mechanism.

5A.4 MULTIPLE BOND FORMATION: SYNTHESIS OF SODIUM AZIDE

Nitrogen-on-nitrogen attack followed by elimination leads to N–N multiple bonds. A good example is the synthesis of sodium azide (NaN_3). Sodium azide has been widely used as the gas-forming agent in automobile airbags; nowadays, however, a milder explosive is generally preferred. The synthesis involves the interaction of nitrous oxide (N_2O) and molten sodium amide ($NaNH_2$) at about 190 °C.

$$2\,NaNH_2 + N_2O \rightarrow NaN_3 + NaOH + NH_3 \tag{5A.16}$$

The structure of the azide anion (N_3^-) indicates that the NH_2^- and N_2O nitrogens must somehow come together and the H and O atoms somehow stripped away. A reasonable working hypothesis is that the amide ion (NH_2^-) acts as a nucleophile and attacks one of the N_2O nitrogens. For NH_2^- attacking the terminal nitrogen of N_2O, the following pathway may be envisioned:

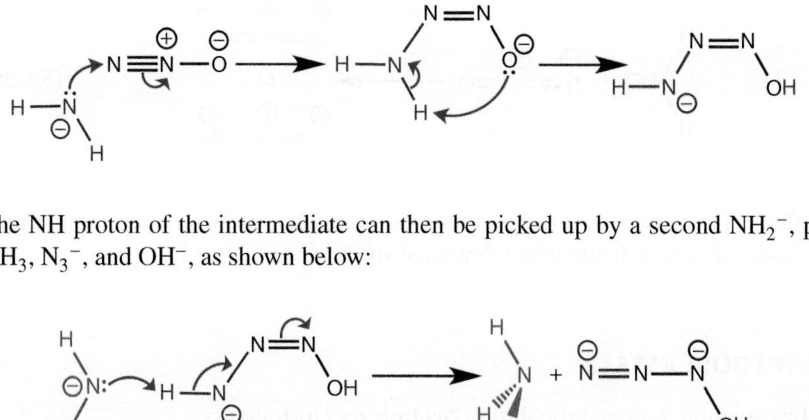

$$\tag{5A.17}$$

The NH proton of the intermediate can then be picked up by a second NH_2^-, producing NH_3, N_3^-, and OH^-, as shown below:

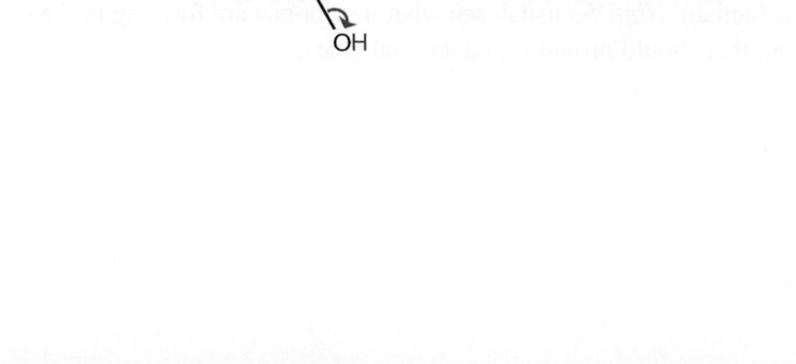

$$\tag{5A.18}$$

Let us now consider a pathway where NH_2^- attacks the "middle" (i.e., nonterminal) nitrogen of N_2O. A couple of the first steps would then be as follows:

$$(5A.19)$$

A second NH_2^- can then pluck off the NH proton from the intermediate, producing, as before, NH_3, N_3^-, and OH^-:

$$(5A.20)$$

Thus, both pathways outlined above appear to be capable of producing NaN_3 from $NaNH_2$ and N_2O. Which of them might correspond to reality? Arrow pushing alone cannot help us here. Selective labeling of the N_2O with ^{15}N (shown in red below), however, might be expected to shed light on the issue:

$$(5A.21)$$

The two different pathways lead to different isotopic arrangements in the azide anion, which should be readily distinguishable with vibrational spectroscopy.

REVIEW PROBLEM 5A.6

In alkaline solution, nitrogen trichloride "hydrolyzes" as follows:

$$2\,NCl_3 + 6\,OH^- \rightarrow N_2 + 3\,ClO^- + 3\,Cl^- + 3\,H_2O$$

Suggest a mechanism. *Hint*: As usual, see what new bonds are forming in the course of the reaction; that should provide some crucial clues.

REVIEW PROBLEM 5A.7

The N–N-bonded hyponitrite anion $[O–N=N–O]^{2-}$ can be accessed via the following solution-phase synthesis:

$$RONO + NH_2OH + 2\ EtONa \rightarrow Na_2N_2O_2 + ROH + 2\ EtOH$$

Suggest a mechanism for the process. (The hyponitrite anion can exist as either *cis* or *trans* isomers. The *cis* isomer serves as a chelating ligand for metal ions.)

5A.5   THERMAL DECOMPOSITION OF NH$_4$NO$_2$ AND NH$_4$NO$_3$

When heated, ammonium nitrite (NH$_4$NO$_2$) and ammonium nitrate (NH$_4$NO$_3$) decompose as follows:

$$NH_4NO_2 \rightarrow N_2 + 2\ H_2O \tag{5A.22}$$

$$NH_4NO_3 \rightarrow N_2O + 2\ H_2O \tag{5A.23}$$

Although both these reactions are common in undergraduate laboratories, where they are typically carried out with just a few milligrams of the salts, the solids are explosive and appropriate precautions are essential. Both reactions involve N–N bond formation, so we must identify a nitrogen nucleophile and a nitrogen electrophile in each case. In the case of NH$_4$NO$_2$, proton transfer from NH$_4^+$ to NO$_2^-$ yields NH$_3$, a potential nucleophile, and HNO$_2$, a potential electrophile, which should link up as follows:

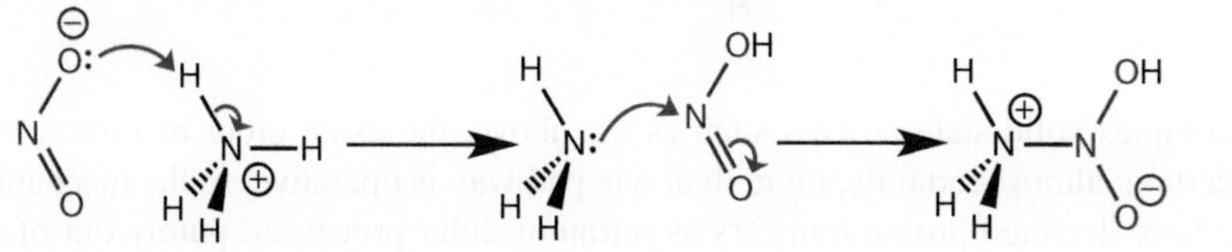

$$\tag{5A.24}$$

Observe that a N–N bond has been created. Additional proton transfers followed by elimination of water lead to a N–N double bond, as shown below.

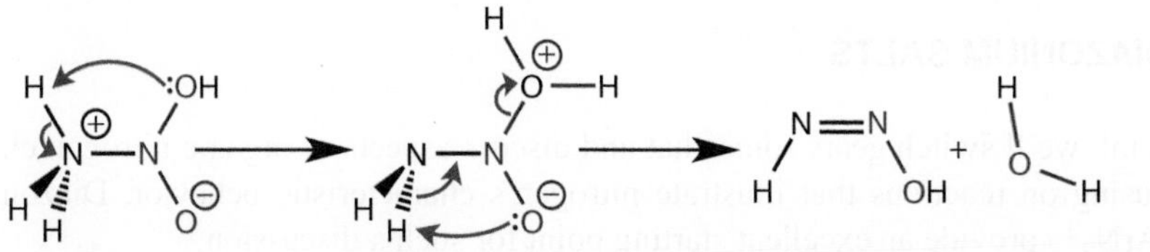

$$\tag{5A.25}$$

Elimination of a second molecule of water then leads to a N–N triple bond, that is, N$_2$:

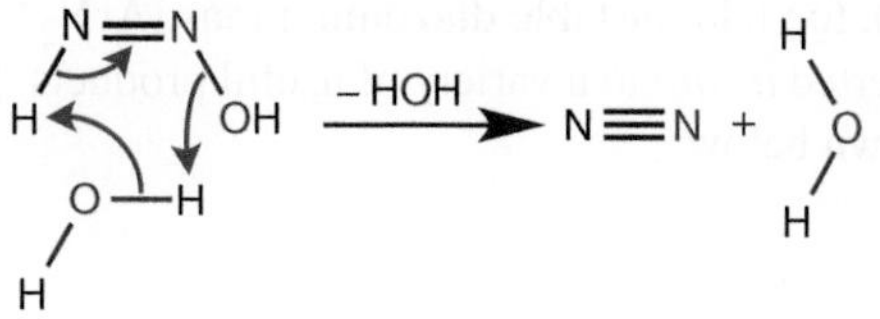

$$\tag{5A.26}$$

The mechanism for the decomposition of NH_4NO_3 is similar. Once again, the nucleophile ammonia is generated via proton transfer from NH_4^+ to NO_3^-. It may seem strange that nitrate, an extremely weak base, should pick up a proton from NH_4^+. Remember, however, that this is a heat-induced solid-state reaction; reactivity patterns are expected to be different under such conditions, relative to lower temperature solution-phase chemistry. Once the N–N bond has formed, we need to do a few proton transfers and eliminate two water molecules to arrive at the final product N_2O, as depicted below:

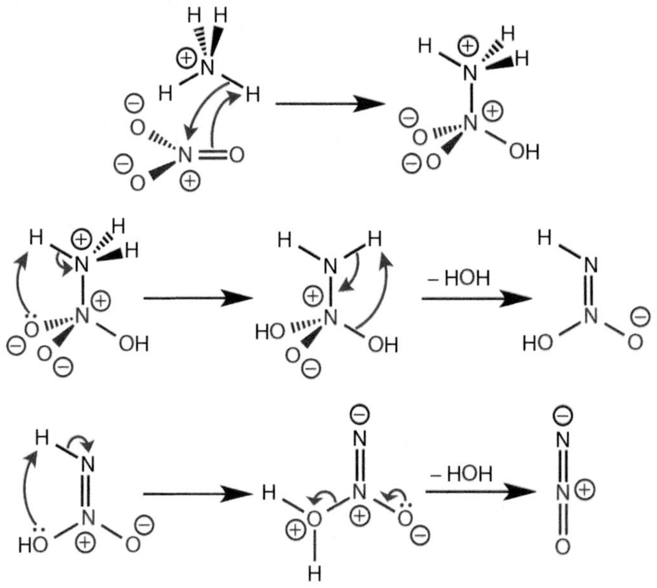

$$(5A.27)$$

For a complex solid-state process such as the above, the exact order of proton transfers is uncertain; almost certainly, more than one pathway is operative. In the mechanisms above, we have depicted proton transfers as intramolecular processes, purely out of convenience and to conserve space. Intermolecular proton transfers almost certainly occur as well.

5A.6 DIAZONIUM SALTS

At this point, we'll switch gears somewhat and discuss aspects of organic nitrogen chemistry, focusing on reactions that illustrate nitrogen's characteristic behavior. Diazonium cations (ArN_2^+) provide an excellent starting point for such a discussion.

Aniline ($PhNH_2$) and other aromatic amines ($ArNH_2$) react with aqueous nitrous acid, or more accurately with NO^+ (which is always present in aqueous solutions of HNO_2, as shown in reaction 5A.8), to yield unstable diazonium ions (ArN_2^+). They are almost never isolated but rather converted *in situ* to a variety of useful products. Some examples of these transformations are shown below:

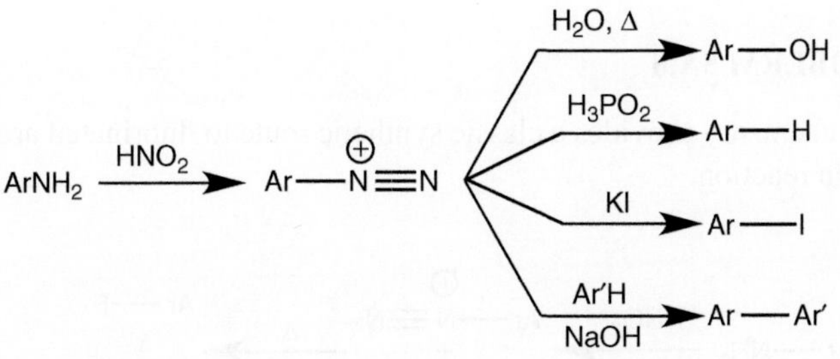

$$\text{ArNH}_2 \xrightarrow{\text{HNO}_2} \text{Ar}-\overset{\oplus}{\text{N}}\equiv\text{N} \begin{cases} \xrightarrow{\text{H}_2\text{O},\ \Delta} \text{Ar}-\text{OH} \\ \xrightarrow{\text{H}_3\text{PO}_2} \text{Ar}-\text{H} \\ \xrightarrow{\text{KI}} \text{Ar}-\text{I} \\ \xrightarrow[\text{NaOH}]{\text{Ar'H}} \text{Ar}-\text{Ar'} \end{cases} \tag{5A.28}$$

The structure of a diazonium cation probably provides the most valuable clue to the mechanism of its formation. A N–N bond has to form, and for that ArNH_2 is a plausible nucleophile and NO^+ the likely electrophile; there are really not many other choices.

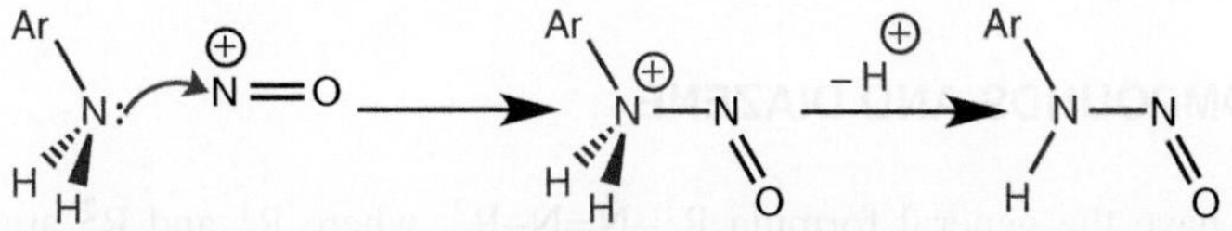

$$(5A.29)$$

Successive protonation of the nitroso (N=O) oxygen and elimination of water then leads to the formation of a diazonium cation, as shown below:

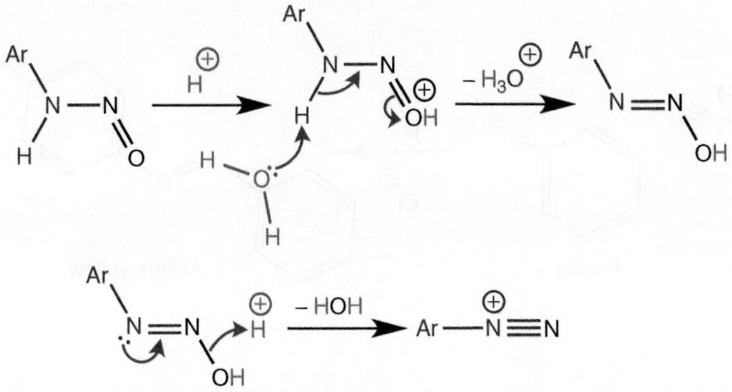

$$(5A.30)$$

The great synthetic utility of diazonium ions stems from the fact that they readily lose N_2, producing highly reactive aryl cations, which can be captured by a variety of nucleophiles (Nu^-):

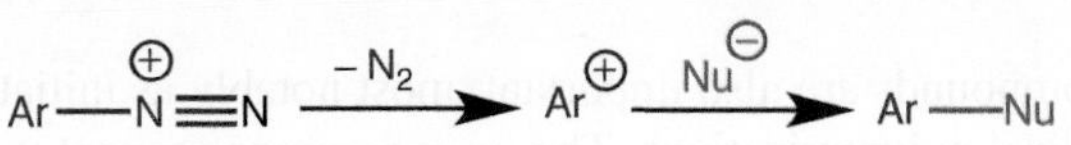

$$\text{Ar}-\overset{\oplus}{\text{N}}\equiv\text{N} \xrightarrow{-\text{N}_2} \text{Ar}^{\oplus} \xrightarrow{\text{Nu}^{\ominus}} \text{Ar}-\text{Nu} \tag{5A.31}$$

Equation 5A.31 provides several examples of this type of transformation.

REVIEW PROBLEM 5A.8

Diazonium salt chemistry provides a classic synthetic route to fluorinated arenes, the Balz–Schiemann reaction:

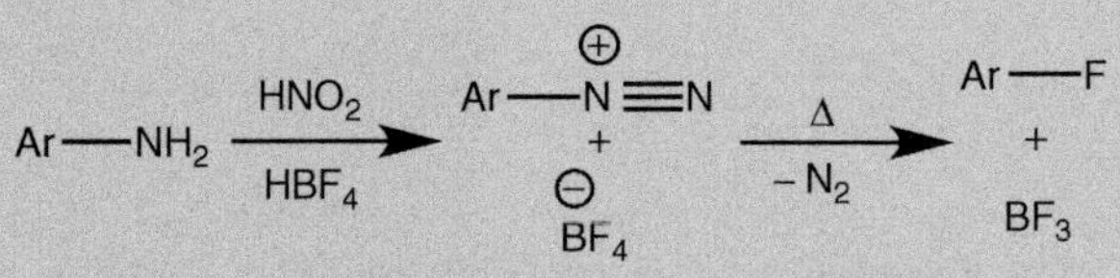

Write out the mechanism for the last step.

5A.7 AZO COMPOUNDS AND DIAZENE

Azo compounds have the general formula R^1–N=N–R^2, where R^1 and R^2 are alkyl or aryl groups. Aromatic azo compounds are generally strongly colored and many are easily prepared via a so-called diazo coupling reaction, in which a diazonium cation is coupled with a relatively "active" (i.e., highly nucleophilic) arene such as a phenol or aniline. The dye called *aniline yellow*, for example, is prepared as follows:

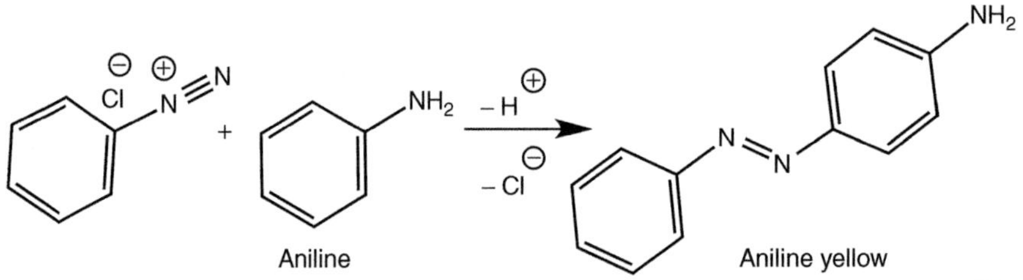

Aniline   Aniline yellow

(5A.32)

REVIEW PROBLEM 5A.9

Suggest a mechanism for the above reaction.

Aliphatic azo compounds are also important, most notably as initiators of radical reactions, including radical polymerizations. The most common reagent used for this purpose is azobisisobutyronitrile (AIBN), which decomposes on gentle heating to N_2 and a pair of 2-cyanoprop-2-yl radicals:

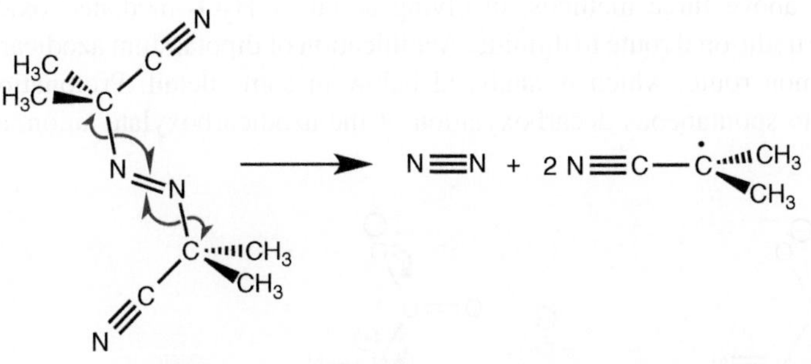

AIBN

$$N{\equiv}N + 2\,N{\equiv}C-\overset{\cdot}{C}\overset{\text{\tiny IIII}CH_3}{\underset{CH_3}{}} \tag{5A.33}$$

The driving force for this decomposition is clearly the formation of N_2. The 2-cyanoprop-2-yl radicals can then abstract a hydrogen from a substrate and set off a radical chain reaction. For example, when heated with a small quantity of AIBN, styrene and maleic anhydride yield a copolymer with almost perfectly alternating monomer units:

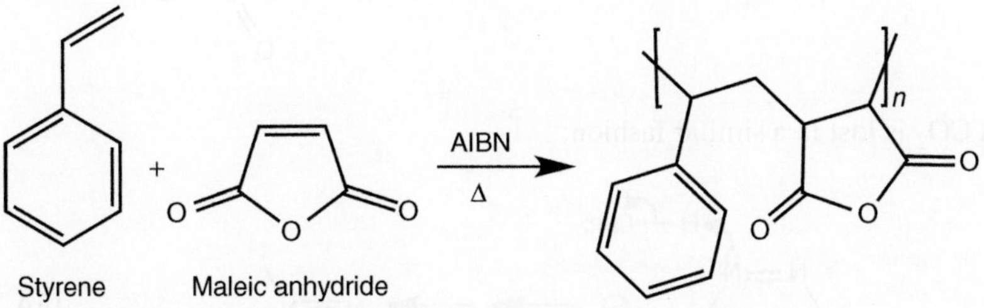

Styrene   Maleic anhydride

$$\xrightarrow[\Delta]{AIBN} \tag{5A.34}$$

REVIEW PROBLEM 5A.10

Suggest a mechanism for the above polymerization.

The parent azo compound HN=NH, called *diimide* or *diazene*, is unstable but is useful as a reducing agent in organic chemistry, where it is typically generated *in situ*. Some standard methods for the generation of diimide are as follows:

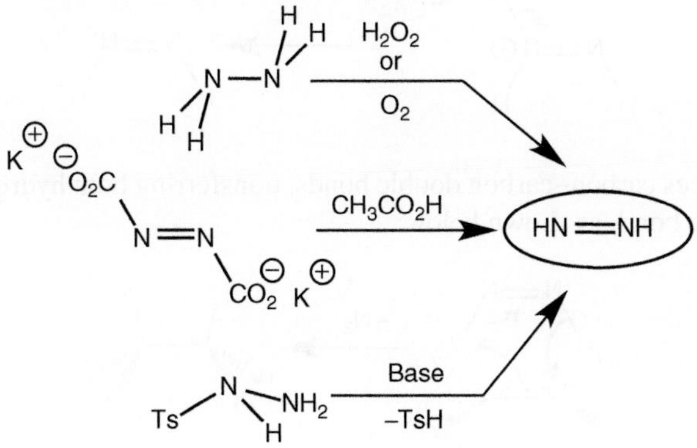

$$\text{HN}{=\!=}\text{NH} \tag{5A.35}$$

The first of the above three methods, involving aerial or H_2O_2-mediated oxidation of hydrazine, is the traditional route to diimide. Acidification of dipotassium azodicarboxylate is another common route, which is analyzed below in some detail. Protonation of the nitrogens leads to spontaneous decarboxylation of the azodicarboxylate anion, as shown below:

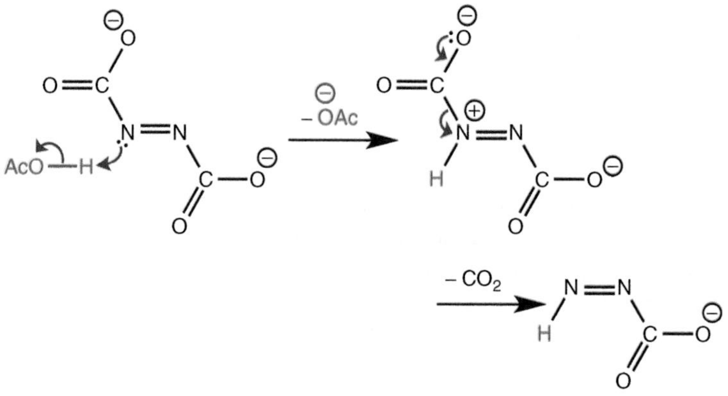

(5A.36)

The second CO_2 is lost in a similar fashion:

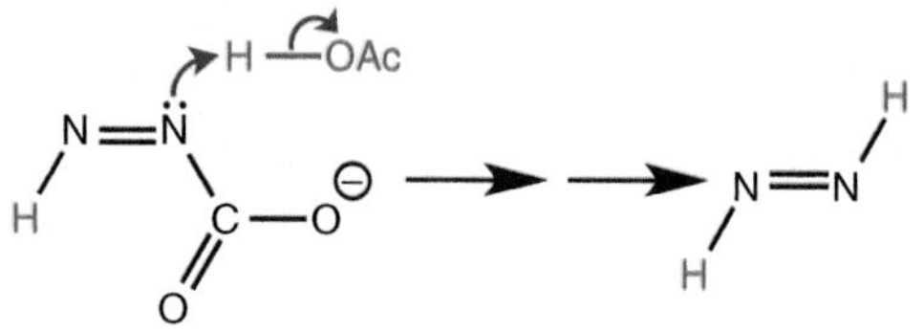

(5A.37)

Like other azo compounds, diimide can exist as either a *cis* or a *trans* isomer. Interconversion between the two is catalyzed by acid, as shown below:

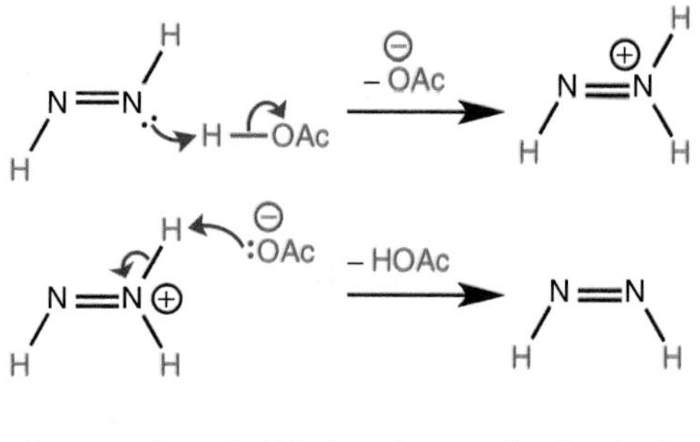

(5A.38)

Cis-diimide reduces carbon–carbon double bonds, transferring both hydrogens to the same face of the double bond, as shown below:

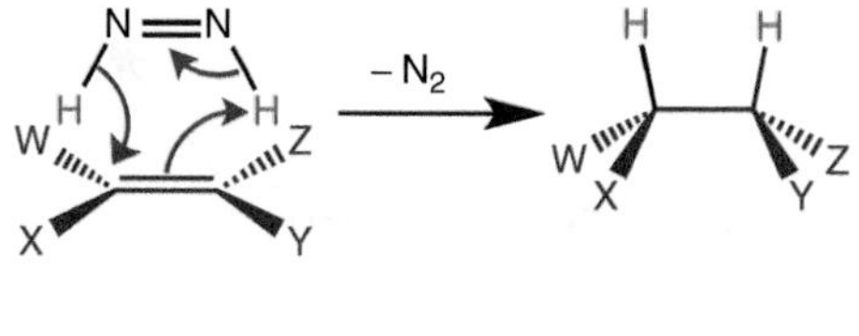

(5A.39)

In this process, diimide reduction resembles catalytic hydrogenation (with metals such as Pd, Pt, and Ni), which also follows the same stereochemical course, that is, delivers H_2 to a given face of the double bond. A disadvantage of diimide reductions is that they tend to be significantly slower than catalytic hydrogenations, especially for tetra-substituted and polarized double bonds. On the positive side, diimide reductions avoid handling of the explosive H_2 gas as well as removal and/or recovery of the transition-metal catalysts.

Diimide reductions typically require an excess of the diimide-generating reagents, because part of the diimide formed is lost to disproportionation:

$$2 N_2H_2 \rightarrow N_2 + N_2H_4 \tag{5A.40}$$

The mechanism is essentially the same as that involved in the reduction of a carbon–carbon double bond:

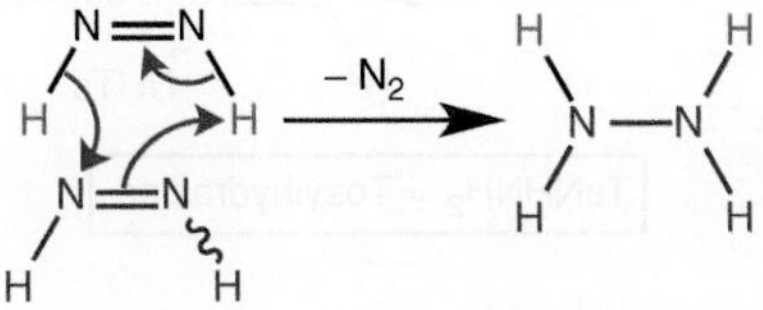

$$\tag{5A.41}$$

Note that for the reaction to occur, one of the diimides (indicated in black above) must be in the *cis* configuration; the other may be either *cis* or *trans*, as indicated by the squiggly N–H bond.

REVIEW PROBLEM 5A.11

A very mild way of generating diimide involves treatment of 2,4,6-triisopropyl-benzenesulfonyl hydrazide (depicted below) with sodium hydrogencarbonate, a mild base:

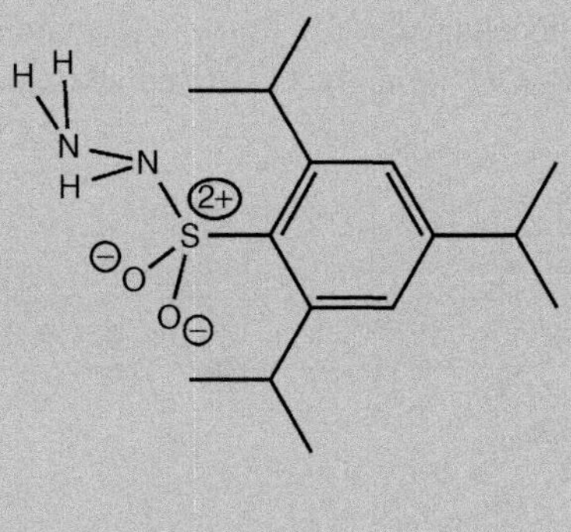

Suggest a mechanism for the process.

5A.8* IMINES AND RELATED FUNCTIONAL GROUPS: THE WOLFF–KISHNER REDUCTION AND THE SHAPIRO REACTION

Carbonyl compounds condense with amines and related functional groups (e.g., hydrazines), with elimination of water, to yield C=N double-bonded products:

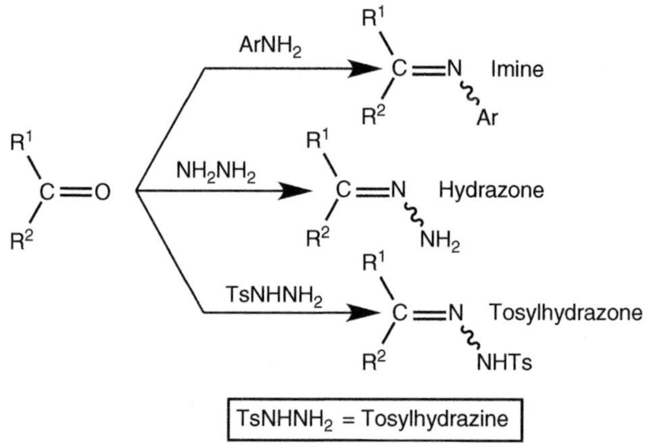

$$(5A.42)$$

With regard to the reaction above (5A.42), recall the abbreviation Ts = tosyl or *p*-toluenesulfonyl (i.e., *p*-CH$_3$-C$_6$H$_4$-SO$_2$-). The *hydrazones* (as the products with hydrazine are called) lend themselves to a number of unique transformations, of which we will discuss the Wolff–Kishner reduction and the Shapiro reaction.

The Wolff–Kishner reduction allows the one-pot transformation of a carbonyl group to a methylene group (CH$_2$) via a hydrazone intermediate, as shown below:

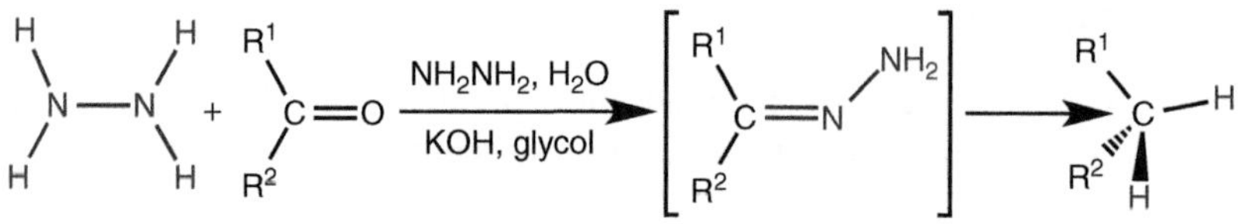

$$(5A.43)$$

We will not go through the mechanism of hydrazone formation. Once formed, the hydrazone may be deprotonated by OH$^-$ or a glycol-derived alkoxide anion:

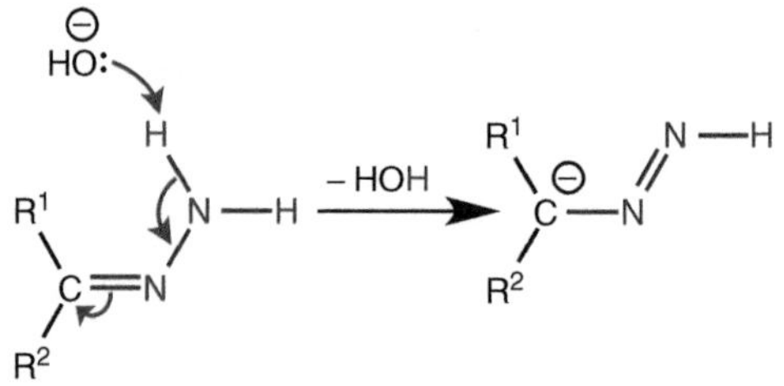

$$(5A.44)$$

The anion so produced is then protonated on carbon, as shown below:

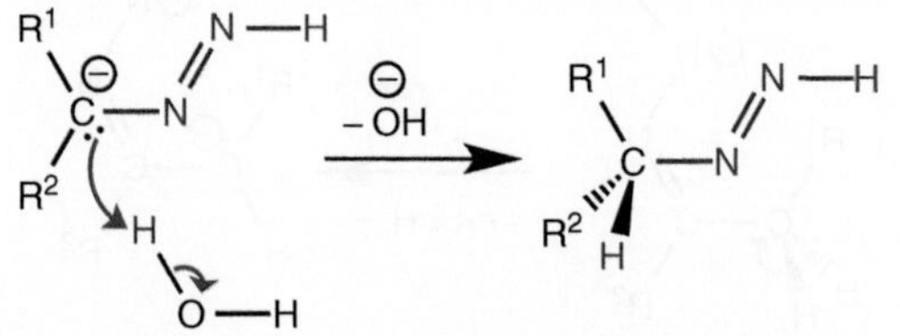

(5A.45)

A second deprotonation then leads to loss of N_2; protonation of the resulting carbanion then produces the final product where the original carbonyl group has been replaced by a CH_2 group.

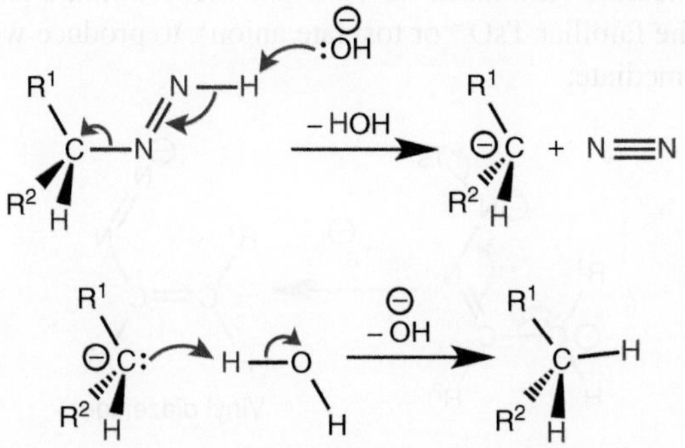

(5A.46)

The Shapiro reaction, on the other hand, allows the transformation of a tosylhydrazone to an alkene, as shown below:

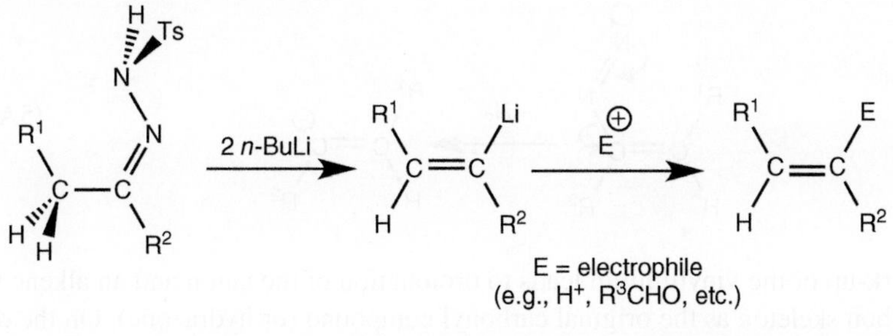

(5A.47)

E = electrophile
(e.g., H^+, R^3CHO, etc.)

Note that the reaction requires (at least) 2 equiv. of a strong base such as *n*-butyllithium. The first equivalent deprotonates the NH proton, and the second equivalent plucks off a proton α with respect to (i.e., adjacent to) the C=N double bond, as shown below:

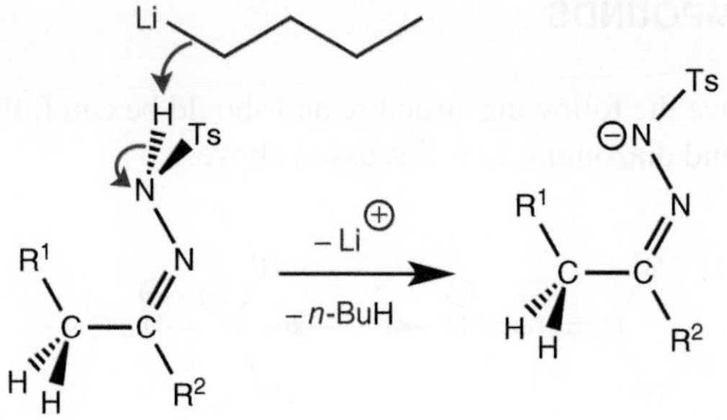

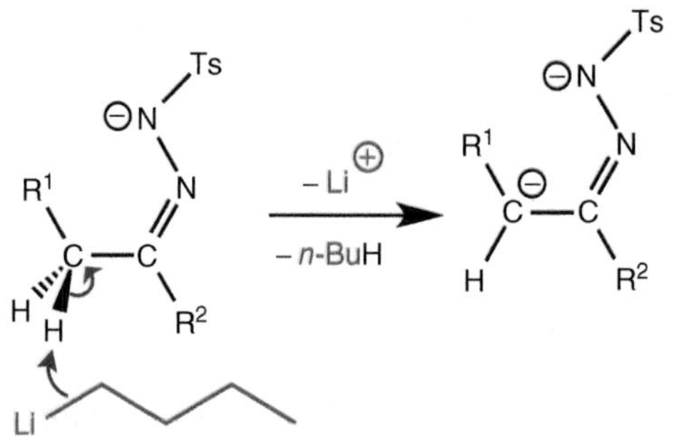

$$\text{(5A.48)}$$

The dianionic intermediate eliminates Ts⁻, the *p*-toluenesulfinate anion (which must be distinguished from the familiar TsO⁻ or tosylate anion), to produce what may be termed a *vinyl diazenide* intermediate:

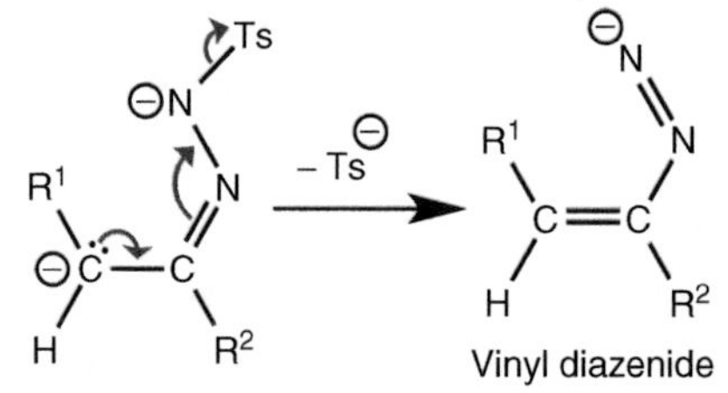

$$\text{(5A.49)}$$

Vinyl diazenide

The vinyl diazenide readily loses N_2 to generate a vinyl anion (or vinyllithium), which may be quenched in various ways:

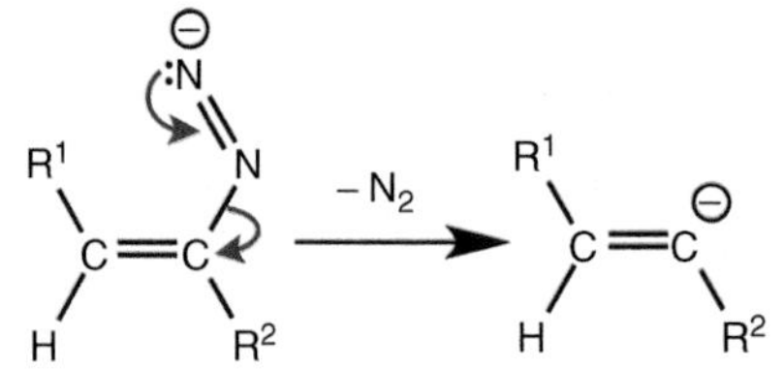

$$\text{(5A.50)}$$

Aqueous work-up of the vinyllithium leads to protonation of the anion and an alkene with the same carbon skeleton as the original carbonyl compound (or hydrazone). On the other hand, quenching with electrophiles such as alkyl halides and aldehydes leads to alkenes with extended carbon skeletons (see reaction 5A.47).

5A.9 DIAZO COMPOUNDS

Diazo compounds have the following structure and should be carefully distinguished from the azo compounds and diazonium ions discussed above:

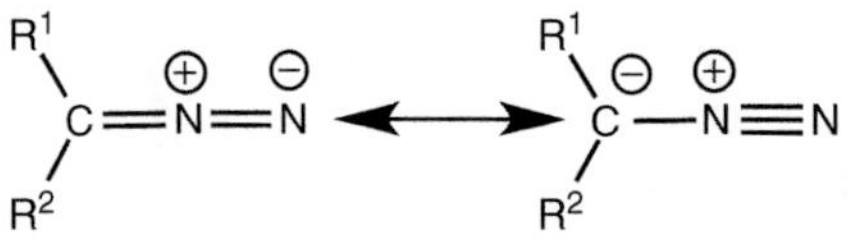

The resonance form with a negative charge on the carbon bearing the N_2^+ group immediately suggests a mild carbanionic character. Diazomethane, thus, may be protonated, which underlies its use as a popular methylating agent in organic chemistry, as shown below for a carboxylic acid:

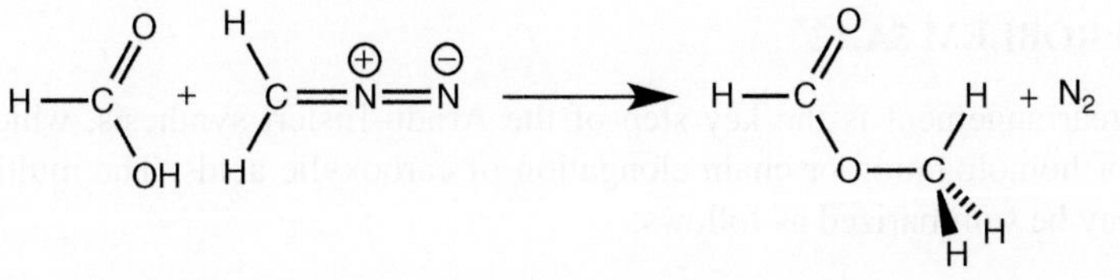

$$\text{(5A.51)}$$

The first step of the mechanism involves protonation of diazomethane:

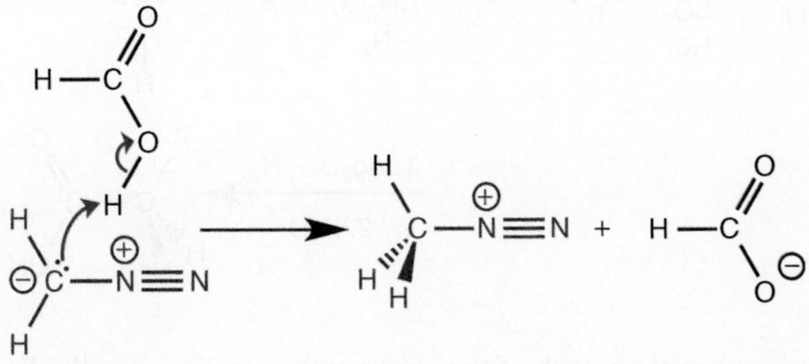

$$\text{(5A.52)}$$

The carboxylate anion then attacks the resulting methyl group, kicking out N_2, which is a superb leaving group:

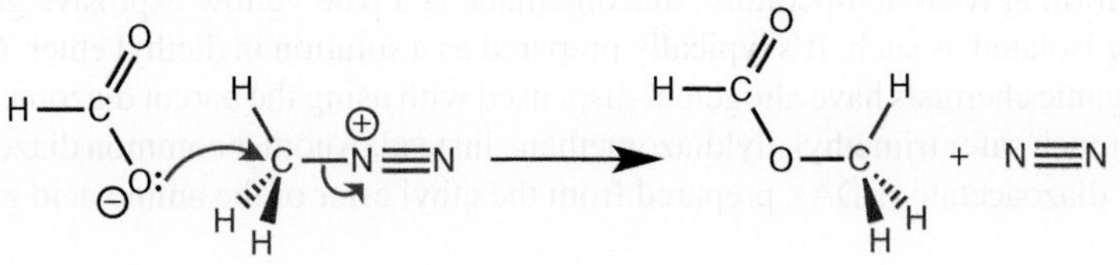

$$\text{(5A.53)}$$

Another classic reaction of diazo compounds is the Wolff rearragement, where an α-diazoketone loses N_2 to form a carbene, which spontaneously rearranges to a ketene:

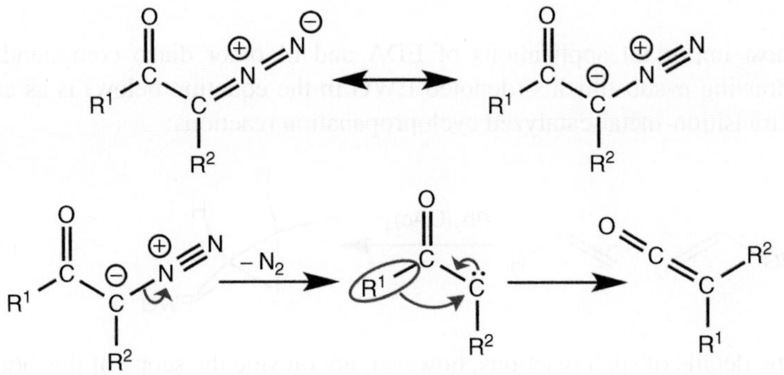

$$\text{(5A.54)}$$

The loss of N_2 is mediated by heat, light, or a transition-metal catalyst such as Ag_2O. Typically, the unstable ketene is not isolated but is trapped by water (to yield a carboxylic acid) or another molecule in the reaction medium.

REVIEW PROBLEM 5A.12*

The Wolff rearrangement is the key step of the Arndt–Eistert synthesis, which is a method for homologation or chain elongation of carboxylic acids. The multistep sequence may be summarized as follows:

As far as you can, try to explain each of the above steps mechanistically. Feel free to consult an organic text.

In its pure form at room temperature, diazomethane is a pale yellow explosive gas and is almost never isolated as such. It's typically prepared as a solution in diethyl ether. Nowadays, many organic chemists have altogether dispensed with using the parent diazomethane, preferring the much safer trimethylsilyldiazomethane instead. Another common diazo compound is ethyl diazoacetate (EDA), prepared from the ethyl ester of the amino acid glycine as follows:

$$(5A.55)$$

One of the most important applications of EDA and of other diazo compounds with electron-withdrawing α-substituents (denoted EWG in the equation below) is as carbene equivalents in transition-metal-catalyzed cyclopropanation reactions:

$$(5A.56)$$

The mechanistic details of such reactions, however, are outside the scope of this book.

Certain diazo compounds, however, do not require a transition-metal catalyst in order to react with carbon–carbon double bonds. Diazomethane itself acts as a 1,3-dipole and undergoes 1,3-dipolar additions with alkenes. The "pyrazoline" intermediate initially produced loses N_2 to produce a cyclopropane:

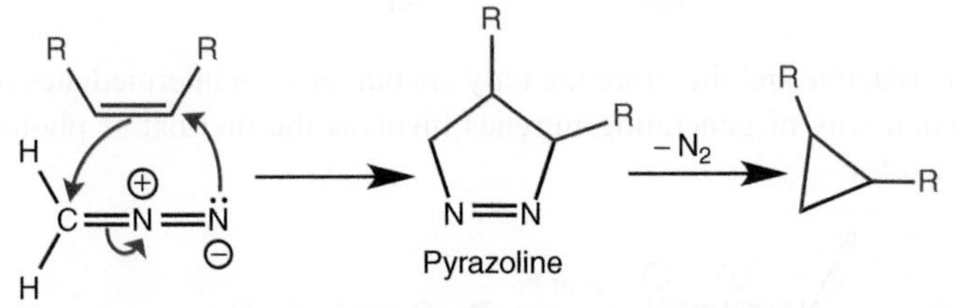

$$(5A.57)$$

Provided it works (which is not always the case), this transition-metal-free reaction may be viewed as a greener alternative to traditional metal-catalyzed cyclopropanation protocols.

REVIEW PROBLEM 5A.13

Trimethylsilyldiazomethane is obtained via the interaction of trimethylsilylmethylmagnesium chloride and diphenyl phosphorazidate (DPPA), as shown below:

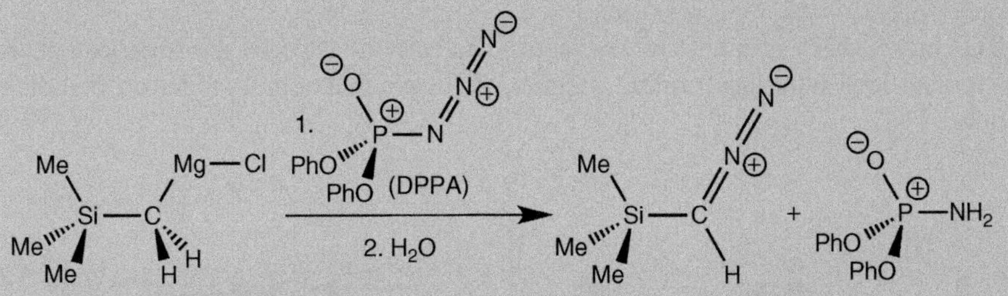

Provide a mechanistic rationale for the synthesis.

Note: The organomagnesium starting material is obtained via a standard Grignard synthesis:

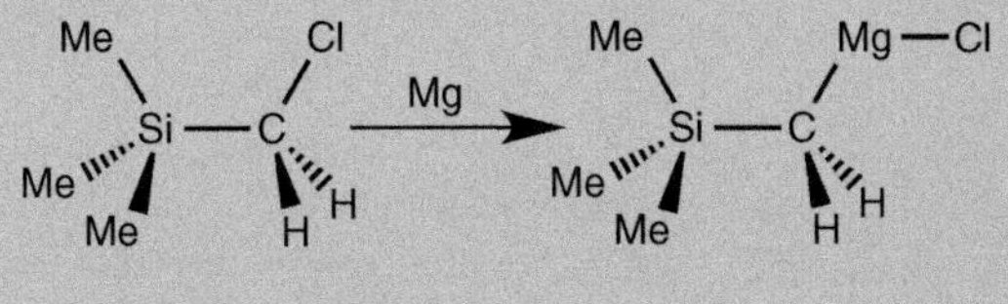

5A.10 NITRENES AND NITRENOIDS: THE CURTIUS REARRANGEMENT

Nitrenes are monovalent nitrogen compounds with a zero formal charge on the nitrogen. They are subvalent in the sense that the nitrene nitrogen has only six electrons in its valence shell in the Lewis structures shown below. They are isoelectronic with carbenes and, like carbenes, also exist in both singlet and triplet states:

$$R\!-\!\overset{\cdot\cdot}{N}\!: \qquad\qquad R\!-\!\overset{\cdot\cdot}{\underset{\cdot}{N}}\cdot$$

Singlet   Triplet

Nitrenes are highly reactive and therefore are only encountered as intermediates in organic reactions. A common way of generating nitrenes involves the thermal or photochemical decomposition of azides:

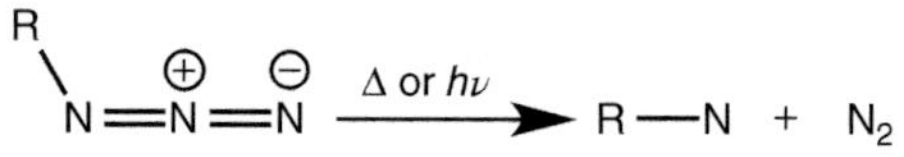

$$\overset{R}{\underset{}{\diagdown}}\,N\!=\!\overset{\oplus}{N}\!=\!\overset{\ominus}{N} \xrightarrow{\;\Delta \text{ or } h\nu\;} R\!-\!N \;+\; N_2 \tag{5A.58}$$

One of the best known reactions involving nitrene intermediates is the Curtius rearrangement of an acyl azide to an isocyanate, discovered by the German chemist Theodor Curtius in 1890:

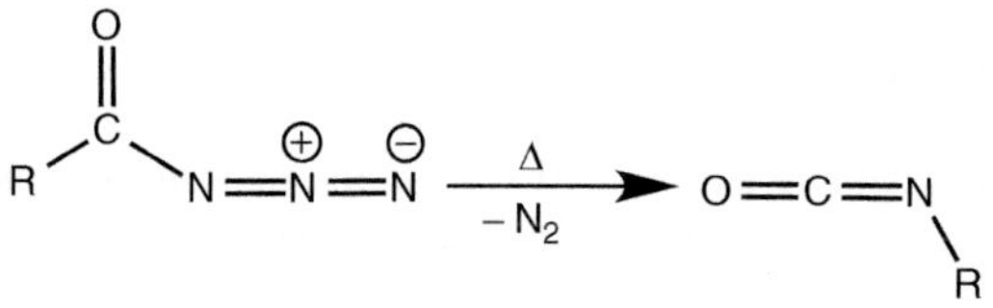

$$\underset{R}{\overset{O}{\overset{\|}{\diagup C \diagdown}}}\,N\!=\!\overset{\oplus}{N}\!=\!\overset{\ominus}{N} \xrightarrow[-N_2]{\;\Delta\;} O\!=\!C\!=\!\underset{R}{N} \tag{5A.59}$$

The intermediacy of a nitrene was invoked as early as 1896 by the American chemist (of German-Jewish heritage) Julius Stieglitz, an insight that appears to have stood the test of time:

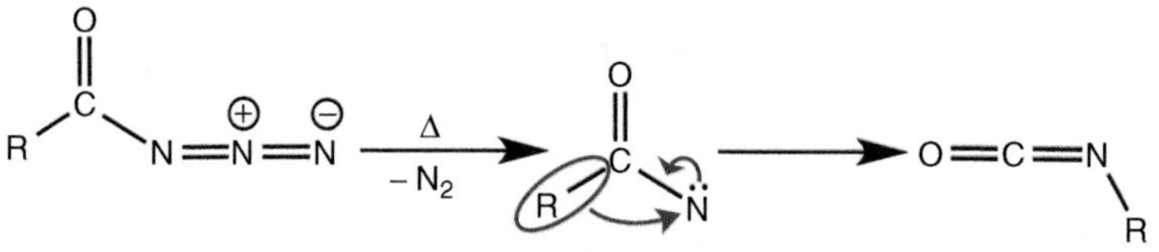

$$\underset{R}{\overset{O}{\overset{\|}{\diagup C \diagdown}}}\,N\!=\!\overset{\oplus}{N}\!=\!\overset{\ominus}{N} \xrightarrow[-N_2]{\;\Delta\;} \;\;\longrightarrow\; O\!=\!C\!=\!\underset{R}{N} \tag{5A.60}$$

In the presence of water, the isocyanate is typically hydrolyzed to an amine, whereas in the presence of an alcohol a carbalkoxy-protected amine is obtained:

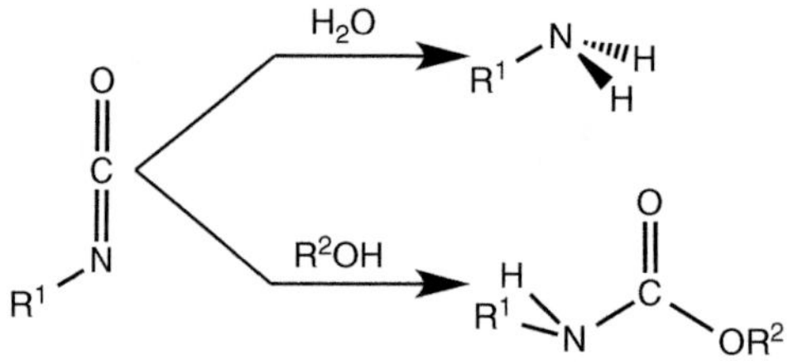

$$\begin{array}{c}
\xrightarrow{\;H_2O\;} R^1\!-\!N\overset{\cdots H}{\underset{H}{}} \\[4pt]
\underset{R^1}{\overset{O}{\overset{\|}{C}}}\!\!=\!\!N \\[4pt]
\xrightarrow{\;R^2OH\;} R^1\!\diagdown_{N}\overset{H}{}\!\diagup\!\overset{O}{\overset{\|}{C}}\!\diagdown OR^2
\end{array} \tag{5A.61}$$

Like carbenes, nitrenes also insert into C–H bonds and add to double bonds. In some of the most useful nitrene addition reactions, however, free nitrenes are not involved; rather a carrier such as a transition-metal complex or an iodine or bromine reagent is used to deliver the nitrene directly to an organic target molecule. We will discuss the role of higher-valent bromine reagents as nitrene delivery agents in the chapter on halogens (Section 7.14).

REVIEW PROBLEM 5A.14*

In Section 3.5, we described a number of small-molecule activation processes mediated by a unique monovalent aluminum β-diketiminate complex. The following are two examples of transformations involving nitrogen-containing functional groups:

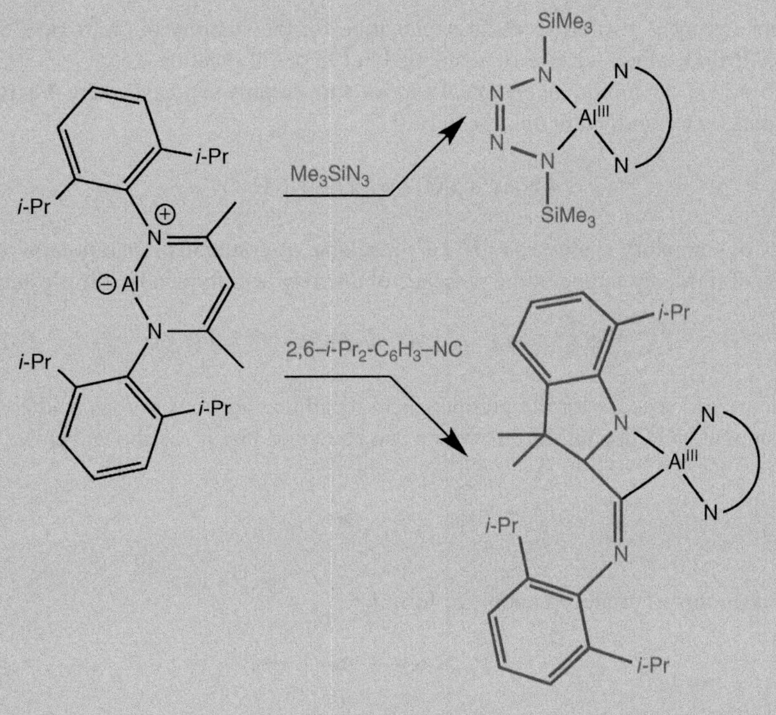

Suggest mechanisms for the two reactions. (Note: For brevity, the β-diketiminate ligand has been indicated as an arc in both the products.)

5A.11 NITRIC OXIDE AND NITROGEN DIOXIDE

We pretty much haven't discussed any radical chemistry in this chapter, and that's all right, given our focus on polar mechanisms in this book. We would be remiss, however, if we didn't briefly mention nitric oxide (NO) and nitrogen dioxide (NO_2), two important and stable nitrogen-based radicals. The stability of both molecules may be attributed to resonance, that is, delocalized bonding, as shown below:

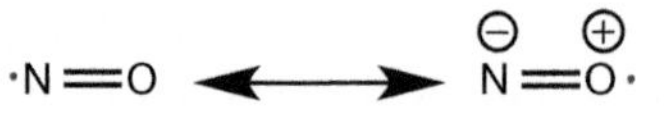

·N═O ⟷ N═O·

Nitric oxide, NO

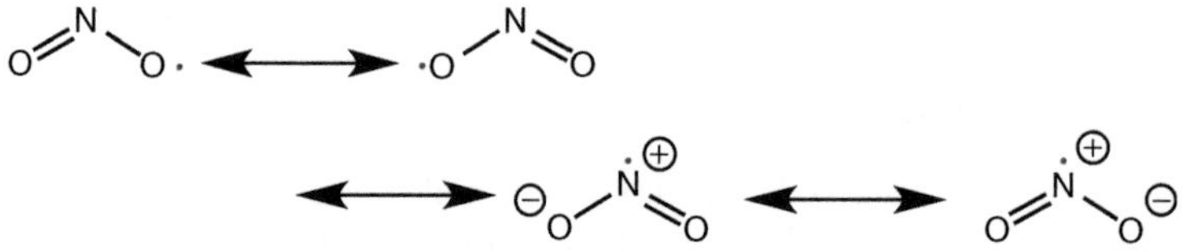

O═N–O· ⟷ ·O–N═O ⟷ O–N═O ⟷ O═N–O

Nitrogen dioxide, NO₂

The same argument also provides an explanation for the stability of "nitroxide" radicals such as TEMPO, which we briefly mentioned in Chapter 2 (Section 2.5).

Nitric oxide is an important chemical and is commercially synthesized by the oxidation of ammonia by O_2 with platinum as catalyst:

$$4\,NH_3 + 5\,O_2 \rightarrow 4\,NO + 6\,H_2O \tag{5A.62}$$

A variety of laboratory syntheses of NO are available. A common preparation involves the reduction of nitrite by iodide in the presence of a strong acid (typically sulfuric acid):

$$2\,NO_2^- + 2\,I^- + 4\,H^+ \rightarrow 2\,NO + I_2 + 2\,H_2O \tag{5A.63}$$

A simple way to account for the products is to recall that aqueous nitrous acid contains a certain amount NO^+ in equilibrium, which can be reduced by I^-, as shown below:

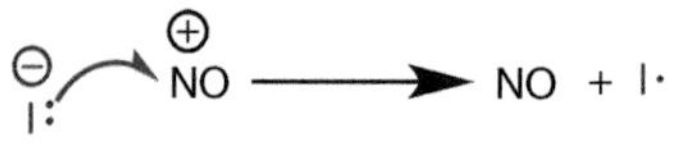

$$I^- \overset{+}{NO} \longrightarrow NO + I· \tag{5A.64}$$

The iodine atoms so produced couple to form I_2:

$$I· \quad ·I \longrightarrow I—I \tag{5A.65}$$

Although nitric oxide is relatively stable under anerobic conditions, it's oxidized by O_2 to nitrogen dioxide (NO_2):

$$2\,NO + O_2 \rightleftharpoons 2\,NO_2 \tag{5A.66}$$

At temperatures around 150 °C, NO_2 reverts back to NO.

As mentioned, nitrogen dioxide is a radical. It's a reddish-brown gas under ordinary conditions and exists in equilibrium with the dimer N_2O_4, a colorless diamagnetic species:

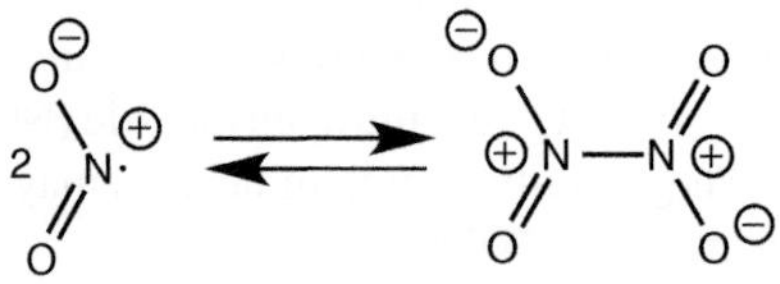

$$2 \; \text{[NO}_2\text{ radical]} \rightleftharpoons \text{[N}_2\text{O}_4\text{ dimer]} \tag{5A.67}$$

The N–N bond in N_2O_4 is long (1.78 Å, compared with 1.49 Å for hydrazine) and weak, and ΔH for the above dimerization is only about –57.2 kJ/mol. Thus, NO_2 is favored at higher temperatures whereas N_2O_4 predominates at lower temperatures.

Nitrogen dioxide reacts with water to yield a mixture of nitrous and nitric acids:

$$2\, NO_2 + H_2O \rightarrow HNO_2 + HNO_3 \tag{5A.68}$$

The reaction is clearly a disproportionation, with NO_2 being simultaneously oxidized to N(V) and reduced to N(III). One possibility is that water, a highly polar solvent, facilitates the transfer of an electron between two NO_2 molecules:

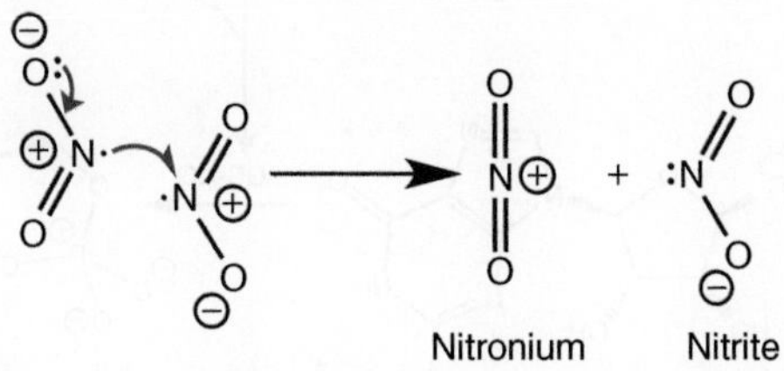

$$\tag{5A.69}$$

The nitronium ion so formed is a powerful electrophile and is immediately hydrolyzed to nitrate, as shown below:

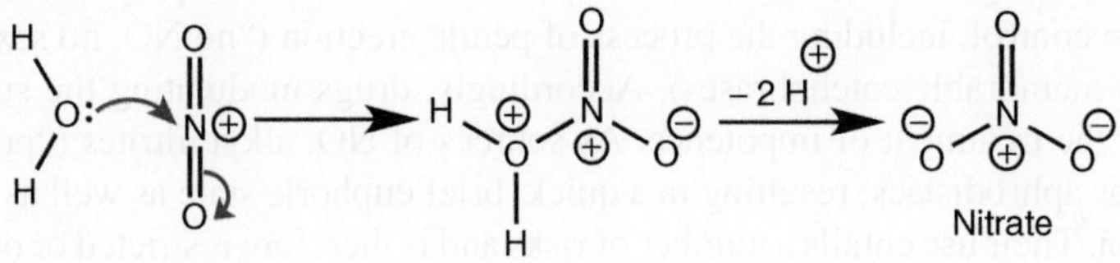

$$\tag{5A.70}$$

Alternatively, water might attack nitrogen dioxide first, with electron transfer occurring in the next step. Feel free to write out the details of such a mechanism.

Both NO and NO_2 are of immense practical importance. Both are byproducts of internal combustion engines and thermal (coal) power plants, and are severe environmental pollutants in many parts of the world. As mentioned, NO is converted to NO_2 by aerial oxidation and also by tropospheric ozone:

$$NO + O_3 \rightarrow NO_2 + O_2 \tag{5A.71}$$

Nitrogen dioxide is toxic by inhalation, leading, among other things, to pulmonary edema. Hydrolysis of NO_2 in cloud droplets, that is, reaction 5A.68, is a contributor to acid rain. Hydroxyl radicals in the troposphere also react with NO_2 to form HNO_3:

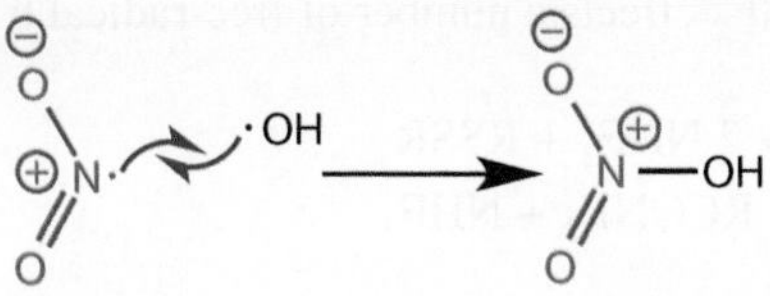

$$\tag{5A.72}$$

Considering its role as a pollutant, it's ironic that NO is one of the most important signaling molecules in biology, affecting nearly every facet of physiological function.

Its biological source is the essential amino acid arginine, from which it is synthesized by the heme-containing enzyme NO synthase. Unfortunately, the mechanism is rather complex, involving a good deal of iron (heme) chemistry, so we are obliged to skip it in this book. Among NO's myriad signaling roles, perhaps the best known is in vasodilation in mammalian systems. NO is sensed by the enzyme-soluble guanylate cyclase (sGC), another heme protein, which has a histidine ligand on the iron. NO binding to the heme iron results in a loosening of the iron–histidine bond; the unligated histidine plays a key role in the transformation of guanosine triphosphate (GTP) to cyclic guanosine monophosphate (cGMP):

$$(5A.73)$$

cGMP sets off a cascade of reactions which ultimately leads to vasodilation; cGMP thus acts as a second messenger (NO being the first signal). Nitric oxide thus plays a critical role in blood pressure control, including the process of penile erection ("no NO, no sex" being one of the more memorable catchphrases). Accordingly, drugs modulating the supply of NO are used for the treatment of impotence. As sources of NO, alkyl nitrites ("poppers") have been used as aphrodisiacs, resulting in a quick, brief euphoric state as well as smooth muscle relaxation. Their use entails a number of risks and is therefore restricted or outlawed in many countries.

REVIEW PROBLEM 5A.15

Unlike the other nitrogen halides, nitrogen trifluoride (NF_3) is a remarkably stable molecule, which is unaffected by water and other common reagents at room temperature. With copper metal at high temperature, it forms tetrafluorohydrazine (N_2F_4). A gas under ordinary conditions, N_2F_4 dissociates readily into NF_2 radicals:

Accordingly, N_2F_4 or NF_2 effects a number of free-radical transformations:

1. $2\,NF_2 + 2\,RSH \rightarrow 2\,NHF_2 + RSSR$
2. $RCHO + N_2F_4 \rightarrow RCONF_2 + NHF_2$

Suggest mechanisms for these two reactions. Can you think of a rationale for the relative stability of the NF_2 radical?

5A.12   SUMMARY

1. In many ways, nitrogen may be viewed as a paradigm of the lighter p-block elements. It is justly famous for its diverse oxidation states, which include every integral value between -3 (e.g., NH_3) and $+5$ (e.g., HNO_3). It forms stable single, double, and triple bonds with other period-2 p-block elements. Nitrogen-based nucleophiles and electrophiles are ubiquitous in both organic and inorganic chemistry.

2. Amines, both aliphatic and aromatic, are widely employed as nucleophiles and as mild bases. Amide anions (NR_2^-) are used as strong bases in organic chemistry.

3. Common nitrogen-based electrophiles include the nitrosonium (NO^+) and nitronium (NO_2^+) ions, as well as nitrogen halides.

4. Elimination of dinitrogen (N_2) is a powerful driving force in chemical reactions. A number of important organic transformations hinge around the elimination of N_2.

5. Nitrenes, which are monovalent nitrogen species, are unstable but are useful reactive intermediates in organic chemistry. We'll encounter them again in Sections 7.11 and 7.15 in the form of iodine- and bromine-based nitrene transfer agents.

6. The two radical species NO and NO_2 are of great practical importance. Nitrogen dioxide (NO_2) is a major pollutant, contributing to acid rain as well as to respiratory illnesses. Nitric oxide (NO) is a ubiquitous signaling molecule in biology.

FURTHER READING

Besides the texts listed in Appendices 1 and 2, the following books cover some of the special topics discussed in this chapter.

1. Stoltzenberg, D. *Fritz Haber: Chemist, Nobel Laureate, German, Jew: A Biography*; Chemical Heritage Foundation, 2005; 336 pp. An excellent biography.
2. Bailey, P. D.; Morgan, K. M. *Organonitrogen Chemistry*; Oxford University Press: Oxford, 1996; 96 pp.
3. Favery, D. E.; Gudmundsdottir, A. D., eds. *Nitrenes and Nitrenium Ions*; John Wiley & Sons, Inc.: Hoboken, NJ, 2013; 586 pp. *A major, up-to-date survey of an important field.*
4. Gilchrist, C. L.; Rees, C. W. *Carbenes, Nitrenes and Arynes*; Nelson: London, 1969; 131 pp. *A short classic.*
5. Butler, A. R.; Nicholson, R. *Life, Death and Nitric Oxide*; Royal Soc.: Cambridge, 2003; 140 pp. *A readable account of the biological role of NO.*
6. Ignarro, L. J., ed. *Nitric Oxide: Biology and Pathobiology*, 2nd ed. Academic: Waltham, MA, 2009; 845 pp. *A definitive volume with contributions by leading researchers.*

5B

The Heavier Pnictogens

Lady Astor: If you were my husband, I'd put arsenic in your coffee.

Churchill: Madam, if I were your husband, I'd drink it!

The chemical properties of the group 15 elements or pnictogens (for which we have occasionally used the symbol Pn) span a wide range but perhaps not quite to the same extent as groups 13 and 14. The chemistry of nitrogen and phosphorus thus do not differ as sharply as that of carbon and silicon. To illustrate, both N and P are classic nonmetals and form molecular acidic oxides. As with other p-block groups, metallic character increases down the group; As and Sb are considered metalloids and Bi is a metal. Thus, As_4O_6 and Sb_4O_6 are molecular species, isostructural with P_4O_6; Bi_2O_3, by contrast, consists of an ionic lattice of Bi^{3+} and O^{2-} ions. Some additional group trends are as follows:

- As for other "first-row" elements (B–F), the coordination number of nitrogen cannot exceed 4, even though a valence of 5 is quite common for nitrogen (e.g., NH_4^+ and NO_3^-). This is believed to be simply a reflection of nitrogen's small atomic size. By contrast, coordination numbers of 5 and 6 are common for all the other group 15 elements; for Bi, even 9-coordination has been documented.

- Unlike nitrogen, the heavier pnictogens do not form true multiple bonds with ease. Thus, molecules such as $POCl_3$ are better viewed as $Cl_3P^+–O^-$, as opposed to $Cl_3P=O$. In this newer view, the shortness of the P–O linkage (1.58 Å) is attributed to its ionic rather than double-bond character.

- For p-block elements, phosphorus exhibits above-average catenating properties. Thus, although linear, single-bonded chains tend to be unstable, elemental phosphorus forms a variety of single-bonded cage structures.

- Valences of 3 and 5 are well established for all the elements, although the relative stability of the tri- and penta-valent states varies considerably among the elements. The propensity to adopt pentavalent states is probably the strongest for phosphorus, especially when molecules containing the highly stable $P^+–O^-$ unit are involved. The bond dissociation energy (BDE) for $R_3P^+–O^-$ ($R_3P^+–O^- \rightarrow R_3P + O$) averages a

Arrow Pushing in Inorganic Chemistry: A Logical Approach to the Chemistry of the Main-Group Elements, First Edition. Abhik Ghosh and Steffen Berg.
© 2014 John Wiley & Sons, Inc. Published 2014 by John Wiley & Sons, Inc.

whopping 544 kJ/mol, which isn't far below that for a C=C double bond (602 kJ/mol). Accordingly, the following isomerization is ubiquitous in much of phosphorus chemistry:

$$ \tag{5B.1} $$

Thus, phosphorus acid, H_3PO_3, is not $P(OH)_3$, but rather $HP^+(OH)_2(O^-)$. Arsenous acid, H_3AsO_3, by contrast, is $As(OH)_3$:

$$ \tag{5B.2} $$

Phosphorous acid   Arsenous acid

- Arsenic exhibits a marked reluctance to adopt a pentavalent state. For example, unlike PCl_5 and $SbCl_5$, which are stable and commercially available substances, $AsCl_5$ decomposes at $-50\,°C$. It was first prepared only in 1976 by UV irradiation of $AsCl_3$ in liquid chlorine at $-105\,°C$. The anomalous instability of $AsCl_5$ is generally ascribed to a phenomenon called *d-block contraction*, which stabilizes the 4s lone pairs of the elements following the first transition series. Unfortunately, limitations of space do not permit a proper discussion of the phenomenon.

- Pentavalent bismuth compounds are typically somewhat unstable and oxidizing. Quite a few are useful as oxidants in various organic transformations. The instability and oxidizing power of pentavalent bismuth is generally ascribed to the inert pair effect (Section 1.27), which we also encountered for thallium and lead.

- The trivalent state can be either nucleophilic or electrophilic depending on substituents and reaction conditions. Thus, Ar_3Pn species are typically used as nucleophiles; Ph_3P is one of the most important reagents in organic chemistry and Ph_3Bi too is used as a precursor to a number of Bi-based reagents. The trihalides (PnX_3) may be nucleophilic, such as in their reactions with molecular halogens, but are often also used as electrophiles.

- The pentafluorides (PnF_5, Pn = P, As, and Sb) are strong Lewis acids and fluoride ion acceptors.

With these general observations in place, we are now in a position to dig deeper into group 15 chemistry.

REVIEW PROBLEM 5B.1

Phosphite esters, $P(OR)_3$, are well known to inorganic chemists as transition-metal ligands. Trimethyl phosphite, however, is particularly reactive and spontaneously rearranges to the dimethyl ester of methylphosphonic acid:

Suggest a mechanism for the reaction.

REVIEW PROBLEM 5B.2

Suggest a mechanism and a hard and soft Lewis acids and bases (HSAB) rationale for the following ligand exchange (metathesis) reaction:

$$PCl_3 + AsF_3 \rightarrow PF_3 + AsCl_3$$

5B.1   OXIDES

The oxides Pn_4O_6 (Pn = P, As, Sb) and Pn_4O_{10} (Pn = P, As) illustrate the preponderance of single bonds in the heavier p-block elements. These structures may be contrasted with N_2O_3 and N_2O_5, which contain N=O double bonds. We urge you to spend a few minutes to learn to draw the Pn_4O_6 and Pn_4O_{10} structures reasonably well. These structures are based on cyclohexane-like rings in the chair conformation arranged in the form of the organic molecule adamantane ($C_{10}H_{16}$):

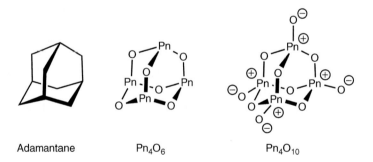

Adamantane Pn_4O_6 Pn_4O_{10}

P_4O_{10}, commonly known by its historic name "phosphorus pentoxide," is prepared by burning elemental phosphorus in a plentiful supply of air:

$$P_4 + 5\,O_2 \rightarrow P_4O_{10} \tag{5B.3}$$

P_4O_{10} is best known as a powerful dehydrating agent. Not only does it react avidly with water but it also extracts the elements of water from various substances. In Section 7.7, we'll go through a remarkable reaction in which P_4O_{10} dehydrates perchloric acid ($HClO_4$), one of the strongest common acids, to Cl_2O_7. A typical acidic oxide, P_4O_{10} reacts with water, producing phosphoric acid (H_3PO_4):

$$P_4O_{10} + 6\,H_2O \;\rightarrow\; 4\,H_3PO_4 \tag{5B.4}$$

The mechanism may be viewed as a series of nucleophilic attacks by water on the P^+ centers with the oxo bridges in the P–O–P linkages as leaving groups, as shown below:

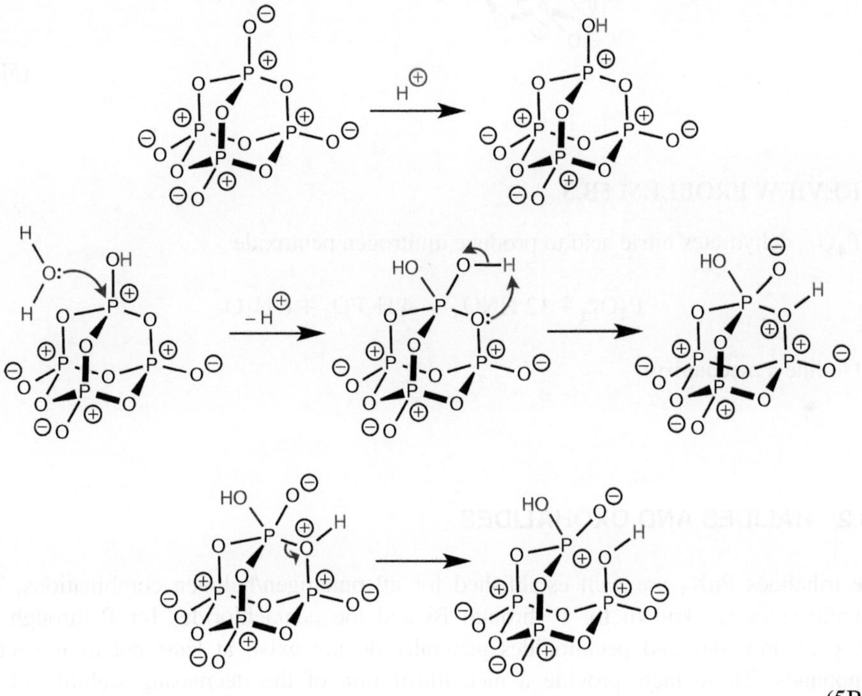

$$(5B.5)$$

A couple of points are worth noting: First, observe the role of protons. Protonation on an adjacent oxygen makes a given P^+ center a better electrophile. Second, protonation of a bridging oxygen also makes it a better leaving group; thus, H_3PO_4 is a better leaving group than $H_2PO_4{}^-$. Second, we have tacitly assumed an S_N2-Si mechanism, that is, the nucleophilic attack and the departure of the leaving group are not shown to be concerted.

Attack by water and opening up of the P–O–P bridges continue until the entire P_4O_{10} skeleton has broken down to phosphoric acid:

(5B.6)

REVIEW PROBLEM 5B.3

P_4O_{10} dehydrates nitric acid to produce dinitrogen pentoxide:

$$P_4O_{10} + 12\,HNO_3 \rightarrow 4\,H_3PO_4 + 6\,N_2O_5$$

Outline a mechanism.

5B.2 HALIDES AND OXOHALIDES

The trihalides PnX_3 are well established for all pnictogen/halogen combinations. The pentafluorides are known for P through Bi and the pentachlorides for P through Sb. The pentabromides and pentaiodides generally do not exist, at least not as molecular compounds. These facts provide a nice illustration of the decreasing stability of the pentavalent state from phosphorus downward.

Phosphorus trichloride (PCl_3) is an industrial chemical that serves as a precursor to other important chemicals such as PCl_5, $POCl_3$, and $PSCl_3$, which in turn serve as starting materials in the synthesis of insecticides, herbicides, plasticizers, and flame retardants. Hundreds of thousands of tons of PCl_3 are produced annually, worldwide, based on direct combination of the elements:

$$P_4 + 6\,Cl_2 \rightarrow 4\,PCl_3 \tag{5B.7}$$

It's reasonable to envision the mechanism as beginning with an attack by a phosphorus lone pair on Cl_2, with concomitant breakage of the weak Cl–Cl bond. The cationic P^+ center thus produced makes for a good leaving group when Cl^- attacks a neighboring P atom:

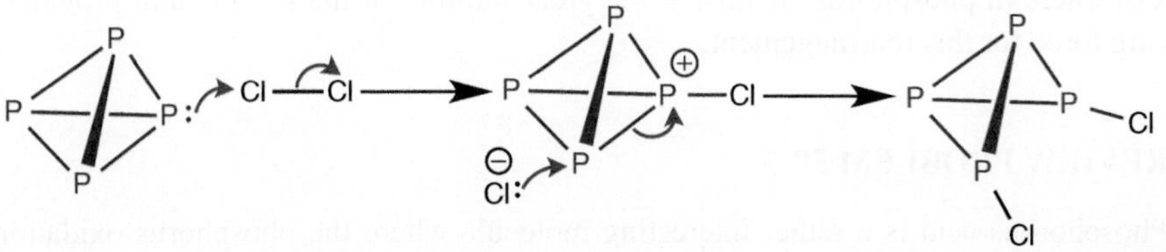

$$(5B.8)$$

The same process essentially continues until the P_4 skeleton is entirely broken down to PCl_3:

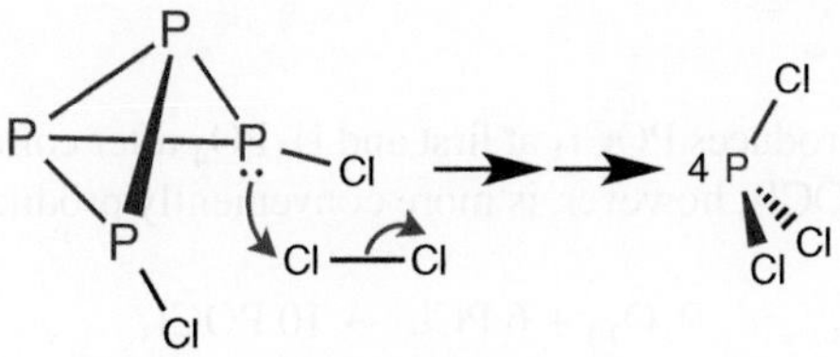

$$(5B.9)$$

REVIEW PROBLEM 5B.4

Phosphorus trichloride is used as a starting material for the important reagent triphenylphosphine. Suggest a reasonable route for this synthesis.

Hydrolysis of PCl_3 leads to phosphorous acid, H_3PO_3:

$$PCl_3 + 3\,H_2O \rightarrow H_3PO_3 + 3\,HCl \qquad (5B.10)$$

The mechanism begins straightforwardly enough with simple nucleophilic displacements of chloride by water:

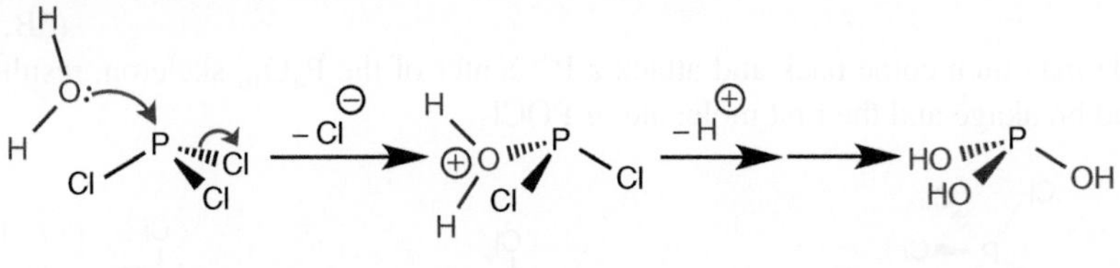

$$(5B.11)$$

As mentioned above, the structure of phosphorous acid is not $P(OH)_3$, so we need a path to attain the actual tetracoordinate structure. Protonation of the trivalent phosphorus and deprotonation of one of the OH groups do the trick, as shown below:

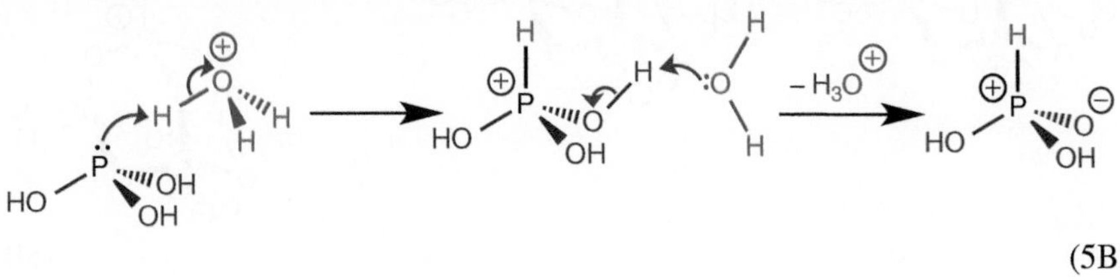

$$(5B.12)$$

As elsewhere in phosphorus chemistry, the great stability of the P^+-O^- unit provides the driving force for this rearrangement.

REVIEW PROBLEM 5B.5

Phosphorous acid is a rather interesting molecule where the phosphorus oxidation state (based on the formula H_3PO_3), coordination number, and valence are all different, +3, 4, and 5, respectively! Explain and comment. If necessary, recall the definitions of the terms from Chapter 1.

Hydrolysis of PCl_5 produces $POCl_3$ at first and H_3PO_4 after complete breakdown. Phosphorous oxochloride, $POCl_3$, however, is more conveniently produced by the interaction of P_4O_{10} and PCl_5:

$$P_4O_{10} + 6\,PCl_5 \rightarrow 10\,POCl_3 \tag{5B.13}$$

With 7 molecules as reactants producing 10 molecules of product, the full mechanism is clearly long and repetitive. Where would one begin in a case like this? Let's start by identifying a likely nucleophile and an electrophilic site it can attack. The anionic O^- groups of P_4O_{10} are plausible nucleophiles, while the pentavalent P in PCl_5 is a plausible electrophile, with Cl^- being a reasonable leaving group:

$$\tag{5B.14}$$

The chloride may then come back and attack a P^+ center of the P_4O_{10} skeleton, resulting in P–O bond breakage and the first molecule of $POCl_3$.

$$\tag{5B.15}$$

Accounting for the full stoichiometry of the reaction (5B.15) is rather tedious, but do work through a few additional steps to make sure that you can handle them. It's worth noting that this reaction is a ligand exchange: oxide and chloride ligands are traded between two pentavalent phosphorus centers. As emphasized in Section 1.19, reactions of this type generally involve bridged intermediates; observe that the migrating oxygen acts as a bridging group in a number of the intermediates above.

REVIEW PROBLEM 5B.6

Suggest a mechanism and an HSAB rationale for the following ligand exchange (metathesis) reaction:

$$PCl_3 + AsF_3 \rightarrow PF_3 + AsCl_3$$

REVIEW PROBLEM 5B.7

Suggest a mechanism for the following reaction:

$$SO_2 + PCl_5 \rightarrow SOCl_2 + POCl_3$$

5B.3 PHOSPHORUS IN BIOLOGY: WHY NATURE CHOSE PHOSPHATE

A chapter on group 15 elements cannot be complete without some discussion of phosphorus in biology. Phosphates in particular are everywhere in biology! In this section, we will be able to present only a few key structures and reactions of biological phosphates, with a little flavor of some of the classic mechanistic work that has taken place in the field. Glycolysis, the 10-step metabolic pathway that breaks down glucose to pyruvic acid, begins with phosphorylation of glucose with adenosine triphosphate (ATP):

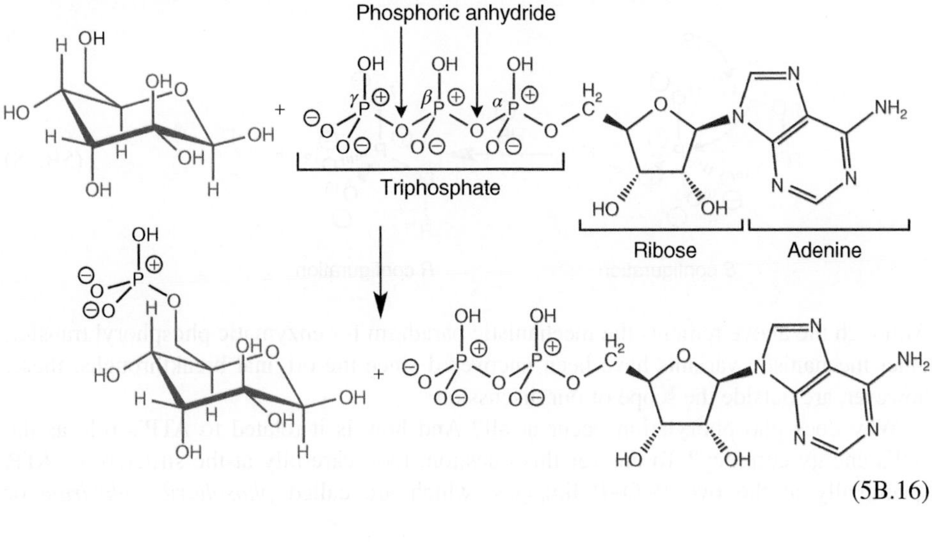

$$(5B.16)$$

The term "phosphorylation" refers to the addition of phosphoryl groups to organic molecules, typically with ATP as the "phosphoryl donor." Such reactions are catalyzed by a class of enzymes called *kinases*. The enzyme catalyzing the above reaction is called *glucokinase*.

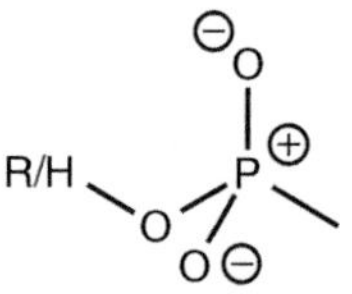

Phosphoryl group

Kinases not only bind ATP and other phosphoryl donors but also enhance the electrophilicity of the migrating phosphate group by coordination with one or two Mg^{2+} ions. The mechanism of phosphoryl transfer is essentially an S_N2 reaction, which we may represent as shown below (B is an active site base). For clarity, we have not included any of the Mg^{2+} ions in our simplified picture:

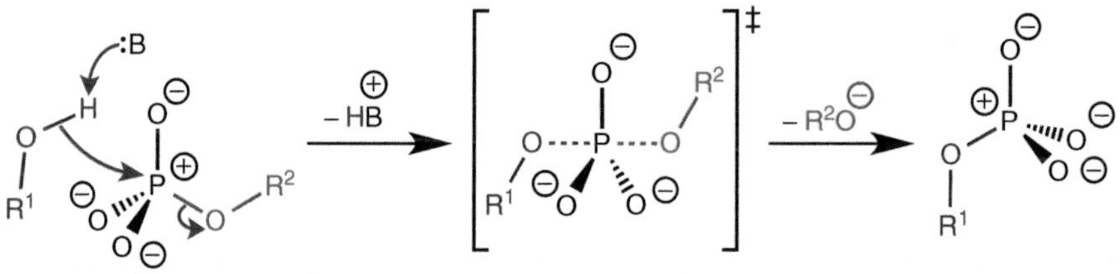

(5B.17)

Note that the mechanism, as indicated above, implies inversion of stereochemistry at the phosphorus. Proving that, however, was no simple matter because an ordinary phosphate group is not stereogenic. In 1978, in an experimental tour de force, Jeremy Knowles and coworkers at Harvard used ^{17}O and ^{18}O isotopes to create organic phosphates with stereogenic phosphate groups in enantiomerically enriched form. Subsequent experimental work confirmed inversion of stereochemistry in the product, providing support for the overall S_N2 nature of the process.

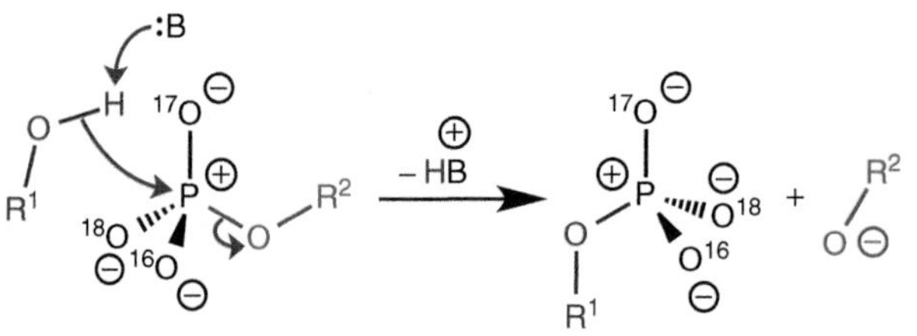

S configuration   *R* configuration

(5B.18)

Although the above remains the mechanistic paradigm for enzymatic phosphoryl transfer, other mechanistic variants have been uncovered since the original breakthroughs; these, however, are outside the scope of our discussion.

Why does phosphorylation occur at all? And how is it related to ATP's role as the cell's energy currency? To answer this question, look carefully at the structure of ATP, specifically at the two P–O–P linkages, which are called *phosphoric anhydride* or *pyrophosphate* linkages. These are the famous high energy "bonds" of ATP. Hydrolysis of these bonds releases a great deal of energy, for much the same reason that hydrolysis of P_4O_{10} does. In phosphorylation, ATP transfers a phosphoryl group to a nucleophile such as an alcohol (glucose in the example above), forming a high energy C–O–phosphoryl linkage; in brief, this is the basis of ATP's role as the cell's energy currency.

Many proteins are inactive in their pristine form and must be "switched on" for activity. Phosphorylation, catalyzed by kinases, is very often that switch. The reverse reaction, dephosphorylation, then is the "off switch" that shuts down the protein's activity; enzymes called *phosphatases* catalyze this process. Reversible phosphorylation is thus an important regulatory mechanism in cells. Phosphorylation typically occurs on amino acids with nucleophilic side chains, namely, serine, threonine, tyrosine, and histidine and, in prokaryotes, even lysine and arginine. Phosphorylation greatly increases the hydrophilicity of a given part of the protein, which may lead to large conformational changes that are necessary for activity. Alternatively, phosphorylation may turn an alcoholic OH group, one of organic chemistry's notoriously poor leaving groups, into phosphate, a much better one.

Another class of phosphate derivatives of paramount importance in biology consists of phosphate diesters or phosphodiesters (to be distinguished from organic diphosphates or pyrophosphates, which are key intermediates in steroid biosynthesis), in which two of the oxygens of phosphoric acid are esterified with alcohols. The sugar–phosphate backbone of DNA is essentially a phosphodiester polymer, where the sugar (deoxyribose) is esterified with phosphoric acid through the 3' and 5' OH groups, as shown below. The bases are attached as side chains at the 1' positions. The primes refer to atom numbering on the sugar, as opposed to the DNA bases.

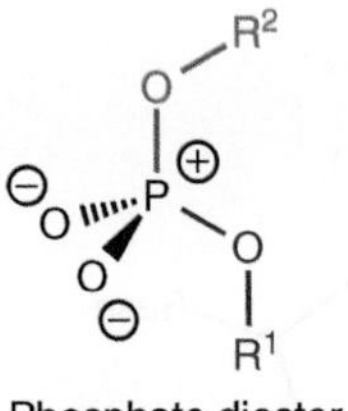

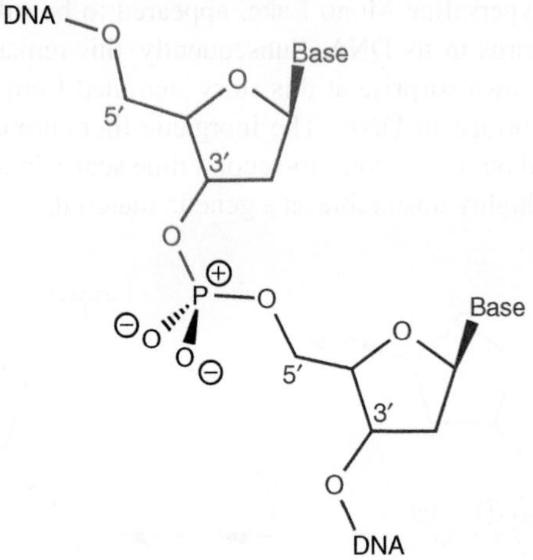

Phosphate diester

On account of their net negative charge, the phosphodiester linkages are highly resistant to hydrolysis. Recall that DNA has been sequenced from woolly mammoths to Egypt's boy-king Tutankhamun and many ancient creatures and people both before and after. Indeed, this ruggedness is part of what makes DNA so perfectly suited as a genetic material. Neither amide (as in proteins) nor ester linkages could have conferred the same degree of stability to a genetic material.

In contrast to DNA, RNA is sensitive to hydrolysis, even at pH 7.0. The reason is that the 2' OH group, which DNA lacks, can attack the 3' phosphate, cleaving the RNA backbone and forming a cyclic ester, which is then hydrolyzed:

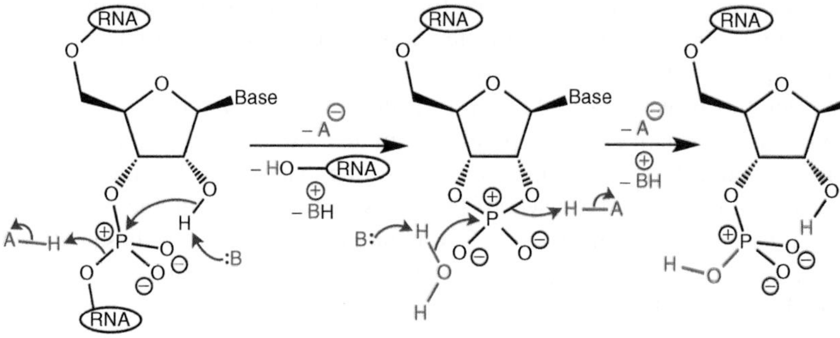

$$(5B.19)$$

Intramolecular five-membered ring formation is generally fast on entropic grounds, and it's so for RNA as well. This instability does not pose a problem for RNA's physiological role, namely, the fast transmission of genetic information rather than long-term information storage. It does, however, pose a significant inconvenience to RNA researchers, who must typically store RNA at low temperatures after freeze-drying or removing excess water by precipitation with ethanol.

5B.4 ARSENIC-BASED DNA

In 2010, a team of NASA researchers reported that GFAJ-1, a halophilic bacterium from California's hypersaline Mono Lake, appeared to be substituting arsenic for a small part of the phosphorus in its DNA. Subsequently, this remarkable claim has turned out to be incorrect. Our own surprise at this story stemmed from the chemical improbability of an arsenodiester bridge in DNA. The inorganic literature clearly shows that arsenate esters are hydrolyzed on millisecond-to-second time scales in aqueous solution, as shown below, making them highly unsuitable as a genetic material:

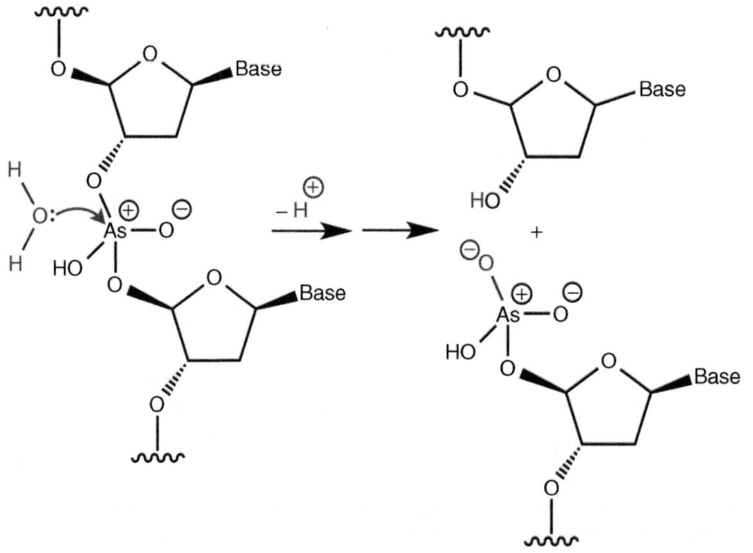

$$(5B.20)$$

Although As-DNA and arsenic-based life do not seem grounded in reality, the episode proved to be a catalyst for important discoveries on arsenic-tolerant organisms. One important question that has been addressed is: how does GFAJ-1 tolerate such high concentrations of arsenate as are found in Mono Lake and how does it discriminate between two such similar species as phosphate and arsenate? A biochemical and crystallographic study on phosphate-binding proteins (PBPs) from GFAJ-1 has shed significant light on this point (Elias, M., *et al. Nature* **2012**, *491*, 124–137). First, the study showed that the PBPs indeed bind phosphate 10^3–10^4 times more strongly than they do arsenate. Second, they found that arsenate binding leads to distortions in the hydrogen-bond network in the binding site, which is finely tuned for phosphate, especially with respect to the geometry of one short hydrogen bond. As far as phosphate/arsenate discrimination is concerned, GFAJ-1 does so very well.

A second chemical reason (besides hydrolytic instability) why an arsenate-based structure would be unsuitable as a genetic material is that it would be vulnerable to reduction by cellular reductants such as the thiol glutathione (GSH):

GSH

We know that, *in vitro*, a variety of reducing agents, including glutathione, can reduce arsenic acid (H_3AsO_4) to arsenous acid. With iodide, for example, the reaction is:

$$H_3AsO_4 + 2\,I^- + 2\,H^+ \rightarrow H_3AsO_3 + I_2 + H_2O \qquad (5B.21)$$

Both polar and radical mechanisms may be envisioned for the process. A polar mechanism might begin with I^-, a good nucleophile, attacking a positively charged and pentavalent As, a good electrophile, to create an I–As bond:

$$(5B.22)$$

The actual reduction step would then involve a second iodide attacking the As-bound I to produce molecular iodine, while kicking out an arsenite anion ($H_2AsO_3^-$) as the leaving group:

$$(5B.23)$$

A radical mechanism, on the other hand, is also conceivable. In this case, iodide would begin by transferring a single electron to arsenic acid, producing a tetravalent As intermediate, which could fall apart in a number of ways, one of which is shown below:

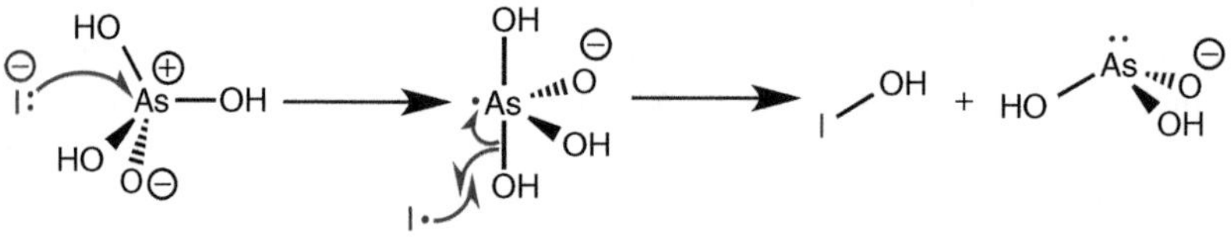

$$(5B.24)$$

The HOI (hypoiodous acid) thus produced could then be scavenged by a second iodide ion to produce molecular iodine; this is a well-known reaction for all the halogens, except fluorine:

$$(5B.25)$$

This chemistry highlights at least two key differences between phosphorus and arsenic. Compared with arsenic acid, phosphoric acid and phosphates are far more resistant to reduction. Second, note once again the tricoordinate structure of arsenous acid (H_3AsO_3) and recall that phosphorous acid (H_3PO_3) has a different tetracoordinate structure.

5B.5   ARSENIC TOXICITY AND BIOMETHYLATION

Arsenic is inextricably linked to poisoning and murder. Kings and emperors, famous and not-so-famous people, have succumbed to arsenic over the ages. In Europe, arsenic as a murder weapon became popular in the Middle Ages and the Renaissance, especially among the ruling classes in Italy. By the nineteenth century, it was recognized as an "element of murder" pretty much everywhere. The so-called arsenic trioxide, As_4O_6, colorless, tasteless, and readily available, was perfectly suited as a poison administered with food. Along with a few other substances used for the same purpose, it acquired the nickname inheritance powder in nineteenth-century France and elsewhere. Arsenic poisoning continues to be widespread today, not so much as a result of human malice but because of groundwater contamination in many parts of the world.

Arsenic sabotages a great many physiological processes, and even a cursory description of them is beyond the scope of this chapter. One of arsenate's most serious biochemical effects is its inhibition of pyruvate dehydrogenase, the enzyme that decarboxylates pyruvate ($CH_3COCOOH$), the end product of glycolysis, to acetyl CoA, the substrate for the ATP-forming citric acid cycle. This and other disruptions ultimately bring about death from multisystem organ failure. Here we will focus on the chemistry of a particular aspect of arsenic metabolism, namely biomethylation, which is best known as a detoxification process in bacteria.

REVIEW PROBLEM 5B.8

Organoarsenic compounds have been used as chemical warfare agents. Chief among was Lewisite, which was used as a vesicant (blister agent) and lung irritant. The main constituent of Lewisite has the structure shown below. During World War II, chemists at Oxford developed an antidote for Lewisite, named *British anti-Lewisite* (BAL) or dimercaprol:

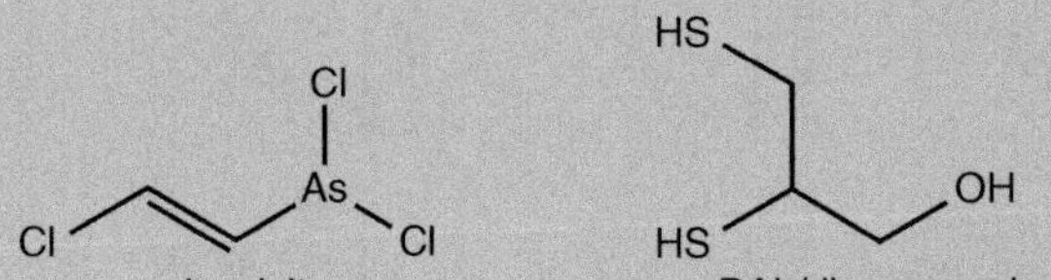

Lewisite   BAL/dimercaprol

Surprisingly enough, BAL is still used today as an antidote for heavy-metal poisoning, particularly for As, but also for Sb, Pb, and Hg. Can you speculate about the chemistry underlying BAL's efficacy?

The discovery of arsenic biomethylation goes back some 200 years, when German physicians noted cases of poisoning attributable to arsenic-containing paints such as Scheele's Green (copper arsenite, below) that were widely used on wallpaper at the time:

It gradually became clear that fungi and bacteria decomposed the arsenic-containing pigments, liberating arsenic-containing gases that affected the inhabitants of the green-painted households. The chief constituent of these gases was ultimately identified as trimethylarsine ($AsMe_3$). By the early part of the twentieth century, other arsenic-containing small organic molecules had also been identified as bacterial and fungal metabolites and a mechanistic model for biomethylation was needed.

Such a model was proposed by Challenger many years ago and it is shown in Figure 5B.1 (Challenger, F. *Chem. Rev.* **1945**, *36*, 315–361). The essentials of the model appear to have stood the test of time.

Impressively, the model hypothesized a "positive methyl group" long before such a methyl donor had been identified. Today we know that *S*-adenosylmethione (SAM, also known as *AdoMet*) is the most important such donor, and glutathione and other cellular thiols are the key reducing agents:

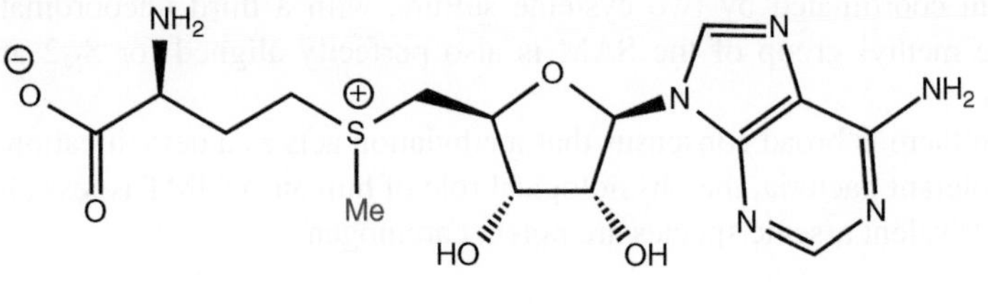

SAM/AdoMet

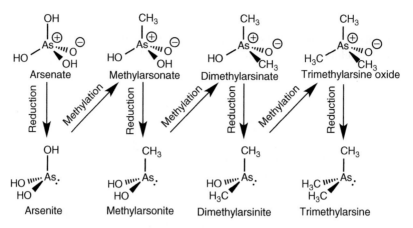

Figure 5B.1 Challenger model for arsenic biomethylation.

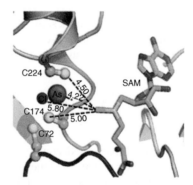

Figure 5B.2 Crystallographic model of a ternary complex of CmArsM, SAM, and trivalent As. Selected distances (Å) from the S-methyl carbon of SAM are shown as dotted lines. (This figure is adapted from Qin, J., et al. Proc. Natl. Acad. Sci. USA **2009**, 106, 5213–5217.)

Enzymes catalyzing the transfer of methyl groups from SAM to trivalent arsenic, called AS3MT, an abbreviation for As(III) SAM methyltransferase, occur in all the kingdoms of life, from bacteria to humans. Crystal structures of such an enzyme (CmArsM) extracted from the thermoacidophilic eukaryotic red alga *Cyanidioschyzon merolae* from Yellowstone National Park have shed significant light on the biomethylation process. A model based on multiple crystal structures of CmArsM (Figure 5B.2) shows an arsenic atom coordinated by two cysteine sulfurs, with a third uncoordinated cysteine nearby. The methyl group of the SAM is also perfectly aligned for S_N2 attack by the trivalent As.

Although there is broad consensus that methylation acts as a detoxification mechanism in arsenic-tolerant bacteria, the physiological role of human AS3MT is less clear, because methylated trivalent arsenic species are potent carcinogens.

REVIEW PROBLEM 5B.9

Using SAM as the methyl donor, use arrow pushing to rationalize a methylation step in the Challenger mechanism.

REVIEW PROBLEM 5B.10

Explain the final step of the Challenger mechanism, namely the reduction of tri-methylarsine oxide to trimethylarsine, using arrow pushing:

$$Me_3AsO + 2 RSH \rightarrow Me_3As + RS-SR + H_2O$$

5B.6 ALKALI-INDUCED DISPROPORTIONATION OF PHOSPHORUS

Switching gears, we'll now turn to a purely chemical topic, with no obvious biological connection. White phosphorus, which consists of P_4 tetrahedra, is the most commonly encountered and used form of phosphorus. It's also the most reactive, and is protected from air by storing under water. Like many nonmetals, white phosphorus reacts with hot alkali, disproportionating to phosphine (PH_3) and sodium hypophosphite (NaH_2PO_2). The reaction is thus useful for the synthesis of these two compounds:

$$P_4 + 3 NaOH + 3 H_2O \rightarrow PH_3 + 3 NaH_2PO_2 \tag{5B.26}$$

Just by looking at the balanced equation, it's difficult to get an insight into the mechanism. But recall the strategy we recommended in Chapter 1: Think like a lone pair! Think about what the likely nucleophile is and where it should attack. You'll quickly see that there aren't many choices. Hydroxide appears to be the only reasonably good nucleophile around and P_4 is a logical electrophile, with strained P–P bonds as promising leaving groups:

$$\tag{5B.27}$$

On ring opening, the departing P^- center picks up a proton from water, producing the first P–H bond. The process repeats itself to create phosphine, PH_3.

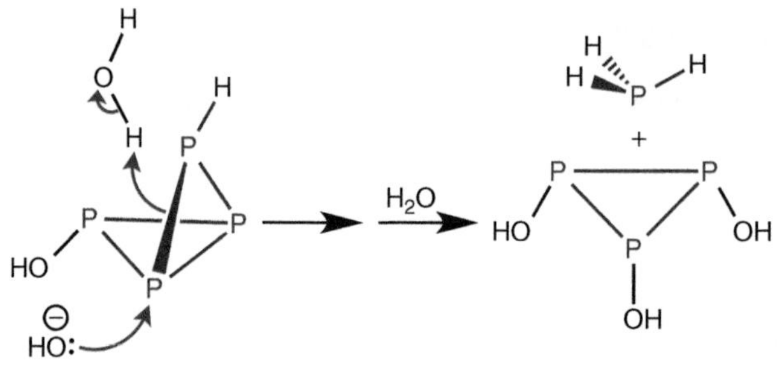

(5B.28)

Hydroxide ions continue to chew on the remainder of the P skeleton, now a hydroxylated P_3 ring, in the same way as depicted above:

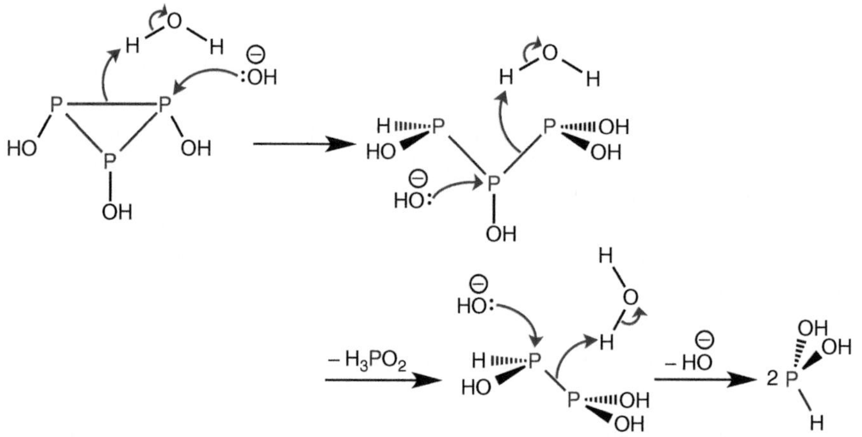

(5B.29)

Note that the product we have obtained above, $HP(OH)_2$, is a protonated form of what we need to get, namely, $H_2PO_2^-$. Like phosphorous acid (H_3PO_3), however, hypophosphorous acid (H_3PO_2) and the hypophosphite anion ($H_2PO_2^-$) both contain tetracoordinate phosphorus, so we still need to move a couple of protons around, which we do below with the help of OH^- and H_2O:

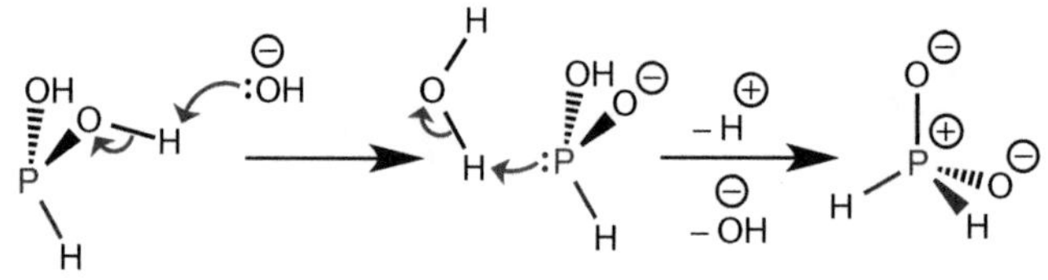

(5B.30)

There we have it: all the products nicely accounted for thanks to arrow pushing! What we have above is clearly speculative, but it's a satisfying way of thinking about what might be happening.

REVIEW PROBLEM 5B.11

Aqueous hypophosphorous acid can be obtained by simple neutralization of the reaction mixture at the end of the above reaction (5B.30) with strong acid:

$$H_2PO_2^- + H^+ \rightarrow H_3PO_2$$

An alternative synthesis of hypophosphorous acid involves oxidation of phosphine with iodine in water:

$$PH_3 + 2\,I_2 + 2\,H_2O \rightarrow H_3PO_2 + 4\,I^- + 4\,H^+$$

Suggest a mechanism for this last reaction.

5B.7 DISPROPORTIONATION OF HYPOPHOSPHOROUS ACID

The thermal decomposition of hypophosphorous acid provides a couple of perfect opportunities for sharpening our arrow-pushing skills:

$$2\,H_3PO_2 \rightarrow PH_3 + H_3PO_4 \qquad\qquad (5B.31)$$

$$3\,H_3PO_2 \rightarrow PH_3 + 2\,H_3PO_3 \qquad\qquad (5B.32)$$

Which one of the two reactions predominates depends on the temperature of the reaction. Mechanistically, both reactions seem rather impenetrable, but the trick is once again to calmly follow the guidelines we espoused in Section 1.23. Look carefully at the products and think about what bonds have been broken and formed in the course of the reaction. In essence, two or three phosphorus atoms have come together and reshuffled the hydrogen and oxygen ligands among them. For each of the above reactions, one phosphorus atom has ended up with no oxygen at all—as PH_3. What kind of process would account for that?

Protonation of the phosphorus atom in H_3PO_2 and dissociation of water as leaving groups would not account for the generation of phosphine, for the simple reason that the P in H_3PO_2 is cationic and does not have a lone pair that can be protonated. The only alternative that comes to mind involves hydride (H^-) transfer from one phosphorus to another. It's an uncommon process in inorganic chemistry, but it's common enough for carbocations (see Section 1.18) so it's not really esoteric. Let's consider the first of the above two reactions in light of this idea.

An oxo-bridged intermediate is a reasonable framework to invoke to preorganize the molecules for hydride transfer:

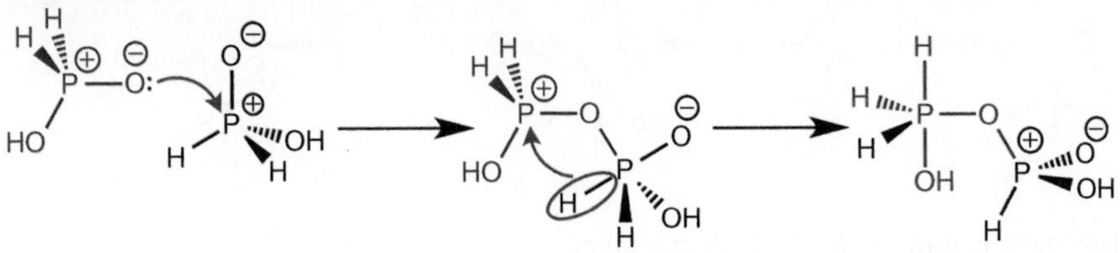

$$(5B.33)$$

So we now have a pentavalent P with three hydrogens and two O ligands. A second hydride transfer seems necessary to kick out one of the oxygens:

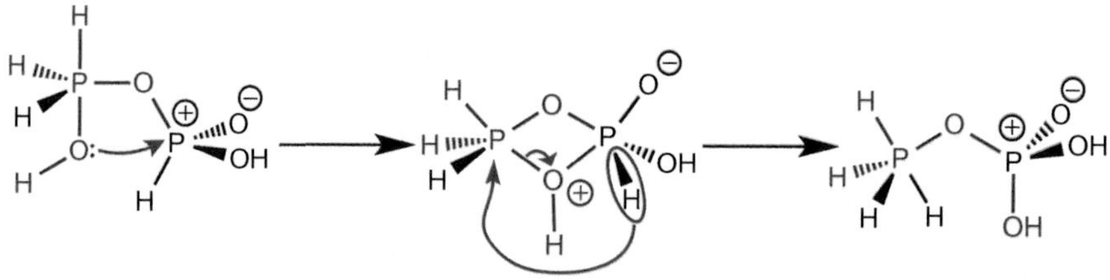

(5B.34)

Deprotonation of the PH_4 group (note that we have reoriented the molecule a bit below) and simultaneous departure of phosphate—effectively a reductive elimination—now accounts for the observed products. Below we have shown an intramolecular path for the process, but an external base could also have accomplished the deprotonation:

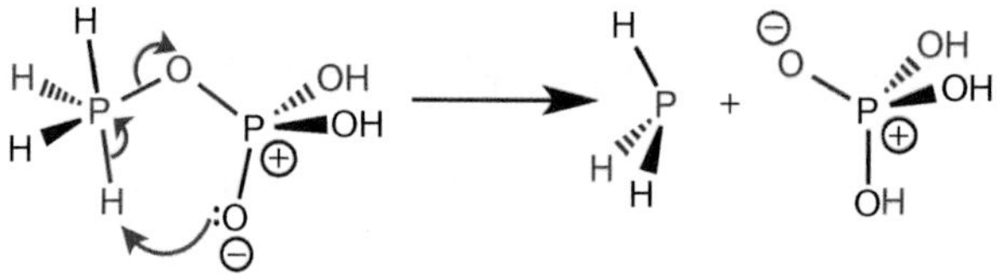

(5B.35)

REVIEW PROBLEM 5B.12

Consider reaction (5B.32):

$$3\ H_3PO_2 \rightarrow PH_3 + 2\ H_3PO_3$$

where three molecules of hypophosphorous acid come together and redistribute their ligands on phosphorus. Suggest a mechanism.

Hint: This is clearly a somewhat complex problem. One solution we have thought of involves an intermediate where three hypophosphorous acid molecules have linked up together, as follows.

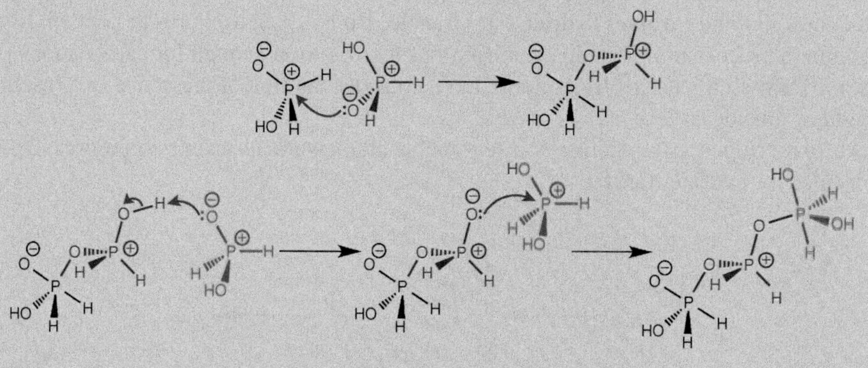

The stage is now set for hydride transfers.

Hypophosphorous acid is an important reducing agent, both in chemistry labs and in industry. Its best known use in organic chemistry is in the reduction of arenediazonium cations (ArN_2^+) to arenes (ArH). In industry, its chief application is "electroless" plating where a metal such as nickel or copper is deposited on a surface by chemical reduction, as opposed to passing an electric current (hence the name).

REVIEW PROBLEM 5B.13

Hypophosphorous acid can reduce elemental iodine to HI, which is the reducing agent of choice for reducing ephedrine or pseudoephedrine to methamphetamine. The relevant reactions are as follows:

$$H_3PO_2 + H_2O + I_2 \rightarrow H_3PO_3 + 2\ HI$$

Ephedrine/pseudoephedrine   Methamphetamine

The sale and use of hypophosphorous acid therefore is strictly controlled in the United States and a number of other countries. Suggest mechanisms for the above two reactions.

5B.8 THE ARBUZOV REACTION

Phosphorus reagents are an integral part of modern organic chemistry, so we'll devote the next few sections to discuss some of the most instructive aspects of organophosphorus chemistry. We'll begin with the Michaelis–Arbuzov reaction, which is a powerful means of creating a carbon–phosphorus bond and hence a key route to organophosphorus chemistry. The reaction involves the interaction of an alkyl halide and a trialkyl phosphite:

$$R^1 - X + P(OR^2)_3 \longrightarrow \qquad + R^2 - X \qquad (5B.36)$$

Here the phosphorus lone pair is a plausible nucleophile and the alkyl halide a plausible electrophile. An S_N2 reaction can thus be envisioned:

$$(5B.37)$$

The phosphonium and halide ions generated now undergo a second S_N2 reaction to generate the final products, as shown:

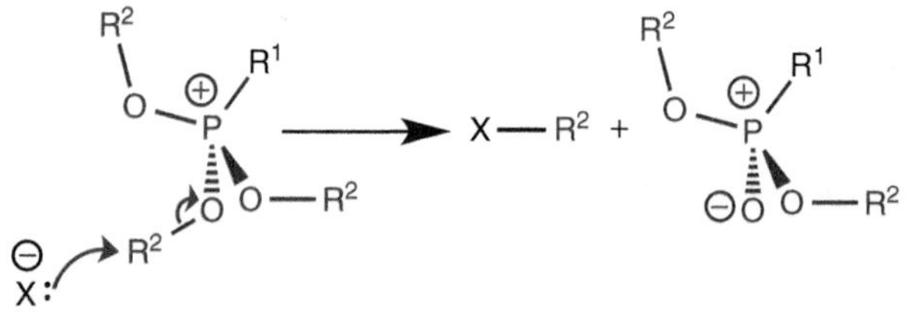

$$(5B.38)$$

As elsewhere in phosphorus chemistry, the driving force for the last step is the creation of the highly stable $P^+–O^-$ linkage. Certain alkylphosphonates (as the products are known) obtained from the Arbuzov reaction serve as starting materials for the Horner–Wadsworth–Emmons reaction, which is described below:

REVIEW PROBLEM 5B.14

Draw a mechanism for the following reaction.

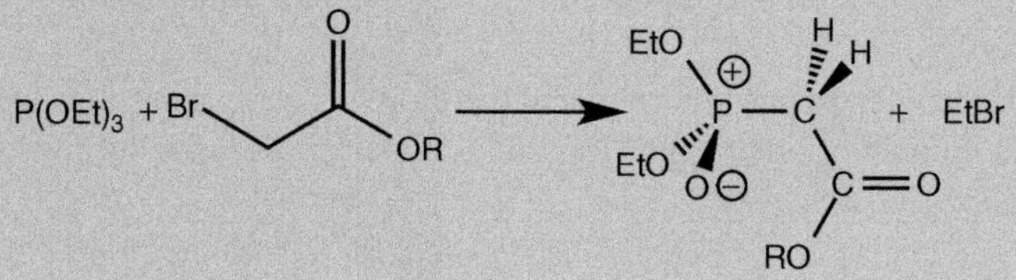

5B.9   THE WITTIG AND RELATED REACTIONS: PHOSPHORUS YLIDES

The Wittig reaction is the reaction of a carbonyl compound with a triphenylphosphonium ylide to yield an olefin and triphenylphosphine oxide. The reaction was discovered by the German chemist Georg Wittig in 1953, for which he received the Nobel Prize in 1979, shared with Herbert C. Brown, the discoverer of hydroboration. The reaction quickly came to rank among the most important synthetic reactions in organic chemistry. The carbonyl compound in the Wittig reaction may be an aldehyde or ketone and the phosphorus ylide may also be variously substituted. For our discussion, we will choose a reaction involving an aldehyde:

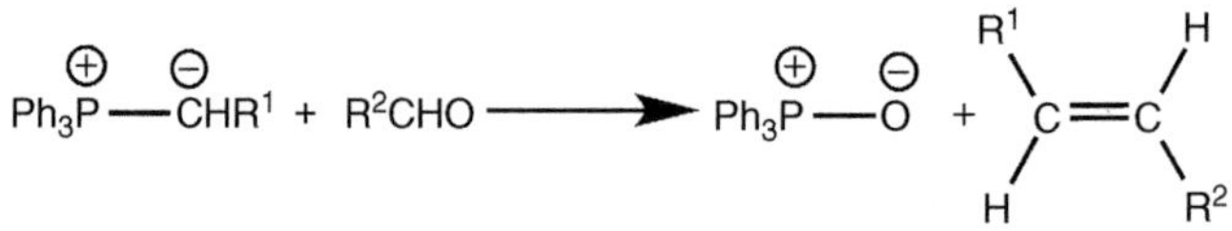

$$(5B.39)$$

The phosphorus ylide starting material is synthesized from the corresponding phosphonium ion, which is deprotonated by a strong base such as *n*-butyllithium:

$$(5B.40)$$

The ylidic carbon may be viewed as a carbanion surrogate and it attacks the carbonyl substrate like a typical strong nucleophile. The zwitterionic product is called a *betaine*, which is defined as a neutral molecule containing a heteroatom-based positively charged group with no attached hydrogens, such as a quaternary ammonium or phosphonium group, and a *nonadjacent* negatively charged center or group, such as carboxylate. The betaine then cyclizes to form a four-membered ring (oxaphosphetane) with a P–O bond:

$$(5B.41)$$

The strained four-membered ring falls apart under the reaction conditions, yielding an alkene and Ph_3PO as the final products.

$$(5B.42)$$

Once again, formation of Ph_3PO is the driving force for the last part of the mechanism. Note the ambiguity in the stereochemistry of the alkene product (as indicated by the squiggly bonds), which results from the fact that the betaine intermediate equilibrates to some extent with both the starting materials and the heterocyclobutane intermediate. With sterically unhindered ylides, a *Z* (or *cis*) stereochemistry is generally the preferred stereochemical outcome.

A major deficiency of the Wittig reaction is the difficulty of separating the desired alkene product from the triphenylphosphine oxide byproduct. The fact that triphenylphosphine is used as a stoichiometric reagent also leads to a poor atom economy, which is defined as:

$$\% \text{ Atom economy} = \frac{\text{Molecular mass of desired product}}{\text{Molecular mass of all reactants}} \times 100\%$$

In other words, on a mass basis, a substantial fraction of the starting materials does not end up as part of the desired product, but is discarded as waste. In principle, a catalytic version of the Wittig reaction is not hard to envision:

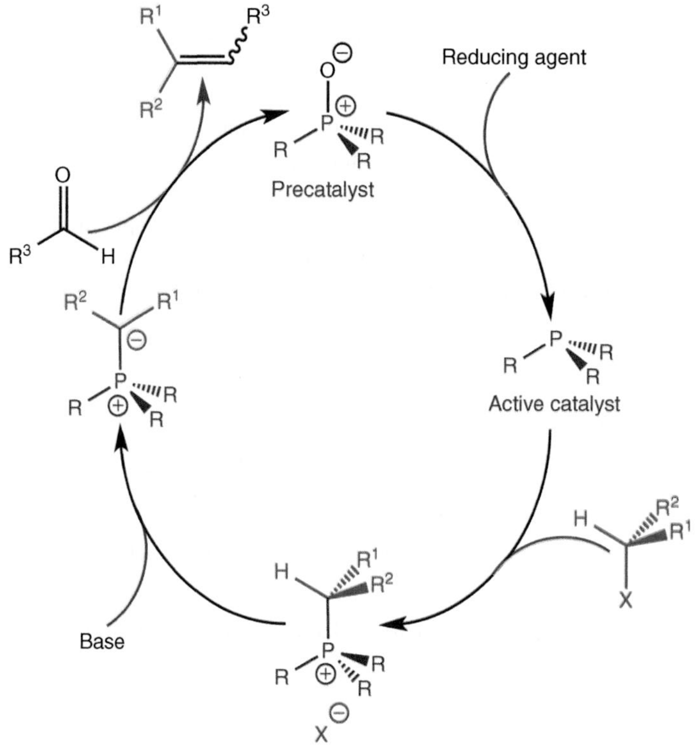

(5B.43)

In practice, however, finding a reducing agent that would reduce the phosphine oxide to phosphine but would not touch aldehydes, ketones, alkenes, and other reducible functional groups is rather a tall order. Recently, University of Nottingham chemist Christopher O'Brien and coworkers found diphenylsilane (Ph_2SiH_2) to be a suitable reducing agent for this purpose (O'Brien, C. J., *et al. Angew. Chem. Int. Ed.* **2009**, *48*, 6836–6839). Some typical conditions reported by the authors are as follows:

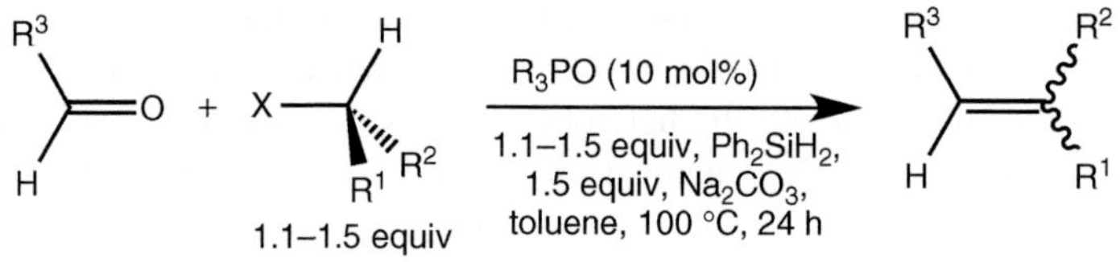

(5B.44)

REVIEW PROBLEM 5B.15

With respect to the catalytic Wittig reaction discussed above, draw a mechanism for the reduction of the phosphine oxide R_3PO by the silane Ph_2SiH_2.

Closely related to the Wittig reaction is the Horner–Wadsworth–Emmons reaction, where a stabilized phosphonate carbanion reacts with an aldehyde or ketone to yield an olefin.

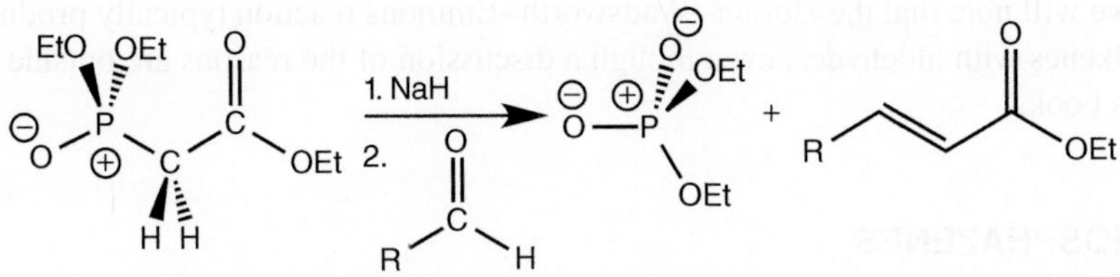

(5B.45)

Sodium hydride deprotonates a C–H unit between the carboxylate and phosphonate ester functionalities to create a carbanion, as shown below. Both ester functionalities help the stabilization of the negative charge of the carbanion.

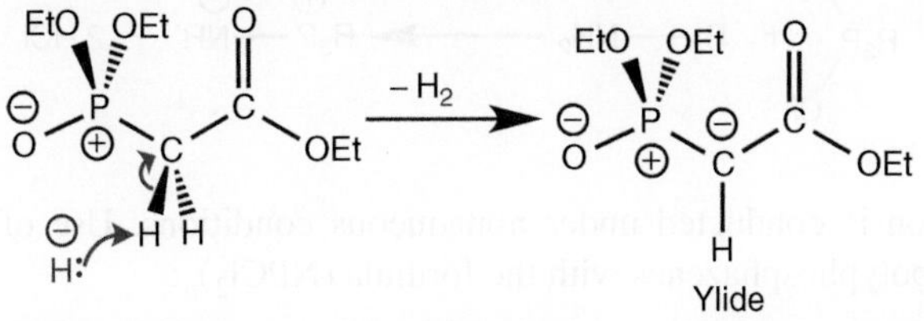

(5B.46)

The carbanion attacks the carbonyl substrate, forming a four-membered heterocyclobutane intermediate, analogous to the one in the Wittig reaction.

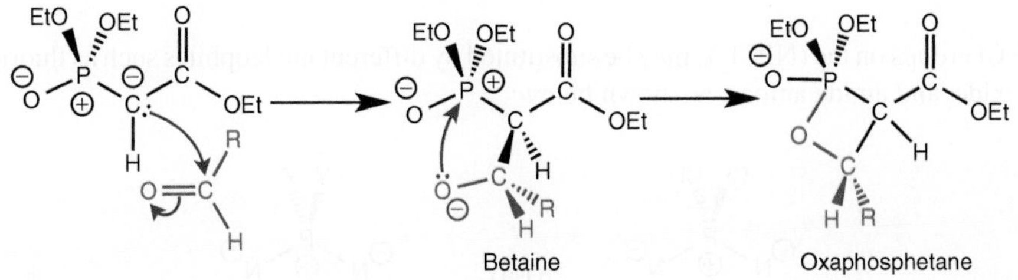

Betaine   Oxaphosphetane

(5B.47)

As in the Wittig reaction, the oxaphosphetane ring falls apart, yielding an alkene as the desired final product:

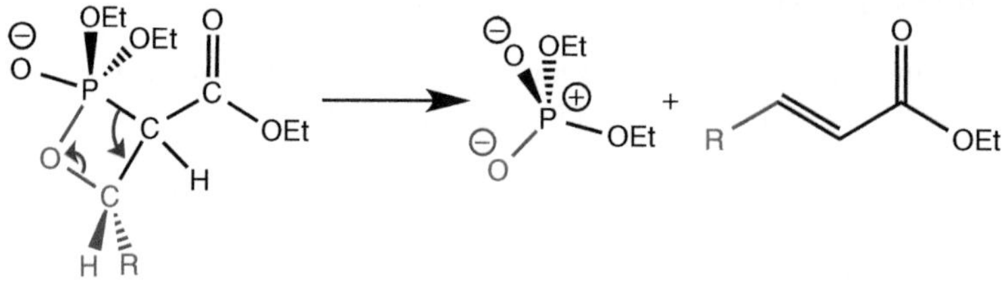

$$(5B.48)$$

As elsewhere in phosphorus chemistry, creation of a new P^+–O^- linkage provides the key driving force for product formation.

Finally, we will note that the Horner–Wadsworth–Emmons reaction typically produces *E* or *trans* alkenes with aldehydes, even though a discussion of the reasons are outside the scope of this book.

5B.10 PHOSPHAZENES

Phosphazenes are a large family of ylide-type phosphorus–nitrogen compounds with the R_3P^+–N^-R' unit, where R and R' are monovalent groups. The following is a general synthesis:

$$(5B.49)$$

where the reaction is conducted under nonaqueous conditions. Use of PCl_5 instead of R_3PCl_2 leads to polyphosphazenes with the formula $(NPCl_2)_n$:

$$PCl_5 + n\,NH_4Cl \xrightarrow[\substack{\text{Chlorobenzene}\\ \text{or}\\ \text{1,2-Dichlorobenzene}}]{\Delta} (NPCl_2)_n + 4n\,HCl$$

$$(5B.50)$$

The Cl groups on the $(NPCl_2)_n$ may be substituted by different nucleophiles such as fluoride, alkoxide, and amide anions, as shown below:

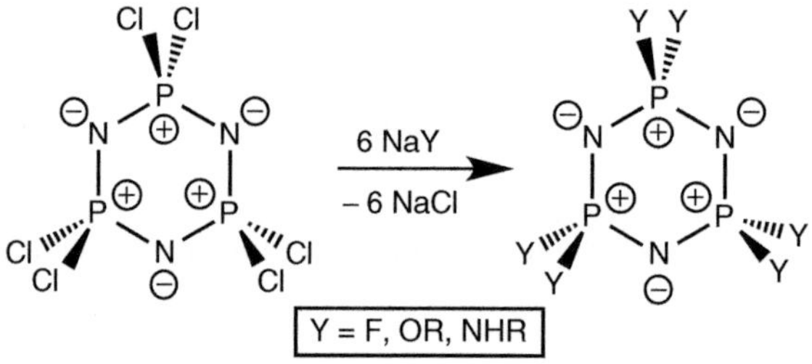

$$(5B.51)$$

Y = F, OR, NHR

REVIEW PROBLEM 5B.16

Suggest a mechanism for the general phosphazene synthesis mentioned above (5B.49).

At present, phosphazenes are probably best known as neutral organic superbases, an application pioneered by Reinhard Schwesinger in the 1980s. They are a big step up from classic amine, amidine, and guanidine bases in terms of basicity. Two commonly used phosphazene bases, BEMP and t-Bu-P4 ("Schwesinger base") are depicted in Figure 5B.3; the pK_a's of their conjugate acids are also compared with those of selected traditional nitrogen bases:

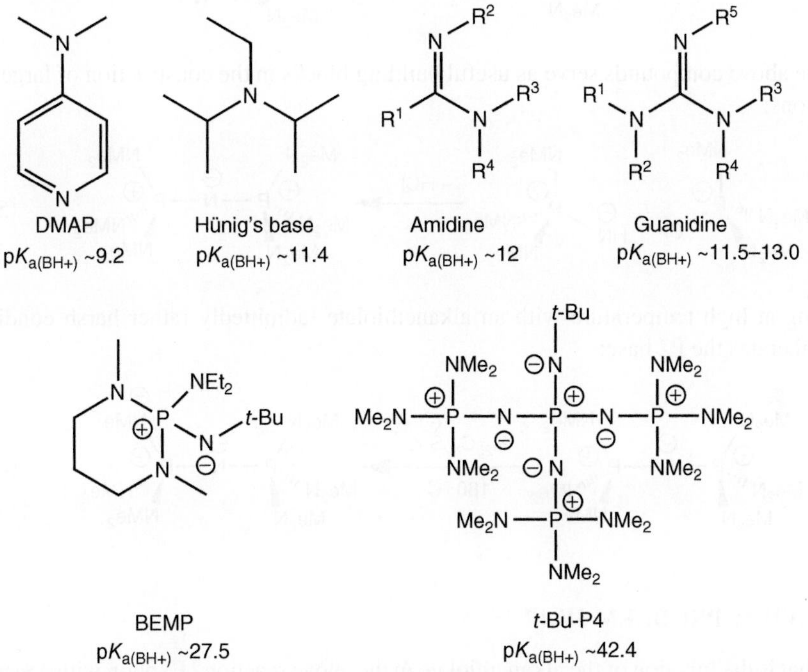

DMAP
$pK_{a(BH+)}$ ~9.2

Hünig's base
$pK_{a(BH+)}$ ~11.4

Amidine
$pK_{a(BH+)}$ ~12

Guanidine
$pK_{a(BH+)}$ ~11.5–13.0

BEMP
$pK_{a(BH+)}$ ~27.5

t-Bu-P4
$pK_{a(BH+)}$ ~42.4

Figure 5B.3 *Selected uncharged nitrogen bases and pK_a's of their protonated forms (the higher the value, the stronger the base).*

The highly basic phosphazenes thus provide an alternative to organolithium bases such as lithium dialkylamides and alkyllithiums. Milder reaction conditions and better solubility are two key advantages associated with phosphazenes. Another important consideration is the lack of a coordinating cation: naked enolates and other anions obtained with phosphazenes often exhibit enhanced reactivity, relative to their lithium salts.

Phosphazene synthesis is conceptually rather simple, as illustrated below for a "P1 base." The P–N skeleton is first put together via a series of nucleophilic displacements (presumably of the S_N2-Si type):

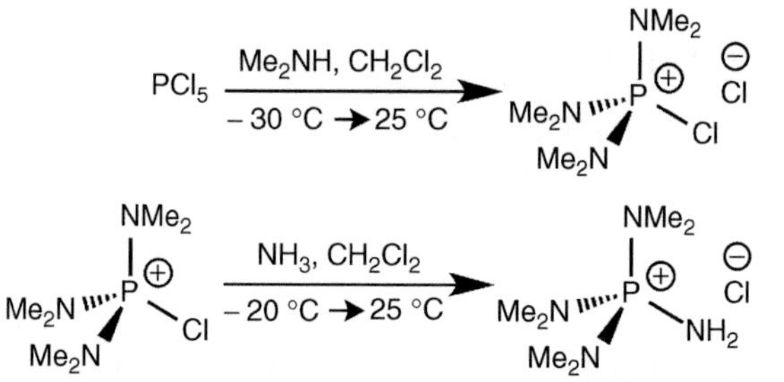

$$ (5B.52) $$

A final deprotonation then generates the P1 base:

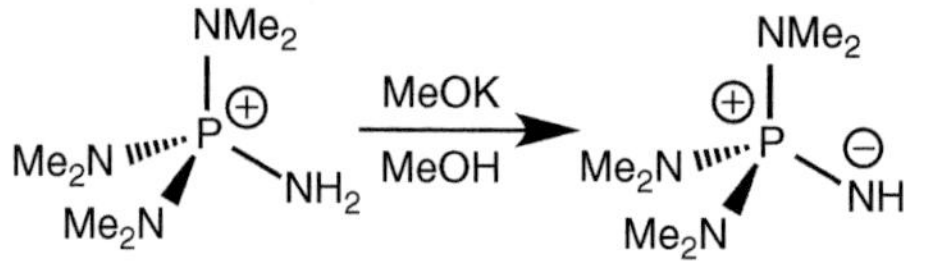

$$ (5B.53) $$

The above compounds serve as useful building blocks in the construction of larger P–N skeletons:

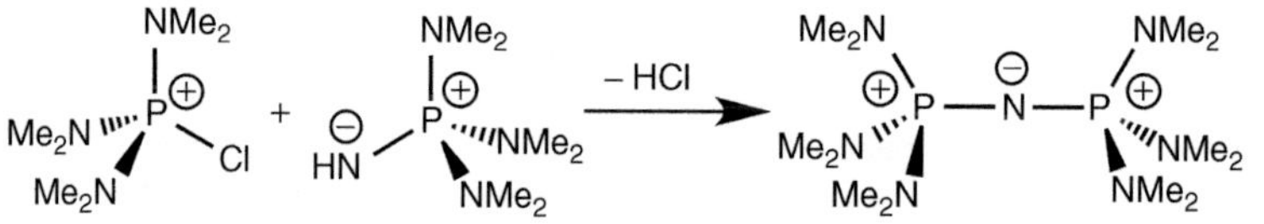

$$ (5B.54) $$

Heating at high temperature with an alkanethiolate (admittedly rather harsh conditions) then liberates the P2 base:

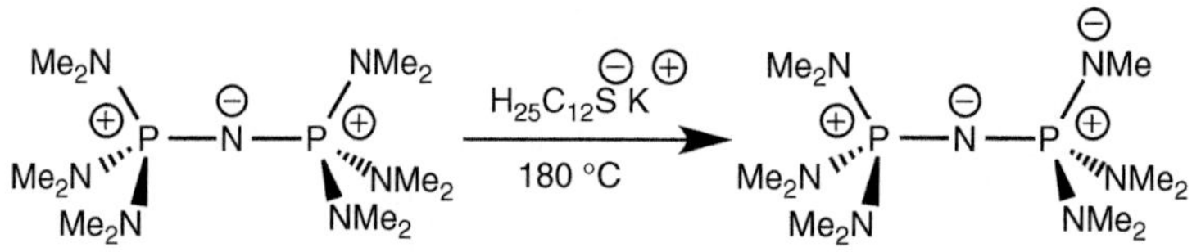

$$ (5B.55) $$

REVIEW PROBLEM 5B.17

What is the function of the alkanethiolate in the above reaction? Explain with a mechanism.

A similar strategy has also been used for *t*-Bu-P4 (Schwesinger base):

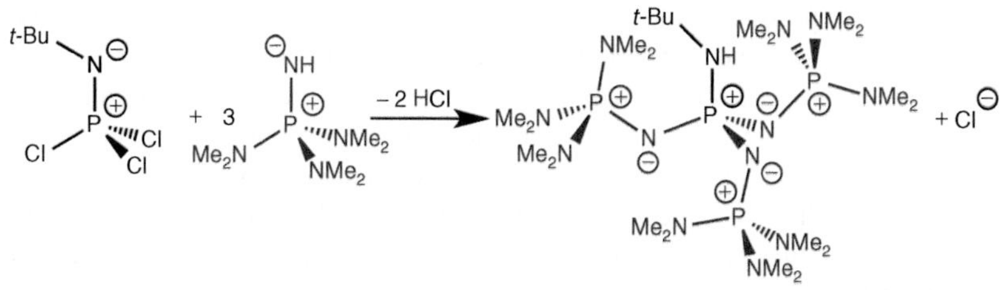

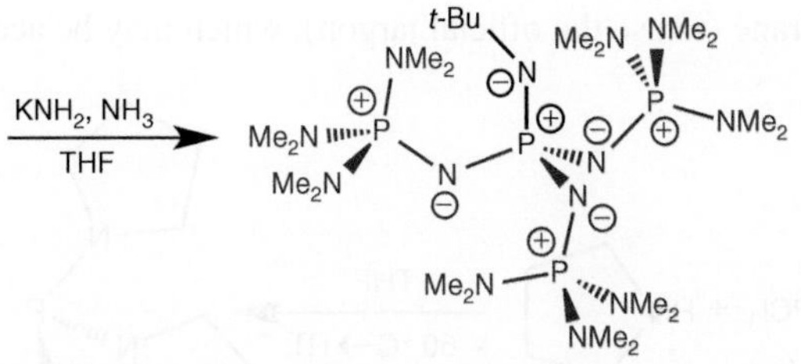

(5B.56)

The Staudinger reaction provides an elegant alternative route to P1 phosphazenes. Normally, the reaction is used as a means to convert alkyl azides to primary amines, as shown below:

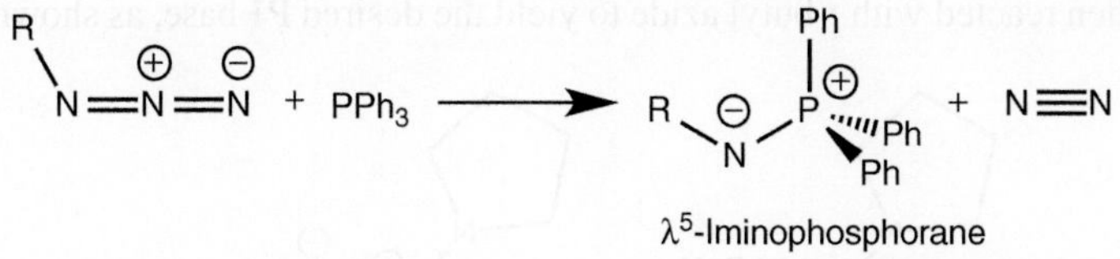

(5B.57)

The λ^5-iminophosphorane produced is then hydrolyzed to a primary amine and triphenylphosphine oxide:

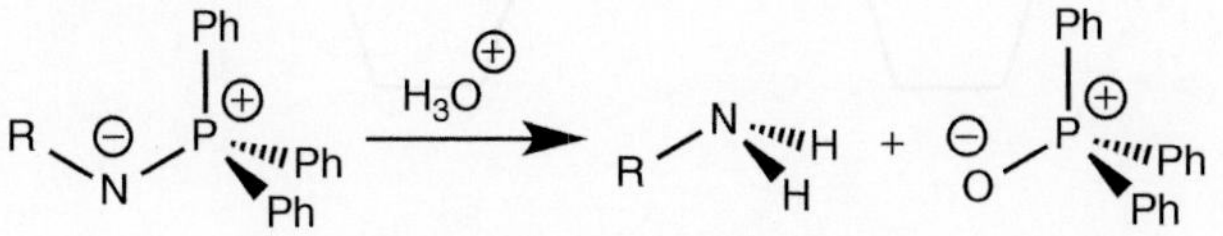

(5B.58)

Mechanistically, the reaction begins with the nucleophilic attack by triphenylphosphine on the terminal nitrogen of the azide group, forming a phosphazide intermediate:

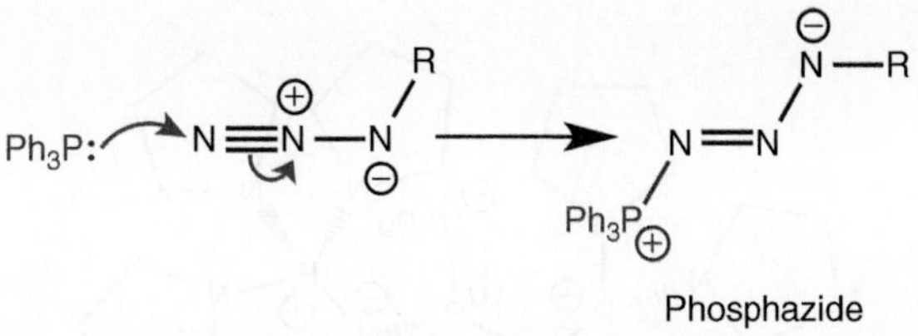

(5B.59)

The phosphazide intermediate is thought to cyclize, forming a cyclic PN_3 intermediate, which spontaneously loses N_2 to form the λ^5-iminophosphorane product:

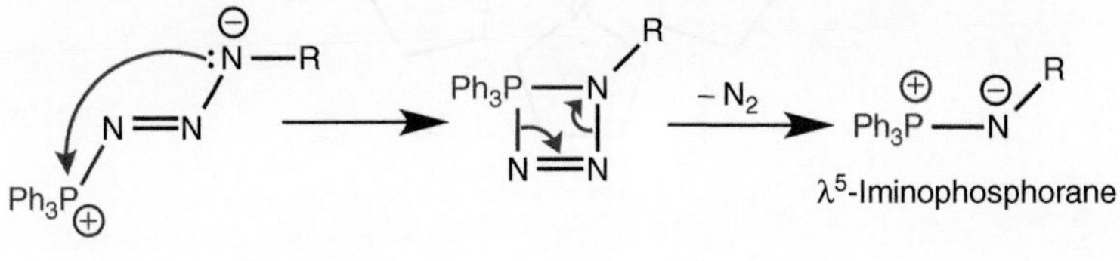

(5B.60)

The Staudinger synthesis of a P1 phosphazene is a simple modification of the above protocol, where triphenylphosphine is substituted with a triaminophosphine—a

λ^3-triaminophosphorane (to use the official jargon), which may be accessed as follows:

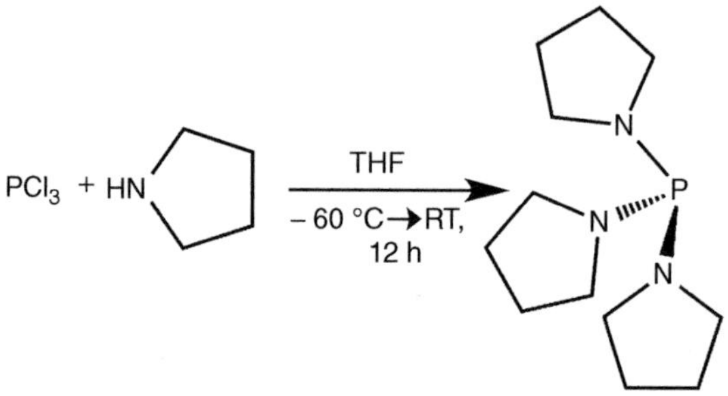
PCl3 + HN
THF
– 60 °C→RT,
12 h
N
P

(5B.61)

The product is then reacted with *t*-butyl azide to yield the desired P1 base, as shown below:

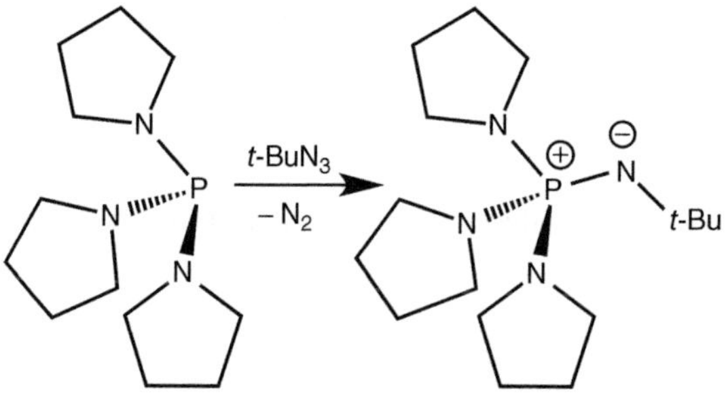
t-BuN3
– N2
N
P
t-Bu

(5B.62)

REVIEW PROBLEM 5B.18

Suggest a synthetic route for the following P4 base:

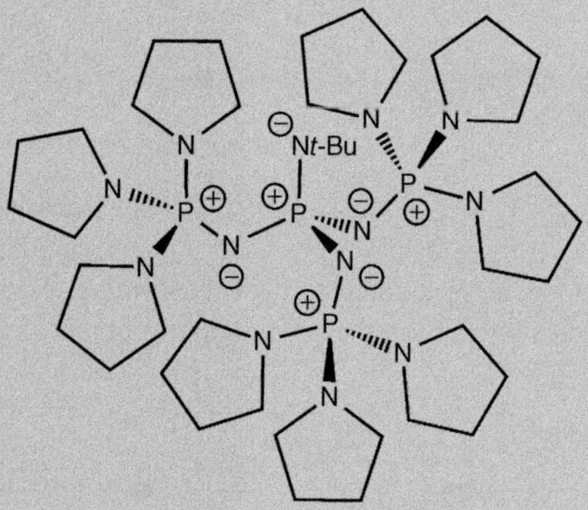
Nt-Bu
N
P

5B.11*   THE COREY–WINTER OLEFINATION

This two-step sequence for converting vicinal diols to olefins is of considerable mechanistic interest. The reagents are thiophosgene ($CSCl_2$) for the first step and a trialkylphosphite, $P(OR)_3$, for the second.

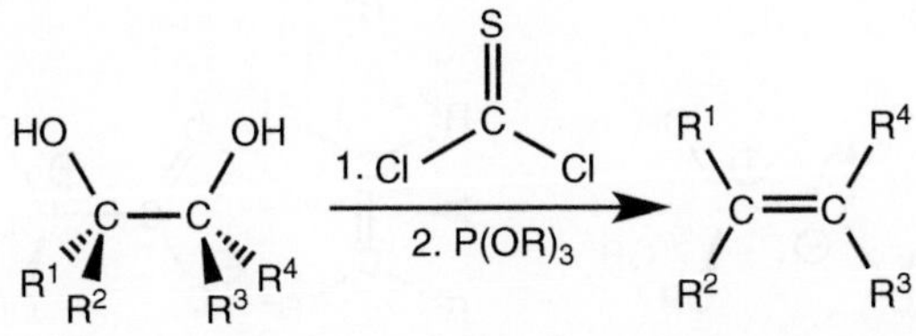

$$(5B.63)$$

The first step is relatively unremarkable, analogous to standard carbonyl chemistry:

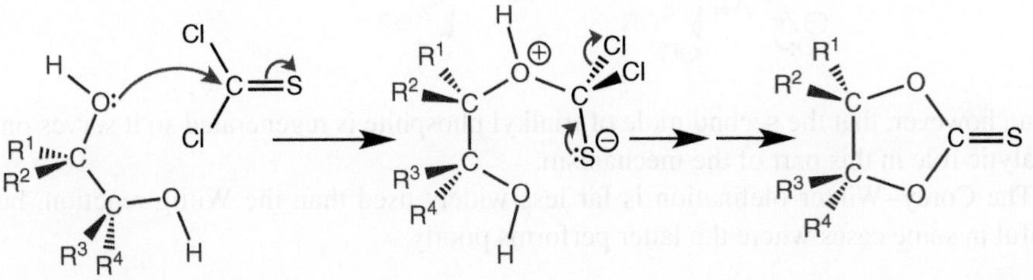

$$(5B.64)$$

The second step, involving the reaction of the cyclic thiocarbonate intermediate formed above with trialkylphosphite, is of great interest. The phosphite P attacks the S of the thiocarbonyl group, generating a cyclic dialkoxycarbene, as shown below:

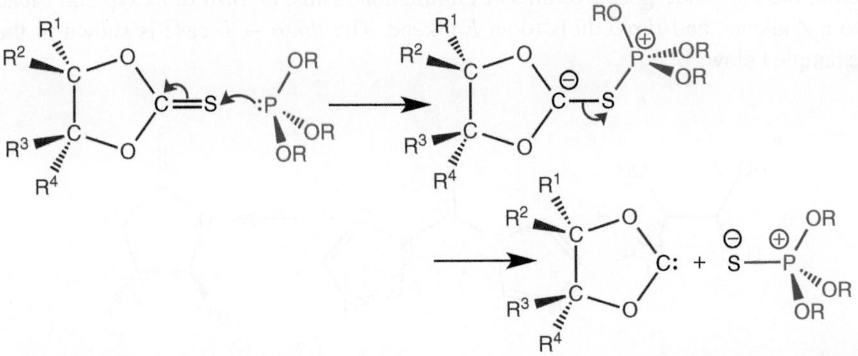

$$(5B.65)$$

Carbene formation appears to be facilitated by at least two significant driving forces: formation of the rather stable P^+–S^- (BDE 335 kJ/mol) linkage, and the electronic stabilization of the carbene by the alkoxy groups.

A second molecule of trialkylphosphite now reacts with the dialkoxycarbene intermediate to generate the olefin product.

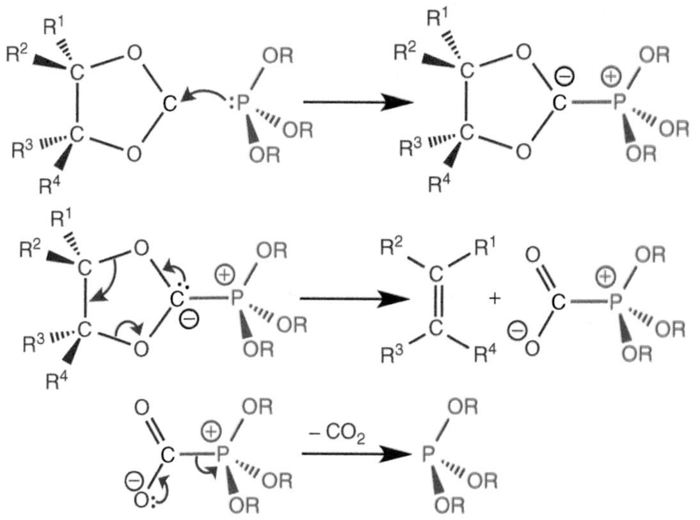

(5B.66)

Note, however, that the second mole of trialkyl phosphite is regenerated so it serves only a catalytic role in this part of the mechanism.

The Corey–Winter olefination is far less widely used than the Wittig reaction, but is useful in some cases where the latter performs poorly.

REVIEW PROBLEM 5B.19[*]

Although the stereochemical course of the Corey–Winter reaction is implicit in the diagrams above, we have not addressed the issue explicitly. The process is stereospecific; the diol undergoes overall *syn* elimination. Thus, *erythro* diols typically lead to a Z alkene, and *threo* diols to an E alkene. The *threo* → E case is shown in the example below:

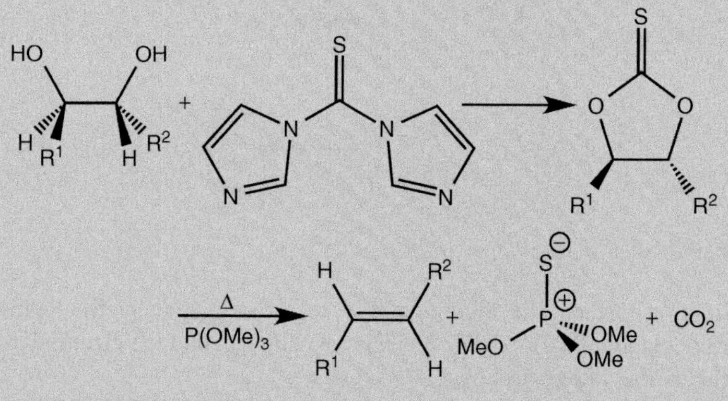

Explain mechanistically the stereochemical course of the reaction. *Note*: Don't be concerned in case you are unfamiliar with the *erythro/threo* nomenclature; we are asking you to simply account for the E or *trans* stereochemistry of the alkene product.

5B.12   TRIPHENYLPHOSPHINE-MEDIATED HALOGENATIONS

Triphenylphosphine is a versatile reagent in organic synthesis. In addition to the applications discussed above, it is also used in a number of halogenation protocols for converting alcohols to alkyl halides. The Appel reaction, shown below for a primary alcohol, is used to prepare both alkyl chlorides and bromides:

$$\text{RCH}_2\text{OH} \xrightarrow[\text{X = Cl, Br}]{\text{CX}_4,\ \text{PPh}_3} \text{RCH}_2\text{X} \tag{5B.67}$$

The first step of the reaction involves PPh_3 attacking a halogen atom in CX_4 to form a phosphonium cation. The fact that a halogen atom of CX_4 acts as the electrophilic center rather than the carbon may seem odd, but it's a reflection of both the nucleophilicity of phosphorus in triphenylphosphine and the sterically hindered nature of the carbon in CX_4 molecules.

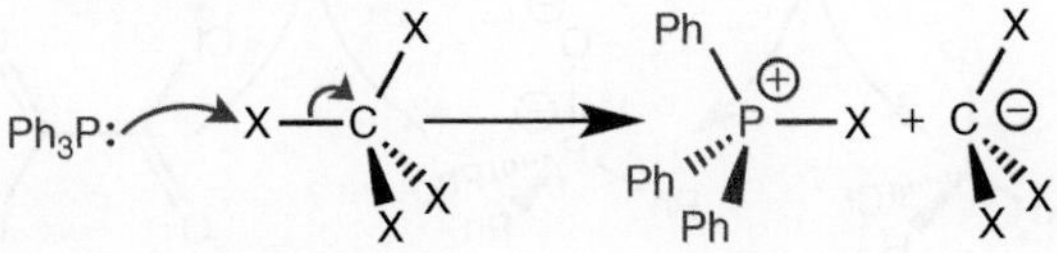

$$\tag{5B.68}$$

The trihalomethide anion then picks up a proton from the alcohol:

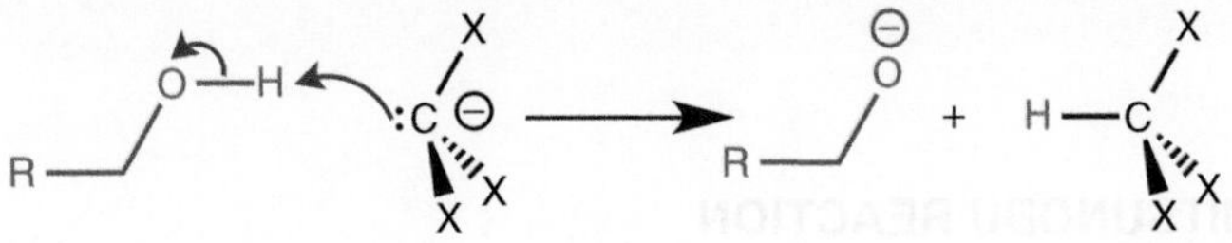

$$\tag{5B.69}$$

The trihalomethane (also called *haloform*) thus formed doesn't play any further role in the mechanism but the alkoxide ion produced reacts with the phosphonium ion to form a pentavalent, pentacoordinate phosphorus intermediate, which falls apart to the alkyl halide and triphenylphosphine oxide.

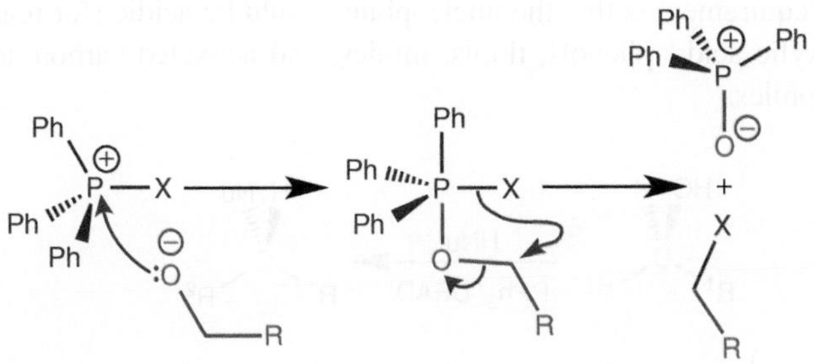

$$\tag{5B.70}$$

Formation of triphenylphosphine oxide, once again, is a key driving force for the reaction.

Traditionally, concentrated HCl or HBr might have been used to convert alcohols to alkyl chlorides or bromides. The Appel reaction avoids the use of strong acids and thus can accommodate acid-sensitive functional groups.

REVIEW PROBLEM 5B.20[*]

As for the Wittig reaction, a catalytic version of the Appel reaction is a worthy goal. British chemists have achieved this goal by substituting CCl_4 with oxalyl chloride as the chlorine source (Denton, R. M. *et al.* **2011**, *76*, 6749–6767) and using a catalytic amount of triphenylphosphine oxide (typically 15 mol%). The overall process may be depicted as follows:

Oxalyl chloride

Explain the role of the oxalyl chloride in mechanistic terms.

5B.13*   THE MITSUNOBU REACTION

The Mitsunobu reaction, discovered by Oyo Mitsunobu (1934–2003) in 1967, is one of the most important among modern synthetic reactions. It allows the replacement of the OH group of primary and secondary alcohols with a variety of nucleophiles, *with clean inversion of stereochemistry* and *under mild conditions*. The key reagents are triphenylphosphine and a dialkyl azodicarboxylate; the latter is very often diethyl azodicarboxylate (DEAD). In addition, a key requirement is that the nucleophile should be acidic (for reasons you'll see below); carboxylic acids, phenols, thiols, imides, and activated carbon acids are all appropriate nucleophiles.

$$(5B.71)$$

How does one come up with a cocktail such as the above—an alcohol, a nucleophile, PPh_3, and DEAD? Insight, a lucky observation or two, and tinkering all surely played a role.

The key objective here is to convert the OH group, normally a lousy leaving group, to an excellent one that can be displaced by the added nucleophile. This is where PPh_3 and

DEAD come in; they react together to form a betaine intermediate, as shown below:

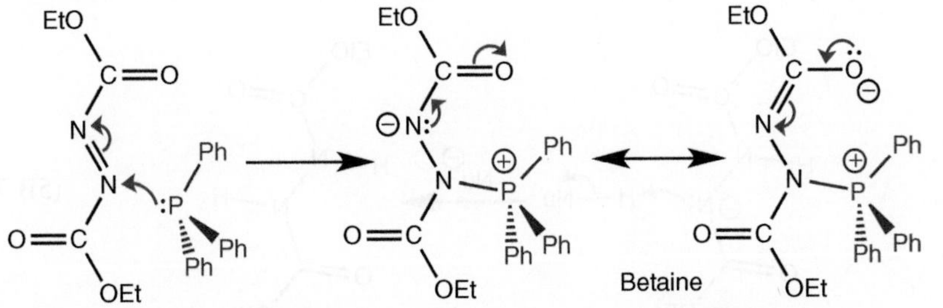

(5B.72)

The betaine then deprotonates the protonated form of the nucleophile:

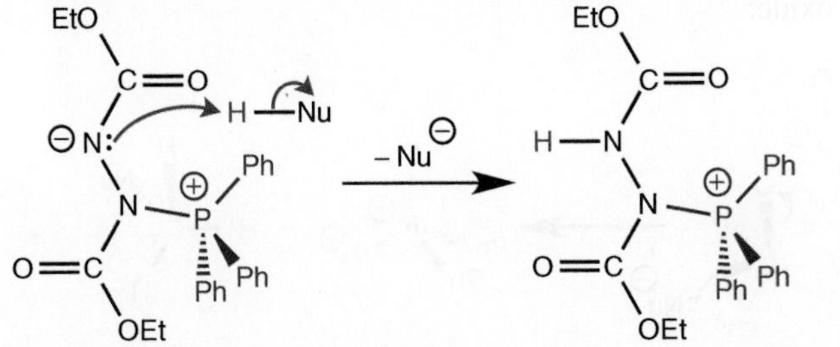

(5B.73)

The cationic phosphorus intermediate then attracts the alcohol, forming a presumptive five-coordinate intermediate, as shown below:

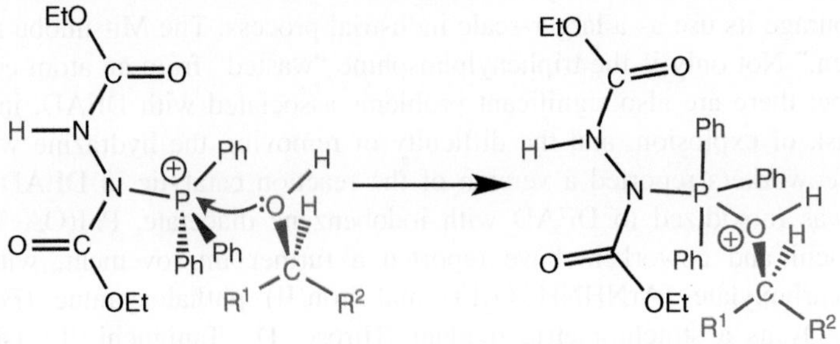

(5B.74)

The coordinated alcohol is now deprotonated by the deprotonated form of the nucleophile, which also leads to the release of the DEAD-H$^-$ anion, slowly setting the stage for the final S$_N$2 displacement of the alcohol oxygen:

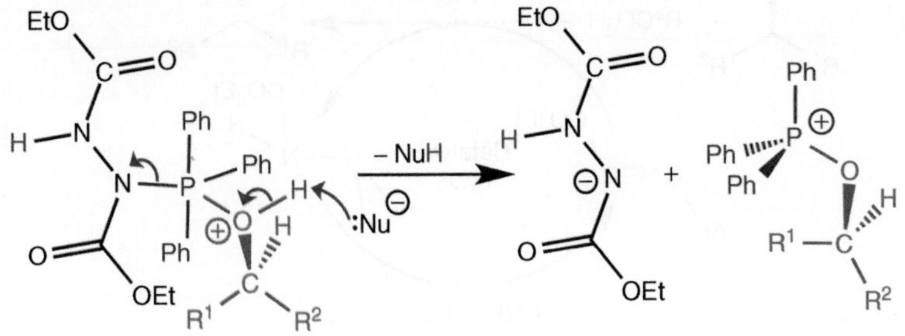

(5B.75)

We, however, need the nucleophile in its anionic form first; the DEAD-H⁻ anion serves as the base for this purpose:

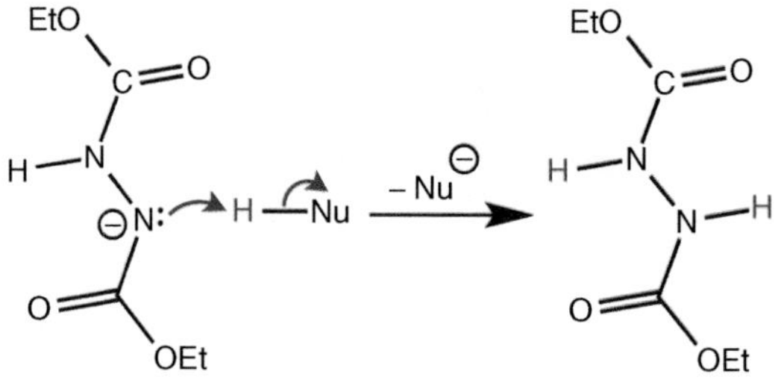

$$(5B.76)$$

We are now set for the grand finale—the clean stereospecific S_N2 displacement of triphenylphosphine oxide:

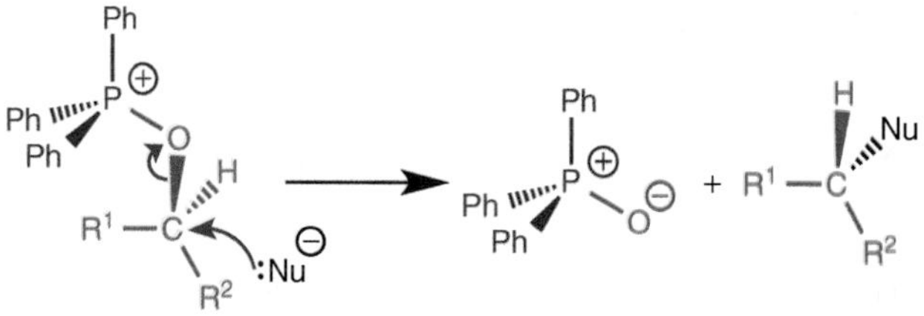

$$(5B.77)$$

The value of the Mitsunobu reaction derives from the mild conditions used and the stereospecificity of the process. It is one of the most important means for inverting a stereogenic center in natural product synthesis. It does, however, have a few serious drawbacks that discourage its use as a larger-scale industrial process. The Mitsunobu reaction is very "un-green." Not only is the triphenylphosphine "wasted" from an atom-economy point of view, but there are also significant problems associated with DEAD, including high toxicity, risk of explosion, and the difficulty of removing the hydrazine waste. In 2006, Toy and coworkers reported a version of the reaction catalytic in DEAD, where the hydrazine was reoxidized to DEAD with iodobenzene diacetate, $PhI(OAc)_2$. Very recently, Taniguchi and coworkers have reported a further improvement, with ethyl 2-arylhydrazinecarboxylate ($ArNHNHCO_2Et$) and iron(II) phthalocyanine (FePc) as cocatalysts and O_2 as a stoichiometric oxidant (Hirose, D.; Taniguchi, T.; Ishibashi, H. *Angew. Chem. Int. Ed.* **2013**, *52*, 4613–4617). The overall concept of a DEAD-free catalytic Mitsunobu reaction is depicted below:

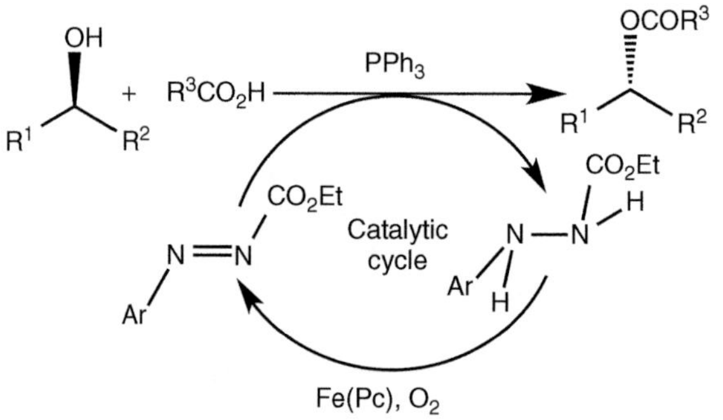

$$(5B.78)$$

The structure of FePc is as follows:

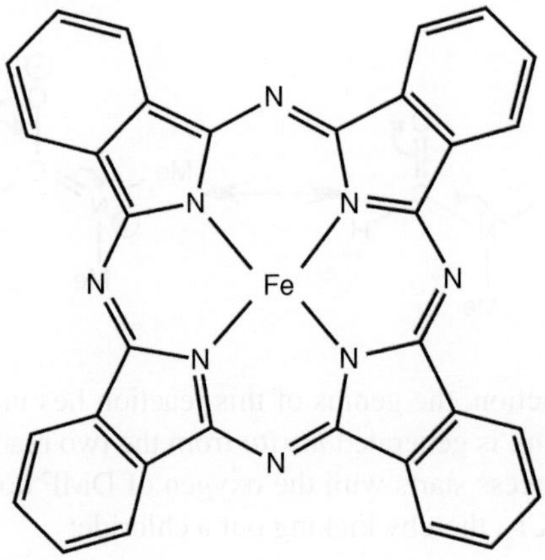

and the best ArNHNHCO₂Et derivative is the following:

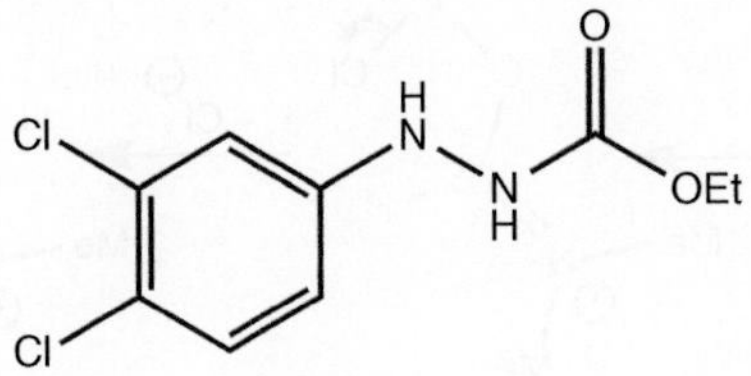

The mechanism of the FePc-mediated oxidation step is outside the scope of this book. It's a fascinating one, however, and may be similar to the famous rebound mechanism of the iron-containing enzyme cytochrome P450; feel free to look it up in a book on bioinorganic chemistry or on the Internet.

REVIEW PROBLEM 5B.21*

Would you describe the Mitsunobu reaction as mechanistically similar to the Appel reaction? Explain.

5B.14* THE VILSMEIER–HAACK REACTION

As our final example of the use of phosphorus in organic synthesis, we will choose the Vilsmeier–Haack reaction, the classic method for electrophilic formylation (addition of an aldehyde functionality) of aromatic rings with $POCl_3$ and dimethylformamide (DMF, $HCONMe_2$).

$$ArH \xrightarrow[\text{2. NaOAc, H}_2\text{O}]{\text{1. DMF, POCl}_3, \Delta} ArCHO \qquad (5B.79)$$

Although DMF is commonly used as a solvent, it is used as a reactant here. Its two main resonance forms are as follows:

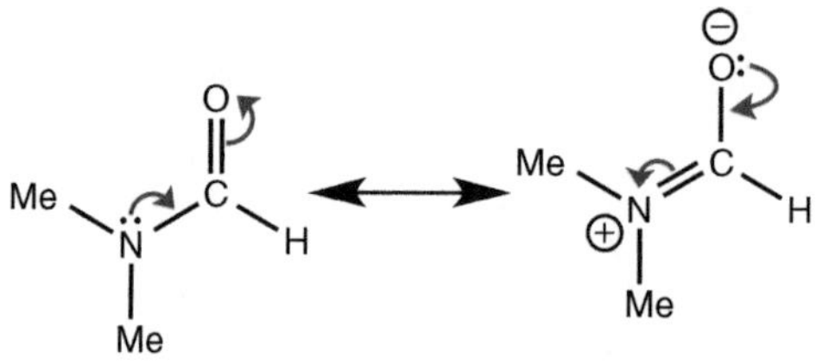

(5B.80)

As for the Mitsunobu reaction, the genius of this reaction lies in that the key reagent, the so-called Vilsmeier reagent, is generated *in situ* from the two readily available compounds $POCl_3$ and DMF. The process starts with the oxygen of DMF doing a nucleophilic attack on the phosphorus of $POCl_3$, thereby kicking out a chloride:

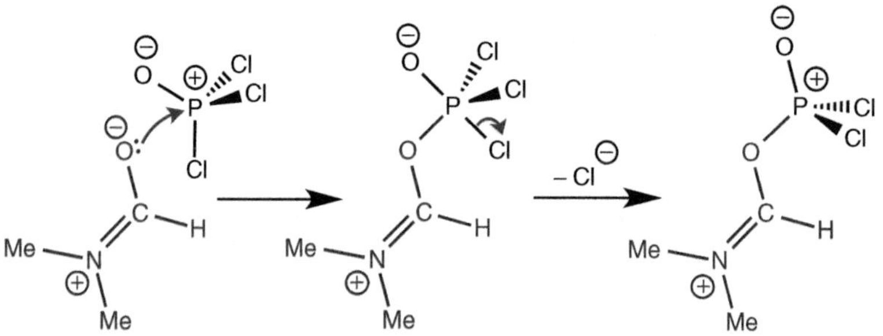

(5B.81)

Chloride then attacks the "iminium ($C=N^+$) carbon," forming a "chloroiminium cation" and the $PO_2Cl_2^-$ anion. The chloroiminium cation (shown in a box below) is the Vilsmeier reagent, the electrophile that reacts with the aromatic rings.

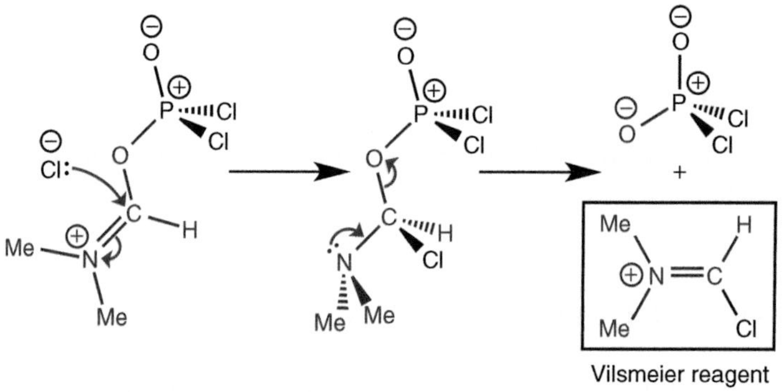

(5B.82)

Vilsmeier reagent

The driving force for formation of the Vilsmeier reagent, once again, is the formation of a P^+–O^- linkage, which by now is a common theme in this chapter.

The rest of the mechanism does not involve any phosphorus chemistry, but having come this far, we might as well complete it. The reaction of the Vilsmeier reagent with benzene is a typical electrophilic aromatic substitution, as shown below:

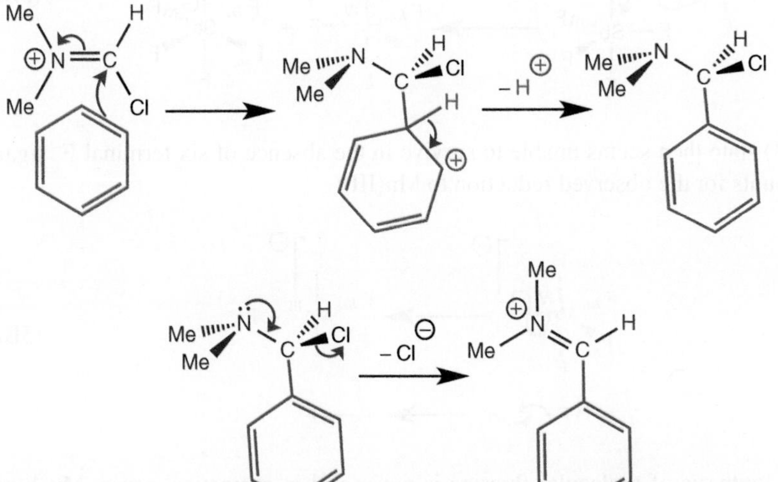

$$(5B.83)$$

The iminium cation product is hydrolyzed under work-up to yield benzaldehyde.

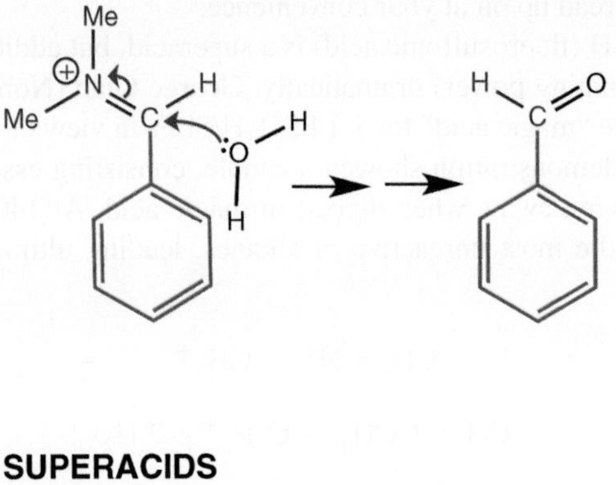

$$(5B.84)$$

5B.15   SbF₅ AND SUPERACIDS

As pointed out previously, the pentafluorides PnF_5 ($Pn = P$, As, Sb, Bi) are powerful Lewis acids and fluoride ion acceptors. While PF_5 and AsF_5 are monomeric, SbF_5 forms F^--bridged oligomers and polymers in the liquid phase. Similarly, solid BiF_5 consists of linear chains of BiF_6 octahedra linked by *trans* fluoride bridges.

Consider the following reaction, the only *chemical* synthesis of F_2:

$$2 K_2MnF_6 + 4 SbF_5 \rightarrow 4 KSbF_6 + 2 MnF_3 + F_2 \qquad (5B.85)$$

The reaction seems clearly driven by SbF_5's hunger for fluoride ions, which it extracts from $[MnF_6]^{2-}$.

$$(5B.86)$$

The Mn(IV) state then seems unable to survive in the absence of six terminal F^- ligands, which accounts for the observed reduction to Mn(III):

$$(5B.87)$$

The above synthesis of molecular fluorine is not useful in a practical sense. Much more important is the application of SbF_5 in the preparation of superacids, which are defined as acids that are stronger (i.e., have greater protonating power) than 100% sulfuric acid. The acidity of such strong acids is typically quantified by the Hammett acidity function, which we encourage you to read up on at your convenience.

Anyhydrous FSO_3H (fluorosulfonic acid) is a superacid, but addition of SbF_5 increases its acidity (i.e., protonating power) dramatically. George Olah (Nobel Prize in chemistry, 1994) coined the name "magic acid" for 1:1 FSO_3H/SbF_5 in view of its ability to protonate alkanes. A dramatic demonstration showed a candle, consisting essentially of long-chain hydrocarbons, dissolving away, when dipped in magic acid. At 140 °C, magic acid even protonates methane, the most unreactive of alkanes, leading ultimately to the relatively stable *t*-butyl cation:

$$CH_4 + H^+ \rightarrow CH_5^+$$

$$CH_5^+ + CH_4 \rightarrow C_2H_5^+ + 2\,H_2$$

$$C_2H_5^+ + 2\,CH_4 \rightarrow (CH_3)_3C^+ + 2\,H_2 \qquad (5B.88)$$

At the cost of a short digression from group 15 chemistry, a brief comment may be warranted on these exotic reactions. The remarkable species CH_5^+ is best thought of as a complex of CH_3^+ and H_2, with a three-center two-electron bond, reminiscent of B–H–B bonding in diborane.

The second of the three reaction steps above (5B.88) can then be viewed as a nucleophilic displacement where a C–H bond of methane acts as the nucleophile to kick H_2 out of CH_5^+:

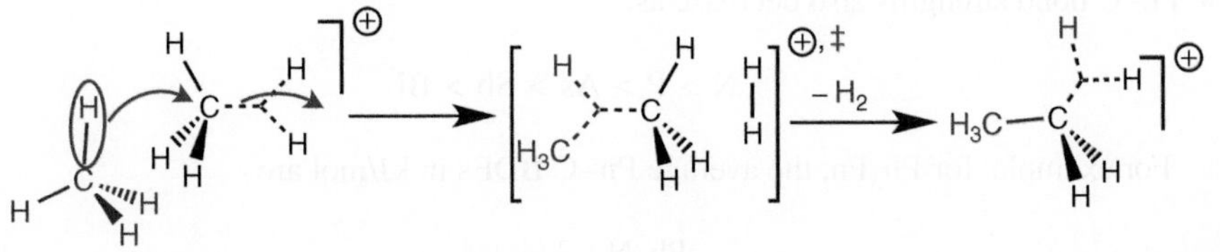

$$(5B.89)$$

The $C_2H_7^+$ cation so produced is expected to be capable of eliminating H_2 to generate the ethyl cation, as shown below:

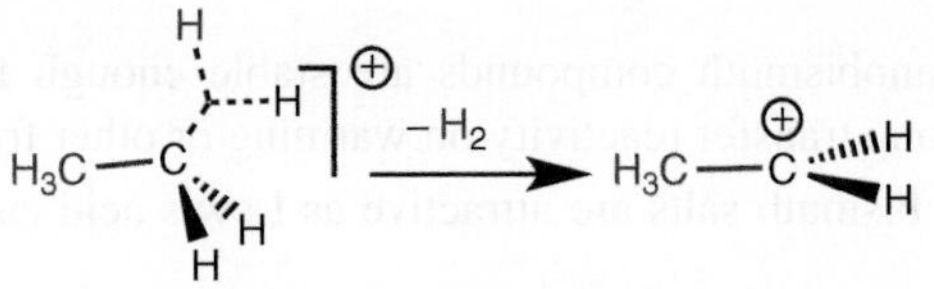

$$(5B.90)$$

Note that we have never invoked the exceedingly unstable methyl cation (CH_3^+), which is known only to occur in the gas phase. Once the ethyl cation is produced, it can react with additional methane to ultimately produce the relatively stable *t*-butyl cation.

Far stronger than magic acid is fluoroantimonic acid, which is a mixture of HF and SbF_5 in various ratios. The 1 : 1 mixture is the strongest superacid known, some 2×10^{19} (20 quintillion) times stronger than 100% sulfuric acid. Its molecular structure may be represented as follows:

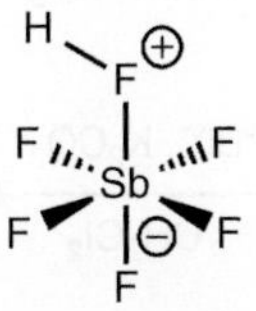

The F–H bond in $HSbF_6$ is clearly exceedingly weak, and the proton as a result is as close to "naked" as is possible in the liquid phase. As might be expected, fluoroantimonic acid is exceedingly corrosive and moisture-sensitive; it can, however, be kept in Teflon containers. Compatible solvents include SO_2ClF, SO_2, and chlorofluorocarbons.

5B.16   BISMUTH IN ORGANIC SYNTHESIS: GREEN CHEMISTRY

Bismuth and its compounds are cheap and nontoxic. Partially hydrolyzed bismuth salicylate, called *pink bismuth*, is widely used as an antacid, anti-inflammatory, and antibacterial in medications such as Pepto-Bismol™. Given these attractive properties, which are unique among heavy metals, bismuth compounds have been surprisingly rarely used by synthetic chemists; in the same spirit, bismuth reagents are conspicuous by their absence in organic texts, even fairly advanced ones. The research literature does document some significant applications of bismuth and we will attempt to provide a good summary. Some general considerations are as follows:

- As mentioned, the inert pair effect is important for bismuth, which means that pentavalent bismuth compounds are fairly strong oxidizing agents.
- Pn–C bond strengths also decrease as:

$$N > P > As > Sb > Bi$$

For example, for Ph_3Pn, the average Pn–C BDEs in kJ/mol are

$$Ph_3N : 374 \pm 4$$
$$Ph_3P : 321 \pm 21$$
$$Ph_3Bi : 194 \pm 11$$

Thus, many organobismuth compounds are stable enough for easy handling, but exhibit useful group transfer reactivity on warming or other forms of activation.

- Finally, trivalent bismuth salts are attractive as Lewis acid catalysts in a number of reactions.

The starting point for most organobismuth chemistry is typically a triarylbismuth, prepared from $BiCl_3$ and a Grignard reagent, as shown in Figure 5B.4. The triarylbismuth can then be oxidized and elaborated to a variety of pentavalent Bi reagents. Molecular halogens are used to produce triarylbismuth dihalides, Ar_3BiX_2. Peroxo reagents such as *t*-butyl hydroperoxide (TBHP) and sodium perborate ("NaBO_3") are used in conjunction with carboxylic acids to generate triarylbismuth dicarboxylate derivatives.

Pentavalent bismuth reagents are useful for oxidizing primary and secondary alcohols to aldehydes and ketones, respectively, under very mild conditions:

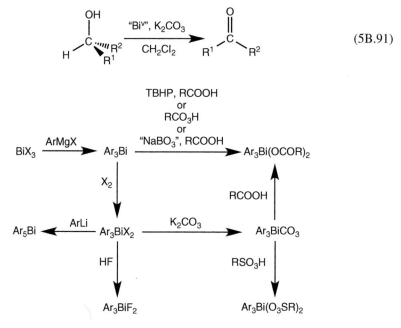

$$(5B.91)$$

Figure 5B.4 *Short summary of organobismuth chemistry.*

Interestingly, two different mechanistic pathways appear to be operative, as shown below. In the first pathway, one of the Bi-bound aryl groups abstracts a proton from the carbon atom carrying the OH group. In the other pathway, an external base (often CO_3^{2-}) plays this role. Regardless of the pathway, this proton abstraction leads to creation of the C=O double bond and reductive elimination of trivalent bismuth:

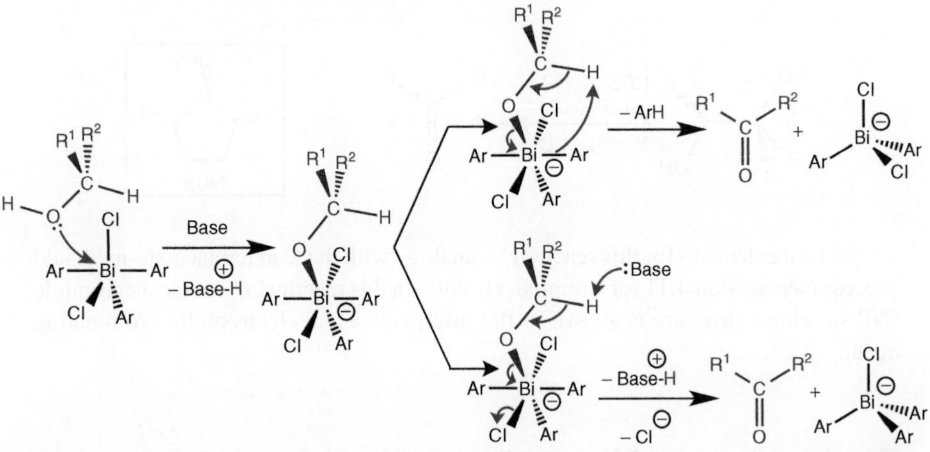

$$(5B.92)$$

REVIEW PROBLEM 5B.22

A remarkable reaction of pentavalent arylbismuth compounds is the arylation of enols and phenols. Three examples are shown below:

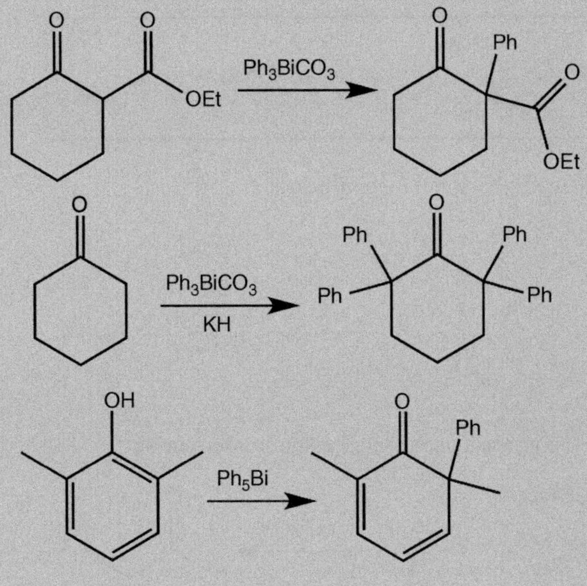

Suggest mechanisms for each of these three reactions.

REVIEW PROBLEM 5B.23[*]

In certain applications, the pentavalent bismuth reagent can be conveniently generated *in situ*, allowing the use of only a catalytic amount of a triarylbismuth reagent. Vicinal diols can thus be cleaved under mild Ph_3Bi-catalyzed conditions:

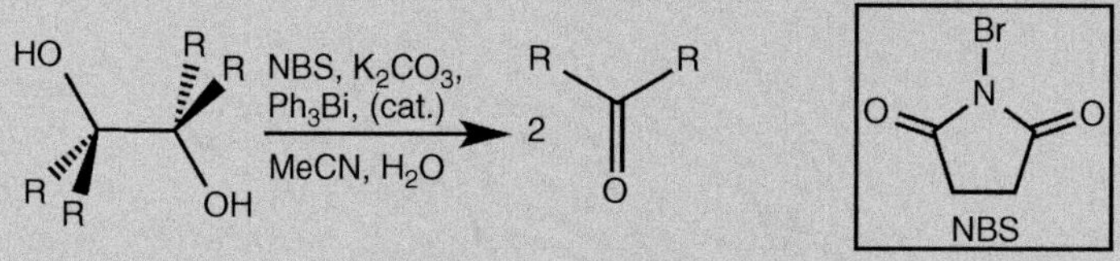

Suggest a mechanism for this reaction by analogy with the lead tetraacetate-mediated process (see Section 4.11 for a reminder). *Note*: In this reaction, *N*-bromosuccinimide (NBS), whose structure is shown in the inset, acts as an electrophilic brominating agent.

Fluoride may be extracted from triarylbismuth difluorides to yield fluorobismuthonium salts, as shown below. A carbon nucleophile, such as an enol ether, an allylsilane, or a vinylstannane, may then displace the Bi-bound fluoride to yield a variety of exceedingly useful triarylbismuthonium intermediates.

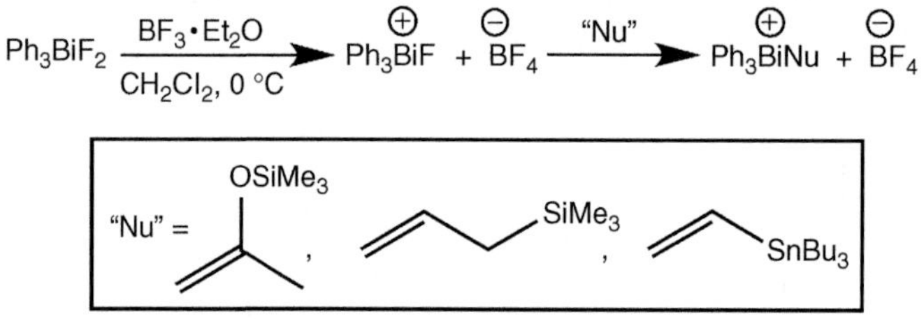

(5B.93)

The reaction with an allylsilane may be illustrated as follows:

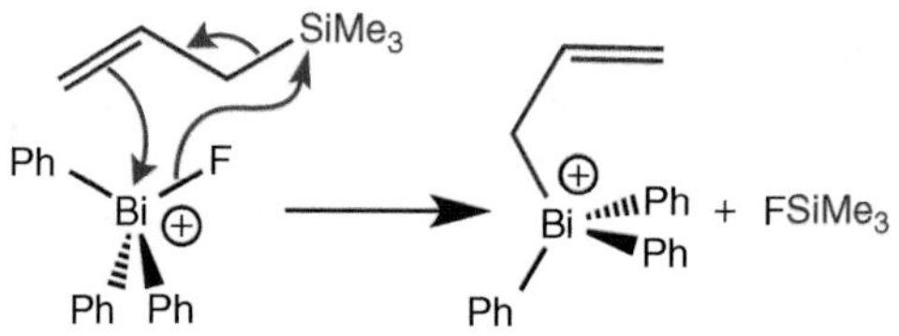

(5B.94)

With enol ethers, the products are α-acylbismuthonium salts:

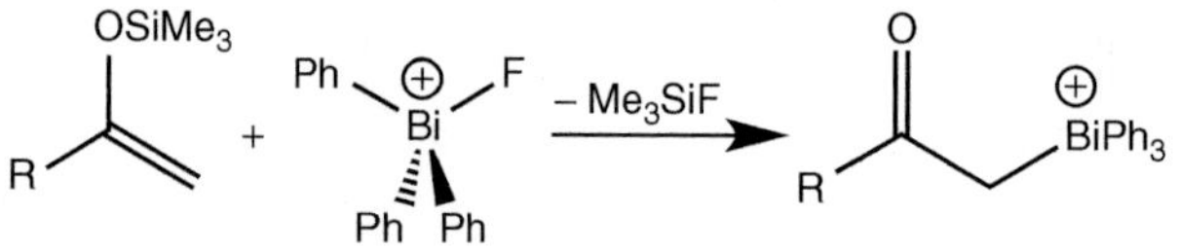

(5B.95)

REVIEW PROBLEM 5B.24

With appropriate assumptions (about work-up, etc.), explain the formation of an α-acylbismuthonium ion in reaction (5B.95).

The α-acylbismuthonium salts thus obtained react with a variety of nucleophiles as shown below, with triphenylbismuth in each case acting as an excellent leaving group:

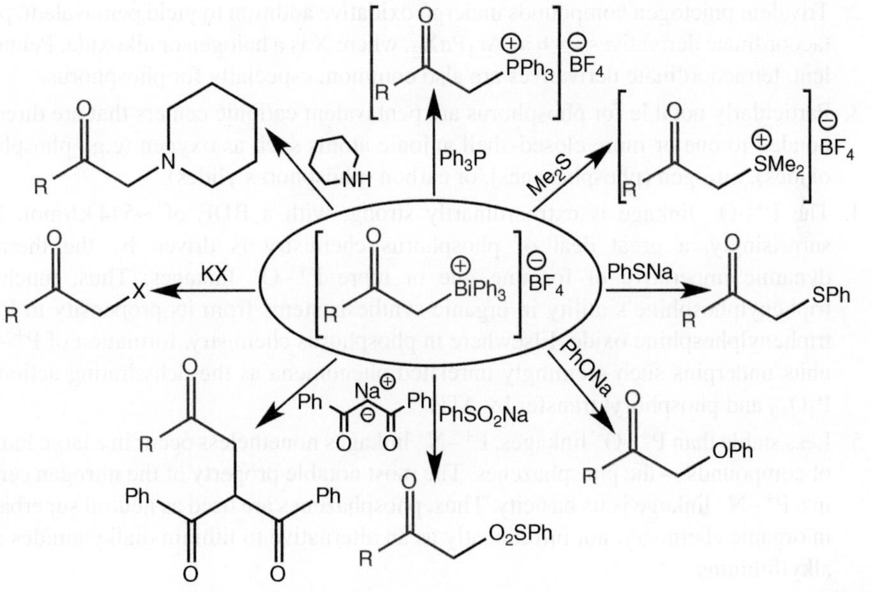

(5B.96)

Allyl- and vinyl-bismuthonium salts are similarly useful and can be elaborated further with a variety of nucleophiles as shown below for the allyl case:

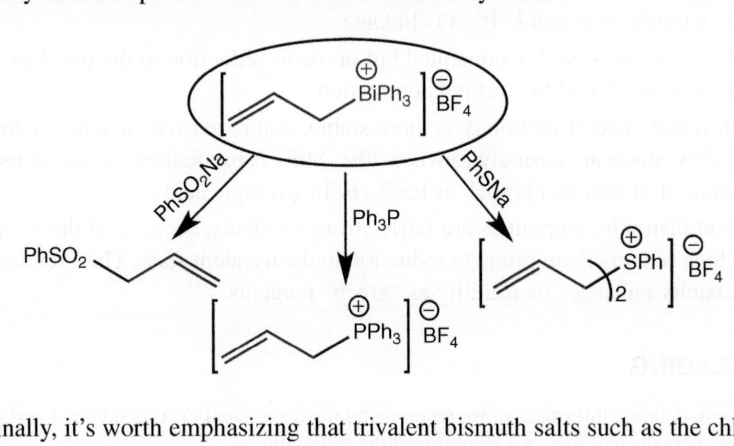

(5B.97)

Finally, it's worth emphasizing that trivalent bismuth salts such as the chloride, nitrate, and trifluoromethanesulfonate (triflate) are extremely attractive as Lewis acid catalysts. Applications include Friedel–Crafts acylations, Diels–Alder reactions, esterifications, acetal and thioacetal deprotections, and addition of trimethylsilyl cyanide to carbonyl groups, among many other reactions.

5B.17 SUMMARY

Some of the main points of our discussion have been the following:

1. Trivalent pnictogen derivatives are typically nucleophilic. A good example is triphenylphosphine, a common nucleophile in organic chemistry.

2. Trivalent pnictogen compounds undergo oxidative addition to yield pentavalent, pentacoordinate derivatives such as Ar_3PnX_2, where X is a halogen or alkoxide. Pentavalent, tetracoordinate derivatives are also common, especially for phosphorus.

3. Particularly notable for phosphorus are pentavalent cationic centers that are directly bonded to one or more closed-shell anionic atoms such as oxygen (e.g., phosphine oxides), nitrogen (phosphazenes), or carbon (phosphorus ylides).

4. The P^+–O^- linkage is extraordinarily strong, with a BDE of ~544 kJ/mol. Not surprisingly, a great deal of phosphorus chemistry is driven by the thermodynamic imperative of forming one or more P^+–O^- linkages. Thus, much of triphenylphosphine's utility in organic synthesis stems from its propensity to form triphenylphosphine oxide. Elsewhere in phosphorus chemistry, formation of P^+–O^- units underpins such seemingly unrelated phenomena as the dehydrating action of P_4O_{10} and phosphoryl transfer by ATP.

5. Less stable than P^+–O^- linkages, P^+–N^- linkages nonetheless occur in a large family of compounds—the phosphazenes. The most notable property of the nitrogen center in a P^+–N^- linkage is its basicity. Thus, phosphazenes are used as neutral superbases in organic chemistry, not infrequently as an alternative to lithium dialkylamides and alkyllithiums.

6. The P^+–C^- linkage is stable enough to exist in isolable synthetic intermediates, the phosphorus ylides, but is quite reactive and functions as essentially a carbanion. In a typical application such as the Wittig reaction, a reactive P^+–C^- linkage is transformed to a much more stable P^+–O^- linkage.

7. Pentavalent arsenic is well documented but prone to reduction to the trivalent state, a phenomenon attributed to d orbital contraction.

8. The pentavalent state of antimony is more stable, compared with arsenic. While all the PnF_5 derivatives are strongly Lewis-acidic, SbF_5 is particularly so and is used in the preparation of superacids such as FSO_3H/SbF_5 (magic acid).

9. Pentavalent bismuth compounds are fairly strong oxidants, because of the inert pair effect, which renders them prone to reduction to the trivalent state. Their remarkable lack of toxicity enhances their utility as "green" reagents.

FURTHER READING

In general, for more detailed information, we recommend the texts listed in Appendices 1 and 2. For some of the more specialized topics, we recommend the following:

1. Emsley, J. *The 13th Element: The Sordid Tale of Murder, Fire, and Phosphorus*; Chemical Heritage Foundation: Philadelphia, 2002; 327 pp. *A popular science book.*

2. Westheimer, F. H. "Why Nature Chose Phosphate," *Science* **1987**, *235*, 1173–1178. *A classic paper on the subject.*

3. Rosen, B. P.; Ajees, A. A.; McDermott, T. R. *Bioessays* **2011**, *33*, 350–357. *A good perspective on the As-DNA controversy.*

4. Whorton, J. C. *The Arsenic Century: How Victorian Britain was Poisoned at Home, Work, and Play*; Oxford University Press: Oxford, 2010; 464 pp. *An entertaining popular book.*

5. Jenkins, R. O. Biomethylation of arsenic, antimony and bismuth, (Ch. 7) *In* Sun, H., ed. *Biological Chemistry of Arsenic, Antimony and Bismuth*; John Wiley & Sons, Inc.: Hoboken, NJ, 2011; 400 pp. *An up-to-date summary of detailed information on biomethylation.*

6. But, T. Y. S.; Toy, P. "The Mitsunobu Reaction: Origin, Mechanism, Improvements, and Applications," *Chem. Asian J.* **2007**, *2*, 1340–1355. *An excellent review article.*

7. Schwesinger, R., *et al.* "Extremely Strong, Uncharged Auxiliary Bases; Monomeric and Polymer-Supported Polyaminophosphazenes (P2-P5)," *Liebigs Ann.* **1996**, 1055–1081. *A review on phosphazenes.*

8. Ishikawa, T. *Superbases for Organic Synthesis: Guanidines, Amidines, Phosphazenes and Related Organocatalysts*; John Wiley & Sons, Inc.: Hoboken, NJ, 2009; 336 pp. *An authoritative reference on uncharged nitrogen bases.*

9. Olah, G. A.; Surya Prakash, G. K.; Sommer, J.; Molner, A. *Superacid Chemistry*, 2nd ed.; John Wiley & Sons, Inc.: Hoboken, NJ, 2009; 850 pp. *A definitive reference.*

10. Ollevier, T., ed. *Bismuth-Mediated Organic Reactions*; *Top. Curr. Chem.* **2012**, vol. *311*. Springer: New York/Heidelberg; 277 pp. *An up-to-date and wide-ranging treatment of the topic.*

11. Lancaster, M. *Green Chemistry: An Introductory Text*, 2nd ed.; Royal Society: Cambridge, 2010; 344 pp. *A readable introductory text.*

12. Grossman, E. *Chasing Molecules: Poisonous Products, Human Health, and the Promise of Green Chemistry*; Shearwater: Washington, DC, 2011; 288 pp. *A popular perspective of green chemistry.*

6

Group 16 Elements: The Chalcogens

Linus Pauling: *Have you ever smelled hydrogen selenide?*
Matthew Meselson: *No, I never have.*
Pauling: *Well, it smells much worse than hydrogen sulfide. …*
Now, hydrogen telluride smells as much worse than hydrogen
selenide as hydrogen selenide does compared to hydrogen
sulfide … some chemists were not careful when working
with tellurium compounds, and they acquired a condition
known as "tellurium breath." As a result, they have become
isolated from society. Some even committed suicide.
Meselson: *Oh.*
Pauling: *But Matt, I'm sure that you would be careful.*
Why don't you think it over and let me know if you would
like to work on the structure of some tellurium compounds?

Interview between Pauling and then prospective graduate
student Matthew Meselson, who during his subsequent
tenure with Pauling did the Meselson–Stahl
experiment, which proved the "semiconservative"
nature of DNA replication.

Two of the group 16 elements—oxygen and sulfur—are among the most familiar ones. Not only is water critical to all life and O_2 to all aerobic life, the ozone layer protects life and civilization on Earth from the sun's harmful ultraviolet rays. The annual worldwide production of sulfuric acid is in the hundreds of millions of metric tons and is the highest by weight for all industrial chemicals. Indeed, the amount of sulfuric acid produced is often a good indicator of a country's state of industrial development and economic status.

Arrow Pushing in Inorganic Chemistry: A Logical Approach to the Chemistry of the Main-Group Elements,
First Edition. Abhik Ghosh and Steffen Berg.
© 2014 John Wiley & Sons, Inc. Published 2014 by John Wiley & Sons, Inc.

GROUP 16 ELEMENTS: THE CHALCOGENS **203**

Interestingly, as a so-called Döbereiner triad (i.e., a triad of elements, whose chemical similarities were recognized by German chemist Johann Wolfgang Döbereiner in the nineteenth century), sulfur, selenium and tellurium played a role in the initial construction of the periodic table.

Many group 16 compounds are malodorous, hydrogen sulfide (H_2S) being probably the most notorious example. Flatulence, bad breath, and skunk spray all contain hydrogen sulfide (H_2S), as well as other volatiles. Hydrogen sulfide is also toxic, so it's important to exercise caution so as to avoid exposure when working with H_2S in the laboratory. Interestingly, against this unattractive backdrop, H_2S has recently been recognized as a gasotransmitter, that is, a gaseous signaling substance in biology, comparable to NO and CO. Unfortunately, we will not be discussing this important biological role of H_2S in this book, partly because the story is still unfolding and key details are still missing.

Dimethyl sulfide (DMS) has a cabbage-like smell, which becomes offensive at higher concentrations. It is, however, an important ingredient of the flavors of several food items such as asparagus, beetroot, and truffles. Along with algal pheromones called *dictyoterpenes*, DMS is a key part of what people call "the smell of the sea." We'll see that DMS is the end product of various dimethysulfoxide-mediated oxidations (e.g., the Swern oxidation of alcohols); washing the glassware with bleach destroys much of the DMS produced.

The group 16 elements or chalcogens (which we will occasionally abbreviate as Ch) exhibit considerable chemical diversity, with the greatest discontinuity in properties between oxygen and sulfur. Sulfur and selenium are chemically rather similar. Tellurium is considered a metalloid and polonium a full-fledged metal. Some general group trends are as follows.

- Oxygen is perhaps best known for its high electronegativity (3.44 on the Pauling scale), which is second only to fluorine (3.98). The electronegativities of sulfur and selenium are much lower—2.58 and 2.55, respectively—essentially the same as that of carbon (2.55). The difference in electronegativity translates to many differences between oxygen and the heavier chalcogens. Thus, O–O bonds are high energy entities, much more fragile than S–S bonds with respect to both homolytic and heterolytic dissociation.

- Elemental oxygen commonly occurs in a couple of different forms that are unique for group 16. The stable form of the element is triplet O_2, which has two unpaired electrons and is paramagnetic. Ozone, O_3, is diamagnetic, a metastable but highly endothermic substance:

$$O_3 \rightarrow 3/2\, O_2 \qquad \Delta H = -142\ \text{kJ/mol} \qquad (6.1)$$

The closed-shell, double-bonded form singlet O_2 (O=O) is an excited state, about 94.3 kJ/mol higher in energy than triplet O_2. In many oxygen-evolving reactions, the O_2 is initially produced in the singlet state, and it subsequently decays to the stable triplet state. Singlet oxygen is highly reactive, aggressively attacking most organic matter.

- As a divalent "first-row" (period 2) element, oxygen forms double bonds with many elements and does so with greater ease compared to the heavier chalcogens.

- Catenation is an important property of the heavier chalcogens. A variety of chains, rings, and cage structures are known for S, Se, and Te.

- One of the most distinctive features of the heavier chalcogens S to Te, relative to oxygen, is their capacity for variable valence. Switching between di, tetra, and hexavalent states is commonplace for S to Te.

- Metallic character increases down the group. This is reflected in the greater prevalence of polymeric as opposed to discrete molecular structures and of anionic complexes such as $SeBr_6{}^{2-}$, $TeBr_6{}^{2-}$, and $PoI_6{}^{2-}$.

- The lower-valent states in the form of H_2Ch, HCh^-, and RCh^- are fairly strongly nucleophilic.

- The tetravalent state is more variable in behavior—nucleophilic and/or reducing in some compounds but electrophilic and/or oxidizing in others. Thus, SO_2 acts as both an oxidizing and a reducing agent, though most commonly as a reducing agent; selenium dioxide, on the other hand, is best known as an oxidant.

- The hexavalent state, best known with O and F as terminal ligands, is often remarkably stable and not particularly prone to reduction.

- Because of its intense radioactivity, polonium has been less studied than the other chalcogens, and the chemical consequences of the inert-pair effect, primarily a sixth-period phenomenon, remain largely unexplored for polonium.

Much of our discussion in this chapter will focus on sulfur, which we'll treat as paradigmatic of the chalcogen group. As for the remarks by Pauling quoted above, selenium and tellurium compounds are indeed toxic. With good laboratory practices, however, they can be safely handled. We will see that selenium reagents allow some unique transformations in organic chemistry, and the same may be said, to a lesser extent, for tellurium.

6.1   THE DIVALENT STATE: FOCUS ON SULFUR

For group 16 elements, the divalent state is probably the most familiar; it's the standard, nonhypervalent state of the elements. Sulfur has a large number of allotropes, and most of these are made up of rings or chains of divalent sulfur; several allotropes are made up of puckered S_8 rings in different packing arrangements. Familiar examples include water, alcohols, and ethers; thiols (RSH) and thioethers (RSR'), including the essential amino acids cysteine (which has a CH_2SH sidechain) and methionine (which has a CH_2SCH_3 sidechain); "catenated" species such as hydrogen peroxide and the sulfanes, H_2S_n; and the chlorides SCl_2 and S_2Cl_2. The hydrides and the chlorides serve as building blocks of many other divalent sulfur species. Consider, for example, the following rational synthesis of cyclohexasulfur.

$$H_2S_4 + S_2Cl_2 \rightarrow \text{cyclo-}S_6 + 2\,HCl \tag{6.2}$$

The SH sulfurs being the negative ends of dipoles are the expected nucleophiles. The SCl sulfurs are good electrophiles, in large measure because of the weakness of the S–Cl bonds (bond dissociation energy, BDE, $\sim$251 kJ/mol) and the resulting efficacy of chloride as a leaving group:

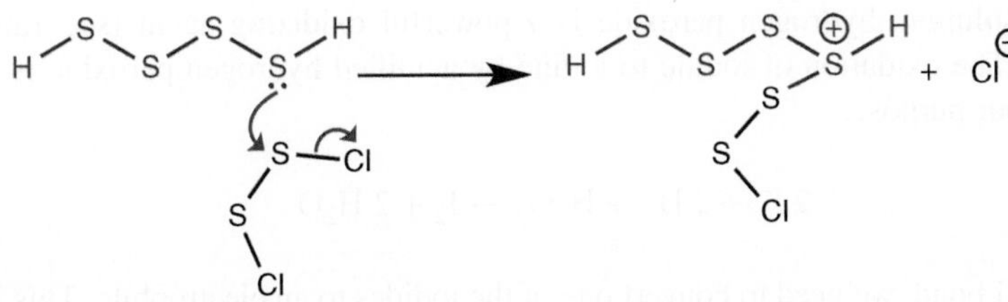

$$(6.3)$$

On leaving, the chloride ions pluck off protons from the cationic sulfurs, as shown below. Another round of the same process then leads to the S_6 product.

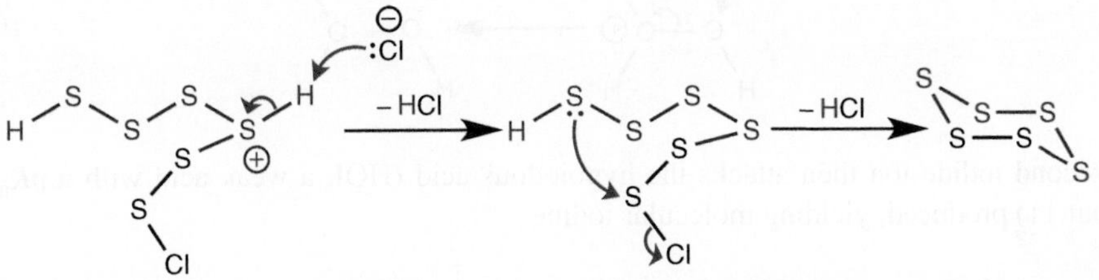

$$(6.4)$$

6.2   THE DIVALENT STATE: HYDROGEN PEROXIDE

Hydrogen peroxide is a relatively stable but reactive molecule. Dihydrogen trioxide or tri-oxidane, HOOOH, is far more reactive, but has been generated in aqueous solution at low temperature. The higher "oxidanes" H_2O_4 and H_2O_5 are also believed to exist but only as fleeting intermediates. In other words, catenation is much more limited for oxygen than it is for the other chalcogens.

With an oxidation state intermediate between those of water and O_2, hydrogen peroxide exhibits a wide range of reactivity. Although it is best known as an oxidant, it reduces stronger oxidants than itself while itself becoming oxidized to O_2. It can act as both a nucleophile and an electrophile.

Hydrogen peroxide oxidizes sulfite to sulfate and organic sulfides to sulfoxides:

$$SO_3{}^{2-} + H_2O_2 \rightarrow SO_4{}^{2-} + H_2O \tag{6.5}$$

$$PhSMe + H_2O_2 \rightarrow Ph\text{-}SO\text{-}Me + H_2O \tag{6.6}$$

The mechanism in each case is likely to involve a nucleophilic attack by sulfur on one of the peroxide oxygens, as shown below for PhSMe:

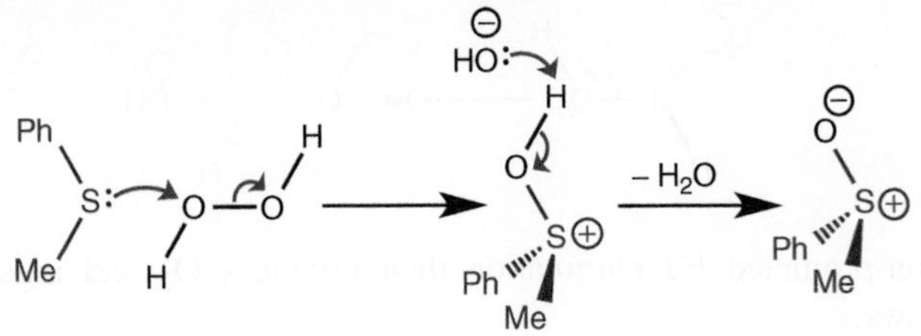

$$(6.7)$$

In acidic solution, hydrogen peroxide is a powerful oxidizing agent (see Table 1.5, Section 1.5). The oxidation of iodide to iodine by acidified hydrogen peroxide is a good example for our purpose:

$$2\,I^- + 2\,H^+ + H_2O_2 \rightarrow I_2 + 2\,H_2O \tag{6.8}$$

To form an I–I bond, we need to convert one of the iodides to an electrophile. This is most simply done by having I^- attack protonated H_2O_2:

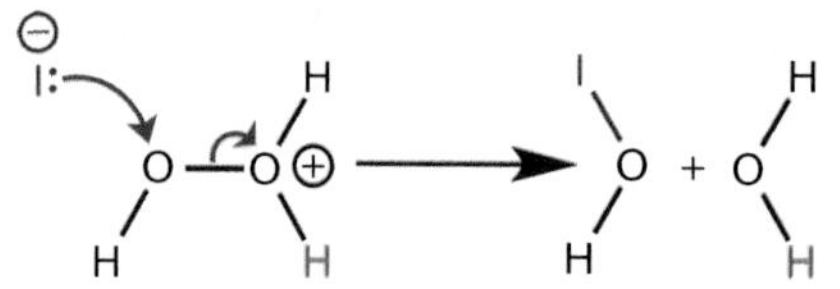

$$\tag{6.9}$$

A second iodide ion then attacks the hypoiodous acid (HOI, a weak acid with a pK_a of about 11) produced, yielding molecular iodine:

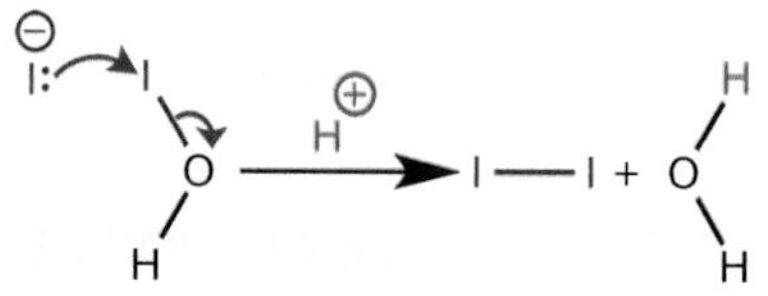

$$\tag{6.10}$$

In neutral or slightly alkaline solution, on the other hand, iodide promotes a different reaction, namely, the catalytic decomposition of hydrogen peroxide:

$$2\,H_2O_2 \rightarrow 2\,H_2O + O_2 \tag{6.11}$$

The first step of the mechanism is still the same as in the above reaction, namely, the production of hypoiodous acid as an intermediate:

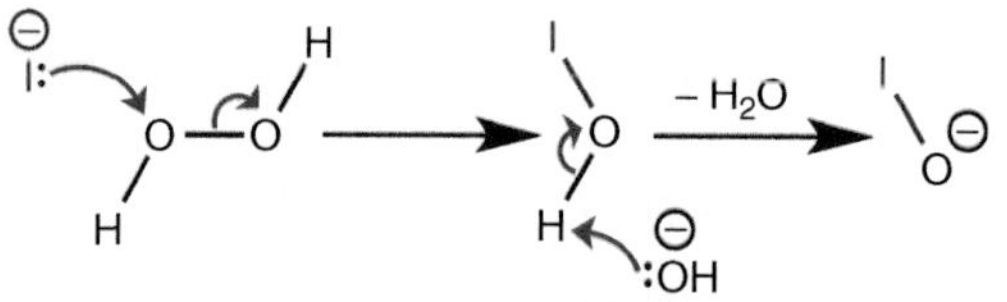

$$\tag{6.12}$$

The hypoiodite anion, acting as a nucleophile, could then attack H_2O_2 to produce the unstable intermediate HOOI:

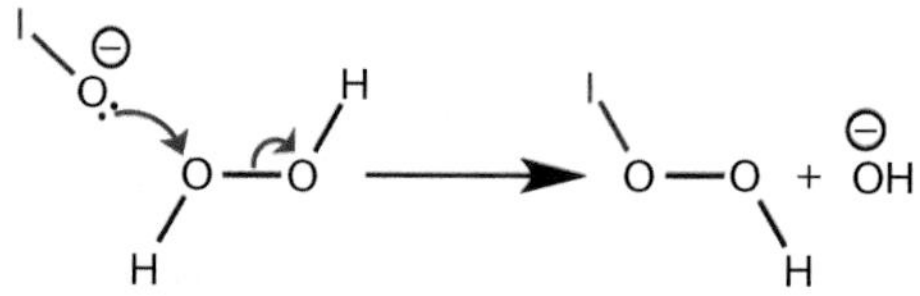

$$\tag{6.13}$$

Hydroxide anion-promoted E2 elimination then produces O_2 and regenerates the I^- catalyst, as follows:

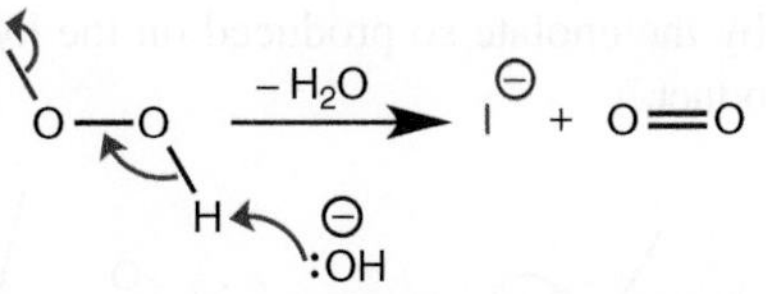

$$\text{(6.14)}$$

This reaction is the basis of the so-called elephant toothpaste demonstration. In a typical experiment, a 2 M KI solution (say, 100 ml) is poured into a graduated cylinder containing 30% H_2O_2 (20 ml), dishwashing liquid (about 5 ml), and some food coloring, the latter two only for dramatic effect. The O_2 produced gets trapped in the colored, soapy water, creating a foamy column—the "elephant toothpaste"—that quickly fills the graduated cylinder and spills out. (*Caution.* Thirty percent H_2O_2 is severely corrosive to the skin, respiratory tract, and eyes. Please therefore consult a full set of instructions before attempting this experiment yourself; this brief description is not intended as a laboratory procedure!)

Some molecular iodine is also produced in this experiment, presumably through the reaction of I^- and HOI:

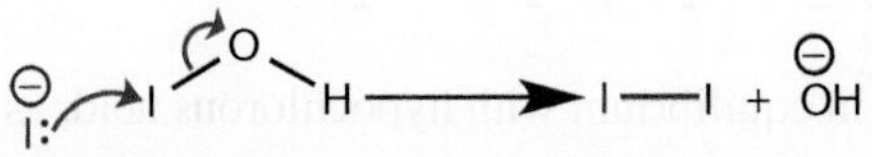

$$\text{(6.15)}$$

Excess iodide converts the I_2 produced to yellow-brown I_3^-:

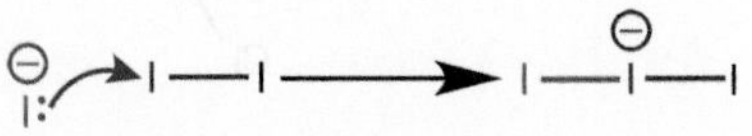

$$\text{(6.16)}$$

Alkaline hydrogen peroxide is widely used for the epoxidation of α,β-unsaturated carbonyl compounds, as shown by the following example:

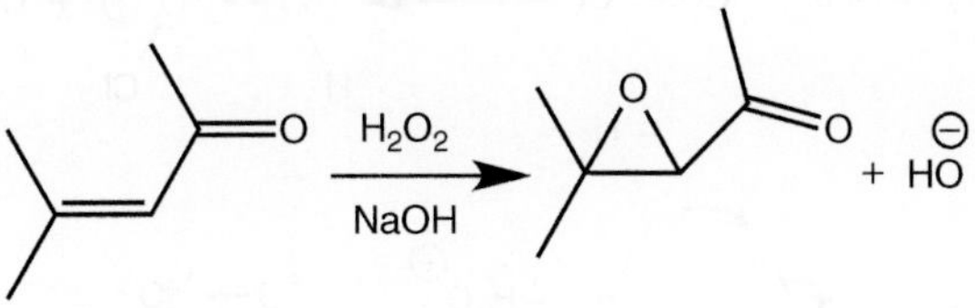

$$\text{(6.17)}$$

In the first step of the mechanism, H_2O_2, or rather its deprotonated form, the hydroperoxide anion, HOO^-, acts as the nucleophile in a Michael addition (for a reminder, see Sections 1.14 and 1.15) to the α,β-unsaturated carbonyl compound, as shown below:

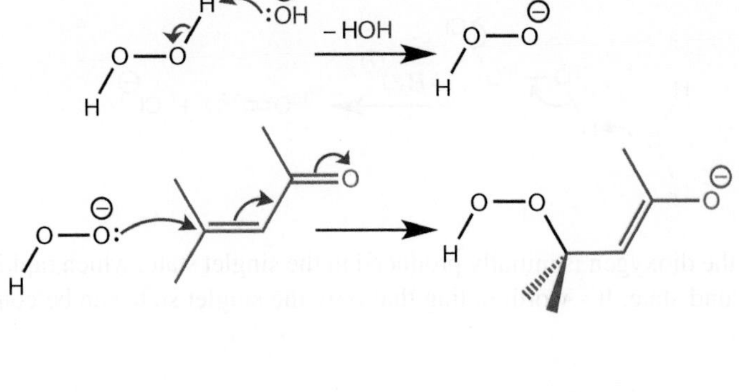

$$\text{(6.18)}$$

An intramolecular attack by the enolate so produced on the hydroperoxide unit then produces the final epoxide product.

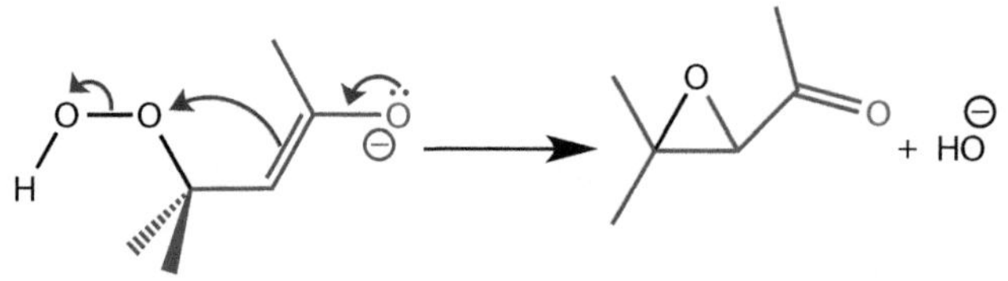

$$\text{(6.19)}$$

The oxidation of H_2O_2 by strong oxidants (Cl_2, MnO_4^-, and Ce^{4+}, among others) has been studied using labeled $H_2{}^{18}O_2$. Analysis of the evolved oxygen proved that both oxygen atoms in the O_2 arise from H_2O_2 and not from water. The same conclusion has also been reached for catalytic decomposition of H_2O_2, for example, the I^--catalyzed process described above. Let us consider the mechanism for H_2O_2 oxidation with Cl_2 as the oxidant:

$$H_2O_2 + Cl_2 + 2\,H_2O \rightarrow O_2 + 2\,H_3O^+ + 2\,Cl^- \qquad (6.20)$$

Chlorine in water exists in equilibrium with hypochlorous acid, as shown below:

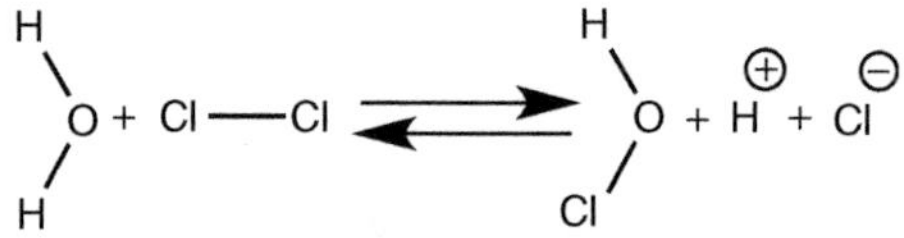

$$\text{(6.21)}$$

Hydrogen peroxide could then do a nucleophilic attack on either Cl_2 or HOCl to produce HOOCl:

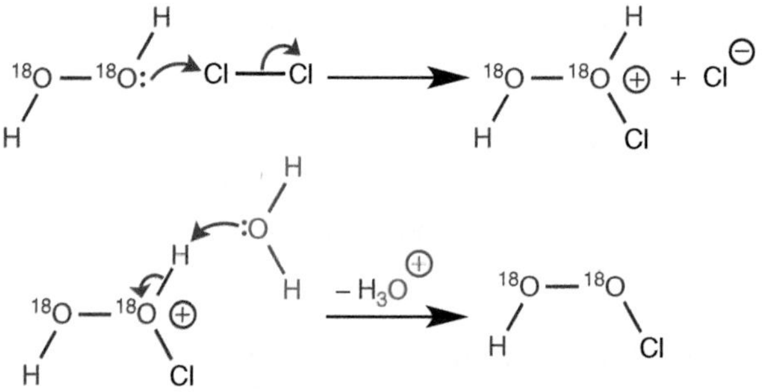

$$\text{(6.22)}$$

Water, acting as the base, might then be expected to decompose the highly unstable intermediate HOOCl via an E2 elimination:

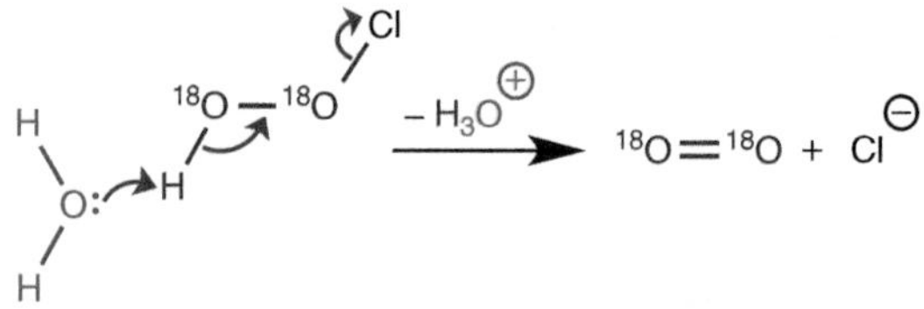

$$\text{(6.23)}$$

In this reaction, the dioxygen is initially produced in the singlet state, which rapidly decays to the triplet ground state. It's worth noting that only the singlet state can be conveniently represented in a Lewis structure, that is, via a simple double-bonded structure (O=O). The triplet ground state requires a molecular orbital description, which is presented in most general and inorganic chemistry textbooks.

Let's reflect for a moment on how the chemistry just described compares with the chemistry of the sulfanes described in the last section. Like the sulfane SH groups, the oxygens of H_2O_2 can act as nucleophiles. Like the sulfur atoms within polysulfide chains, the oxygens in H_2O_2 can also act as electrophiles. One distinct feature of H_2O_2 chemistry is of course its potential for oxidation to O_2. Sulfanes too are prone to oxidation but not to diatomic S_2, which is unstable, but rather to medium-sized S_n rings.

6.3 S_2Cl_2 AND SCl_2

Passing chlorine through molten sulfur results in disulfur dichloride (S_2Cl_2), a fuming orange liquid. Further chlorination leads to sulfur dichloride (SCl_2).

$$S_8 + 4\,Cl_2 \rightarrow 4\,S_2Cl_2 \tag{6.24}$$

$$S_2Cl_2 + Cl_2 \rightarrow 2\,SCl_2 \tag{6.25}$$

Taking the first reaction first, we need to first identify the nucleophile and electrophile, that is, what attacks what. Considering that Cl^- is a much weaker base than HS^- (which is a different way of saying that HCl is a strong acid and H_2S is a weak one), one of the S_8 sulfurs seems likely to be the nucleophile that attacks Cl_2, kicking out a chloride ion:

$$\tag{6.26}$$

The displaced chloride now attacks an adjacent sulfur, creating the linear-chain molecule ClSSSSSSSSCl:

$$\tag{6.27}$$

One of the internal sulfurs of S_8Cl_2 linear chain can now act as the nucleophile and attack a second molecule of Cl_2:

$$\tag{6.28}$$

A chloride can then attack an adjacent sulfur, cleaving the S_8 chain in the process:

$$(6.29)$$

We have thus produced the first molecule of S_2Cl_2; the remaining S_6Cl_2 chain can now be broken down in much the same manner:

$$(6.30)$$

Let us now consider how chlorination of S_2Cl_2 leads to SCl_2. Once again, the first problem is to identify the nucleophile and the electrophile. And again, as the softer atom, sulfur seems most plausible as the initial nucleophile:

$$(6.31)$$

Nucleophilic attack by Cl^- then results in two molecules of SCl_2:

$$(6.32)$$

Sulfur dichloride (SCl_2) is unstable and loses chlorine and reverts back to S_2Cl_2 over several hours:

$$2 \ SCl_2 \rightarrow \ S_2Cl_2 + Cl_2 \tag{6.33}$$

Selenium dibromide ($SeBr_2$) also behaves analogously:

$$2 \ SeBr_2 \rightarrow \ Se_2Br_2 + Br_2 \tag{6.34}$$

REVIEW PROBLEM 6.1

Selenium dichloride, dissolved in acetonitrile, disproportionates via a different pathway, relative to SCl_2 and $SeBr_2$:

$$3 \ SeCl_2 \rightarrow Se_2Cl_2 + SeCl_4$$

Suggest a mechanism and also a possible rationale for the different behavior.

REVIEW PROBLEM 6.2

Selenium dibromide can be prepared from selenium dichloride via interaction with trimethylsilyl bromide in an aprotic solvent such as THF:

$$SeCl_2 + 2 \, Me_3SiBr \rightarrow SeBr_2 + 2 \, Me_3SiCl$$

Suggest a mechanism for this reaction. Also, comment on the overall transformation vis-à-vis the HSAB principle.

Hint. Observe that Se and Si are essentially swapping ligands. As noted in Section 1.19, such ligand exchanges typically involve bridged intermediates. In this case, however, there are at least a couple of options. In our view, the most reasonable way to start the mechanism is to have Cl attack Si, since the latter is the most potent electrophilic center in the system.

6.4 NUCLEOPHILIC BREAKDOWN OF CYCLOPOLYSULFUR RINGS

The S_8 ring is completely broken down by a variety of nucleophiles. With triphenylphosphine, the overall reaction is

$$S_8 + 8 \, Ph_3P \rightarrow 8 \, Ph_3PS \qquad (6.35)$$

The mechanism might not be obvious, but triphenylphosphine is such a superb nucleophile that it's a good bet to have it attack the S_8 ring; the S—S bonds are fairly weak (~213 kJ/mol) and cleave with relative ease:

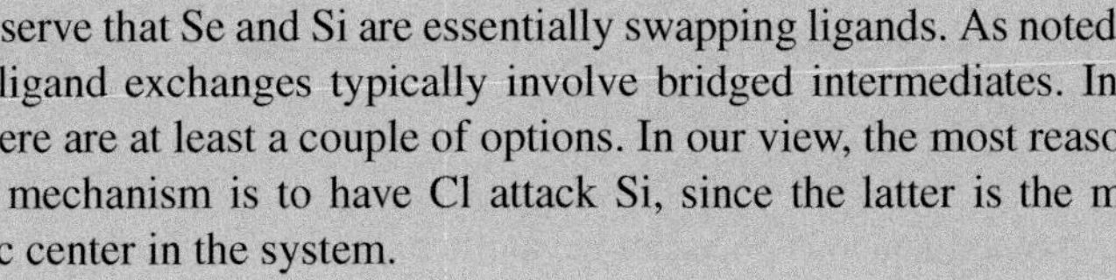

$$(6.36)$$

A Ph_3PS unit is now poised as an excellent leaving group, and attack by a second Ph_3P, as shown below, gives us our first molecule of triphenylphosphine sulfide, Ph_3PS.

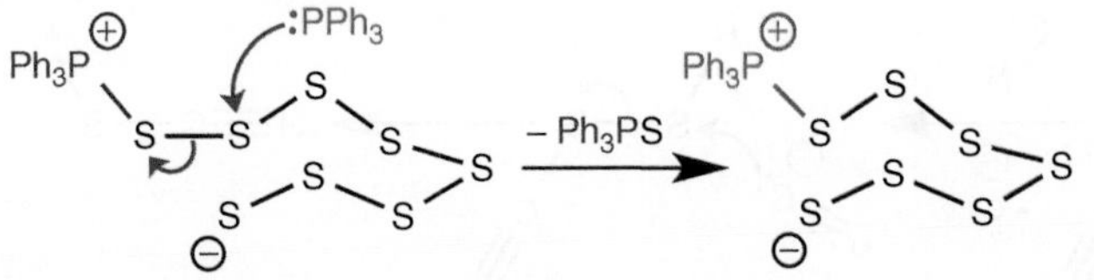

$$(6.37)$$

Additional Ph_3P molecules continue to chew on the shrinking sulfur chain until we are left with just the following stub:

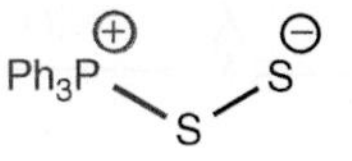

If you don't want to see a triphenylphosphine attack a negatively charged sulfur, you could isomerize this intermediate to a cyclic form with no formal charges. Its breakdown to two molecules of Ph_3PS would then happen as follows:

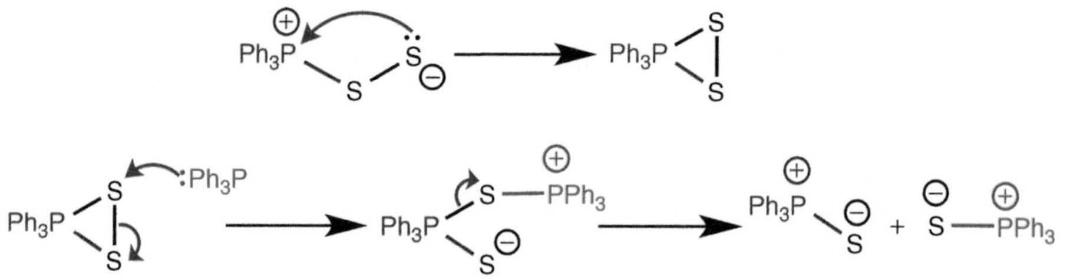

$$(6.38)$$

In the same manner, aqueous potassium cyanide breaks down the S_8 ring to thiocyanate (NCS^-):

$$S_8 + 8\,CN^- \rightarrow 8\,NCS^- \qquad (6.39)$$

The first few steps proceed exactly as for the Ph_3P reaction, that is, reaction 6.35, until we arrive at the following intermediate:

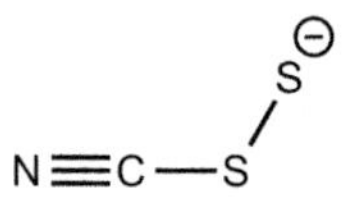

Once again, the puzzle consists of cleaving the last S–S bond and producing the last thiocyanate ions. You probably don't want a cyanide ion attacking a negatively charged sulfur. The negative charge on the terminal sulfur, however, is easily neutralized by protonation, paving the way for the final S–S bond cleavage, as shown below:

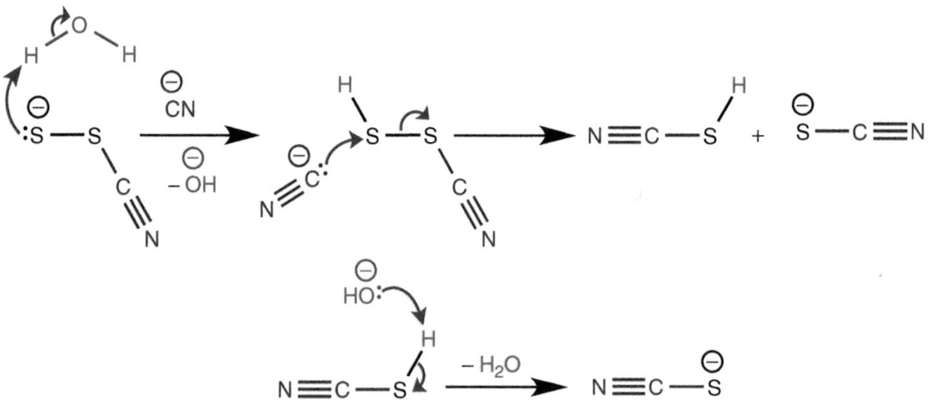

$$(6.40)$$

REVIEW PROBLEM 6.3

Elemental sulfur may be oxidized to a variety of polysulfur cations. Thus, oxidation of S_8 in liquid SO_2 with AsF_5 or SbF_5 leads to S_8^{2+} among other species:

$$S_8 + 3\,AsF_5 \rightarrow [S_8][AsF_6]_2 + AsF_3$$

The bonding in S_8^{2+} is probably fairly subtle but one contributing resonance form is expected to be the following:

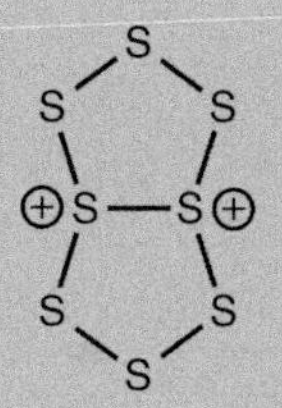

The process might very well involve single-electron transfers from S_8 and various radical intermediates. Here, however, we ask you to write out a polar mechanism for the reaction.

6.5 CYCLOOCTACHALCOGEN RING FORMATION

We'll now consider the reverse of what we discussed above, that is, how a cyclooctachalcogen ring is assembled, a very common process in chalcogen chemistry. Consider, as an example, the following reaction:

$$4\,Ph_3PSe + 4\,SeCl_2 \rightarrow Se_8 + 4\,Ph_3PCl_2 \tag{6.41}$$

An examination of the products indicates that a large number of Se–Se bonds must form, strongly suggesting that Se-on-Se nucleophilic attacks must be taking place. With that insight, the first couple of steps are not difficult to surmise:

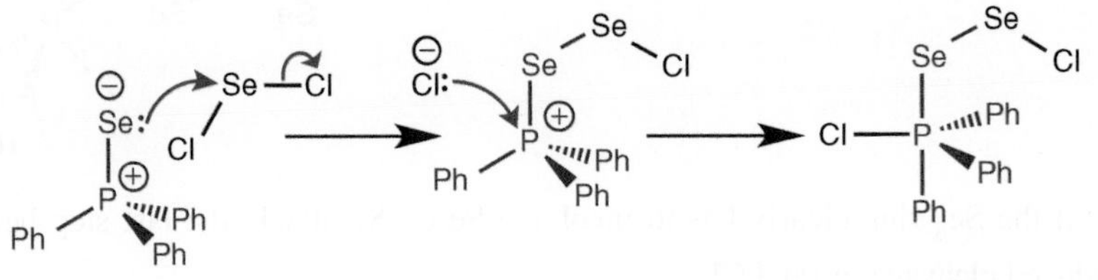

$$(6.42)$$

Observe that we have formed a Se–Se linkage. A second nucleophilic attack by Ph_3PSe on the product above now leads to a Se_3 chain, as shown below:

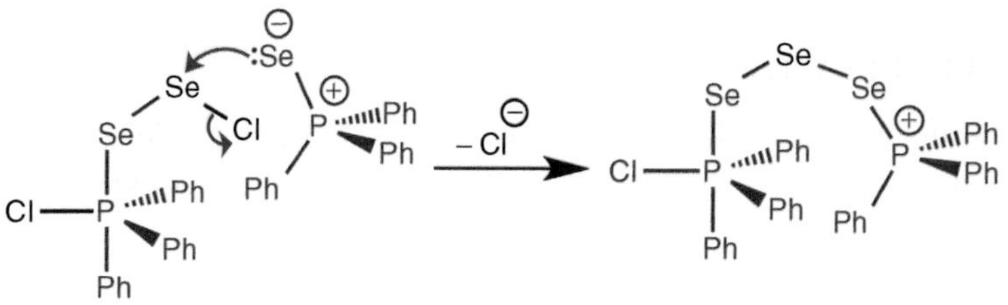

(6.43)

For the Se chain to grow, we need to somehow clip off one of the P's. Observing that the final product is Ph_3PCl_2, a chloride ion seems clearly indicated as the nucleophile needed for this purpose:

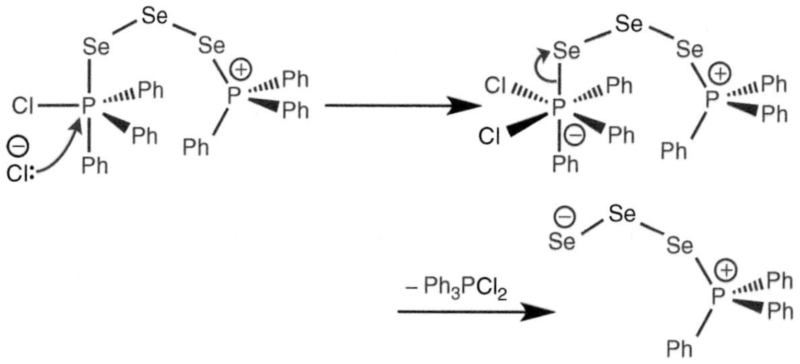

(6.44)

The Se_3 chain now has an anionic Se atom at one end, which can carry out a nucleophilic attack on $SeCl_2$. Proceeding in this manner, we can build up a Se_8 chain, as shown below:

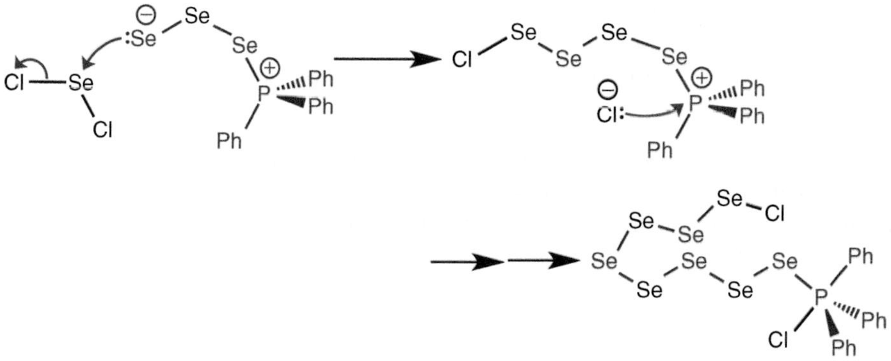

(6.45)

Closure of the Se_8 ring clearly has to involve a Se-on-Se attack, the last step being a chloride-induced cleavage of Ph_3PCl_2:

$$(6.46)$$

In our experience, the mechanisms of cyclooctachalcogen ring formation are perceived as a bit more difficult than nucleophilic breakdown of the same rings. They are certainly quite challenging exercises in arrow pushing. We hope you'll get more comfortable with practice!

6.6 HIGHER-VALENT STATES: OXIDES AND OXOACIDS

The two common oxides of sulfur, SO_2 and SO_3, are typical acidic oxides; they react with water to yield sulfurous acid (H_2SO_3) and sulfuric acid (H_2SO_4), respectively:

$$SO_2 + H_2O \rightarrow H_2SO_3 \qquad (6.47)$$
$$SO_3 + H_2O \rightarrow H_2SO_4 \qquad (6.48)$$

A by-product of fossil fuel combustion, SO_2 is a major pollutant, and the above two reactions, which take place in cloud droplets, play a key role in the processes leading to acid rain.

Sulfur trioxide is a strong Lewis acid and reacts with Lewis bases (indicated below as L) such as pyridine and triphenylphosphine to give stable adducts:

$$(6.49)$$

REVIEW PROBLEM 6.4

Sulfur trioxide reacts with anhydrous HF and HCl to yield the corresponding halosulfonic acids, XSO_3H:

$$HX + SO_3 \rightarrow XSO_3H$$

Suggest a mechanism.

REVIEW PROBLEM 6.5

Unlike SO_2, which is a gas under ordinary conditions, SeO_2 is a white polymeric solid:

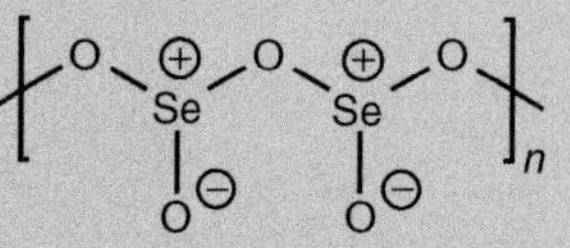

It sublimes, however, at 588 K to yield a gas consisting of discrete SeO_2 molecules. Solid SeO_2 reacts (i) with water to yield selenous acid, H_2SeO_3, an analog of sulfurous acid, and (ii) with aqueous sodium hydroxide to yield sodium selenite, Na_2SeO_3. Write balanced equations for the two reactions and use arrow pushing to account for the products H_2SeO_3 and Na_2SeO_3.

With an intermediate valence of 4 and a sulfur oxidation state of +4, SO_2 can act as both a reducing agent and an oxidizing agent. The reducing properties, however, are more commonly of interest. Thus, aqueous SO_2 (or H_2SO_3) reduces both bromine and iodine.

$$X_2 + HSO_3{}^- + H_2O \rightarrow 2\,HX + HSO_4{}^- \tag{6.50}$$

Although the reaction doesn't look too complicated, a plausible mechanism may not be readily apparent. The fact that sulfur is being oxidized (i.e., acts as an electron donor) while molecular halogen is being reduced suggests that a nucleophilic attack by the hydrogensulfite ($HSO_3{}^-$) sulfur on X_2 may be worth considering:

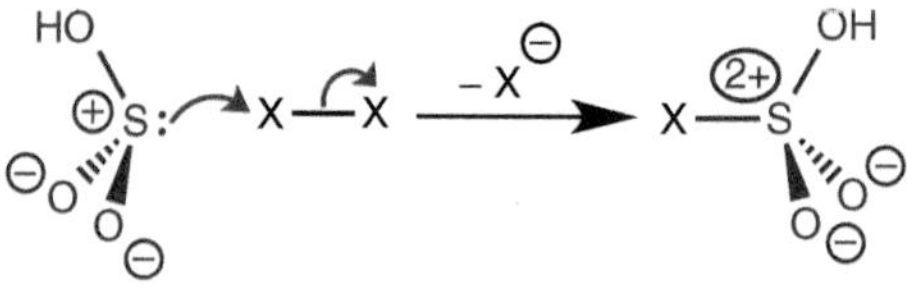

$$\tag{6.51}$$

The product thus obtained is then hydrolyzed to hydrogensulfate:

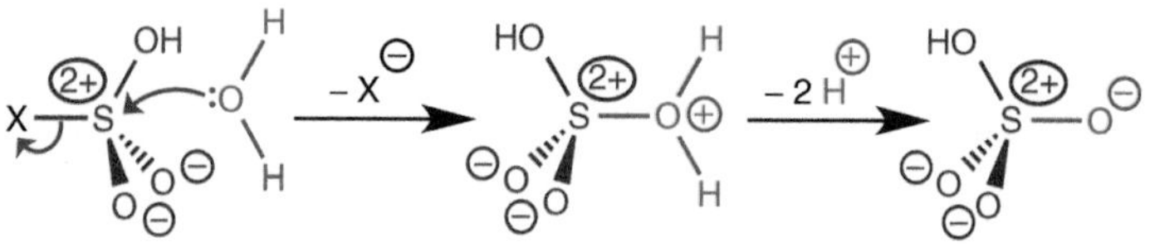

$$\tag{6.52}$$

REVIEW PROBLEM 6.6

Consistent with the nucleophilic character of the sulfur in HSO_3^-, one might expect that the S-protonated tautomer should also be reasonably stable. [17]Oxygen NMR studies strongly suggest the presence of both species in equilibrium in aqueous solutions:

Suggest a mechanism for the interconversion.

A variety of sulfur oxoacids and oxoanions are known that contain more than one sulfur per molecule or ion. Although a systematic account is outside the scope of this book, the thiosulfate anion, $S_2O_3^{2-}$, on account of its long-standing role as a "fixer" in traditional photography, is worth mentioning. The corresponding free acid, thiosulfuric acid, $H_2S_2O_3$, can be obtained by the reaction of chlorosulfonic acid, $ClSO_3H$, and H_2S at low temperature and under anhydrous conditions.

$$ClSO_3H + H_2S \rightarrow H_2S_2O_3 + HCl \qquad (6.53)$$

The structure of the product, $H_2S_2O_3$ (see below), provides the key clue: an S–S bond has been formed. The sulfur in H_2S then is a plausible nucleophile, and the sulfur in $ClSO_3H$, bearing a Cl, a potential leaving group, is a plausible electrophilic site:

$$(6.54)$$

Chloride then picks up a proton, producing HCl and $H_2S_2O_3$ as the final products:

$$(6.55)$$

The thiosulfate anion, a soft base, strongly coordinates Ag^+, a soft Lewis acid through the terminal sulfur atom. Aqueous sodium thiosulfate ("hypo"), $Na_2S_2O_3$, is therefore used to remove unchanged AgBr from exposed photographic film. The process is called *fixing*, as it makes the image permanent and light resistant. In the age of digital photography, this once-crucial process is increasingly of historical interest only.

The thiosulfate anion reacts rapidly with molecular iodine, bleaching the brown/purple color of the latter to produce colorless iodide and tetrathionate, $S_4O_6^{2-}$.

$$(6.56)$$

This reaction is a key part of iodometric titrations, which are used for quantitative determination of oxidants in aqueous samples, such as oxygen saturation in ecological samples or active chlorine in swimming pool water.

The structure of the product, tetrathionate, may not immediately suggest a mechanism. One way of approaching the problem is simply to look for potential nucleophiles and electrophiles. The anionic terminal sulfur of thiosulfate should be a good nucleophile and I_2 a reasonably good electrophile. Their interaction may be expected to be as follows:

$$(6.57)$$

The product of the above step is interesting in that the outer sulfur carries an iodine, a potentially excellent leaving group, which can be displaced by the terminal sulfur of a second thiosulfate:

$$(6.58)$$

And there we have our product!

In hindsight, the structure of the tetrathionate ion should perhaps have afforded a clue to the mechanism. If we ignore the charges, tetrathionate is simply a dimer of thiosulfate. We should perhaps have guessed that the terminal sulfur of one of the thiosulfates somehow had to become electrophilic so that it could join up with a second thiosulfate via a nucleophilic displacement. The take-home message is that, even when the structure of the product(s) doesn't seem to afford immediate mechanistic insights, simply identifying potential nucleophiles and electrophiles and having them interact will often put you on the right track.

6.7 SULFUR OXOCHLORIDES

Sulfur forms two well-known oxochlorides—thionyl chloride ($SOCl_2$) and sulfuryl chloride (SO_2Cl_2). Both are important reagents in organic chemistry and inorganic chemistry. For reasons of space, we will focus here largely on $SOCl_2$. As far as SO_2Cl_2 is concerned, we will note that it is prepared from SO_2 and Cl_2 in the presence of a catalyst such as activated charcoal:

$$SO_2 + Cl_2 \rightarrow SO_2Cl_2 \tag{6.59}$$

In Section 6.15, we will encounter SO_2Cl_2 again, in the role of an oxidant in the synthesis of the remarkable compound S_4N_4.

The main industrial synthesis of $SOCl_2$ is based on the reaction of SO_3 and SCl_2:

$$SO_3 + SCl_2 \rightarrow SOCl_2 + SO_2 \tag{6.60}$$

This is an example of an oxygen atom transfer (OAT) reaction. In other words, the reaction results in the net transfer of one oxygen from SO_3 to SCl_2. The question now is: is it a direct OAT, as shown in the mechanism below?

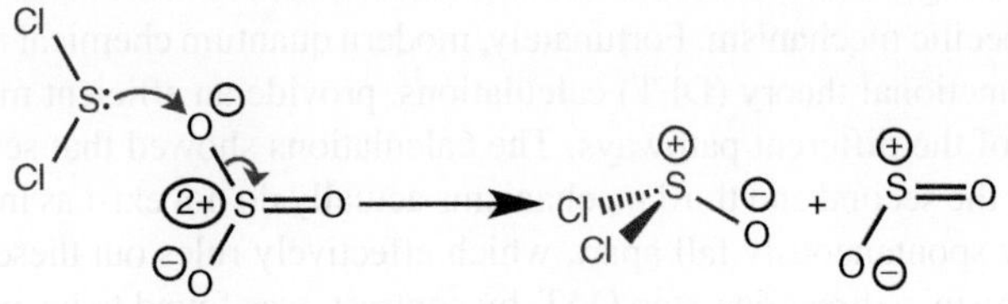

$$\tag{6.61}$$

An awkward feature of this mechanism is that the electrophile is an oxygen atom with a formal charge of −1! We therefore searched around for alternative mechanisms.

A simple way to avoid SCl_2 attack an anionic oxygen is to have it attack the doubly bonded oxygen, as shown below:

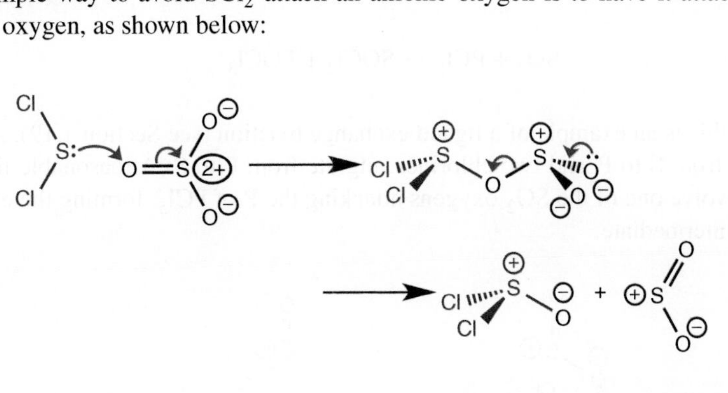

$$\tag{6.62}$$

The key difference between the above two mechanisms is that the second mechanism involves an oxo-bridged intermediate. The first mechanism, by contrast, involves just one step and therefore no intermediates.

There are still other mechanistic possibilities. Consider, for example, the following mechanism, which starts off with a sulfur-on-sulfur attack.

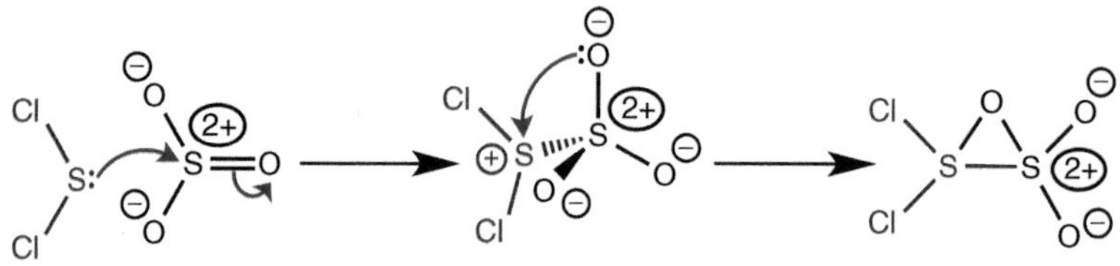

$$(6.63)$$

In this pathway, the three-membered ring formed above allows the oxygen to migrate from one sulfur to the other; the ring falls apart, as shown below, to produce the observed products:

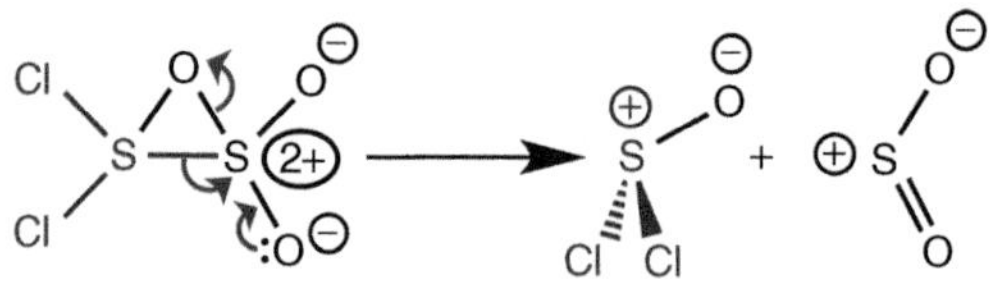

$$(6.64)$$

Which of the above mechanisms is the right one? That's a difficult question to answer, based on arrow pushing alone. The three mechanisms above are all more or less reasonable. Somewhat surprisingly for such a well-known reaction, the research literature did not help us pin down a specific mechanism. Fortunately, modern quantum chemical methods, particularly density functional theory (DFT) calculations, provide an efficient means of judging the plausibility of the different pathways. The calculations showed that several of the key intermediates in the second and third mechanisms actually do not exist as independent entities; that is, they spontaneously fall apart, which effectively rules out these two pathways. The first mechanism, a direct one-step OAT, by contrast, was found to be energetically reasonable; in other words, a transition state with a reasonable energy was found. We will encounter additional examples of OAT in the remainder of the book (see Sections 6.8, 7.2, and 7.7)

Another useful synthesis of thionyl chloride involves the interaction of SO_2 and PCl_5:

$$SO_2 + PCl_5 \rightarrow SOCl_2 + POCl_3 \qquad (6.65)$$

Observe that this is an example of a ligand exchange reaction (see Section 1.19). An oxygen migrates from S to P, and two chlorines migrate from P to S. A reasonable first step then might involve one of the SO_2 oxygens attacking the P of PCl_5, forming the expected oxo-bridged intermediate:

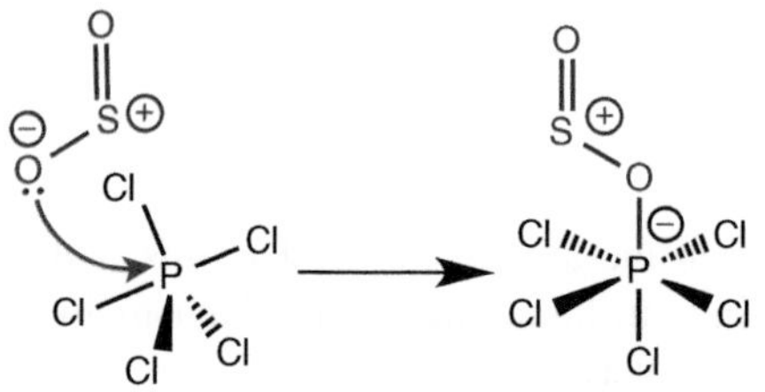

$$(6.66)$$

Two successive chloride ion migrations from P to S would then yield the observed products:

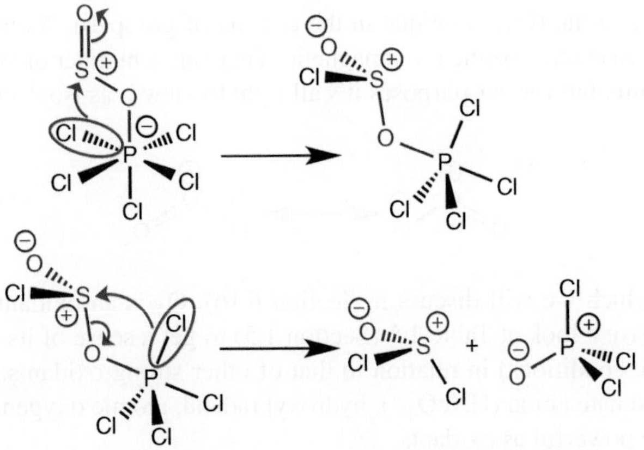

$$(6.67)$$

For compactness, we have depicted the above migrations as 1,3-shifts, but in reality they could also be happening stepwise as a series of D and A reactions, whereby chloride anions would be detaching from the phosphorus and reattaching to the sulfur.

REVIEW PROBLEM 6.7

Thionyl chloride reacts vigorously with water to yield H_2SO_3 and HCl. Suggest a mechanism.

REVIEW PROBLEM 6.8

Another ligand exchange reaction leading to the formation of $SOCl_2$ is the following:

$$SO_2 + COCl_2 \rightarrow SOCl_2 + CO_2$$

Suggest a mechanism.

REVIEW PROBLEM 6.9

SeO_2 and $SeCl_4$ undergo ligand exchange to yield $SeOCl_2$:

$$SeO_2 + SeCl_4 \rightarrow SeOCl_2$$

Suggest a mechanism. *Note*: SeO_2 exists as a discrete molecule in the gas phase, but as a solid it has a polymeric structure. For this exercise, however, you may assume a simple monomeric structure.

6.8   OZONE

The chemistry of ozone (O_3) is unique in the context of group 16. As mentioned, unlike O_2, which is paramagnetic, ozone is diamagnetic. There are a number of subtleties about its electronic structure, but for our purposes it's all right to view it as isoelectronic with SO_2:

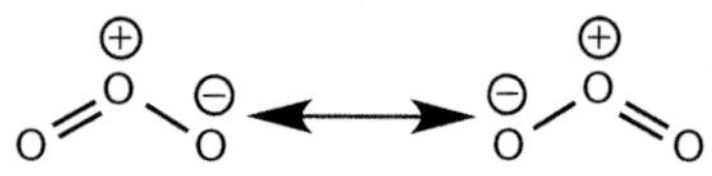

Like SeO_2 (which we will discuss in Section 6.16), O_3 is an oxidant, but a far more powerful one. Have a look at Table 1.5 (Section 1.5) to get a sense of its reduction potential (under acidic conditions) in relation to that of other strong oxidants. Only molecular fluorine, the perxenate anion ($HXeO_4^-$), hydroxyl radical, atomic oxygen, and a few other species are more powerful as oxidants.

In the laboratory, ozone is generally prepared in an ozonizer by silent or sparkless electric discharge through oxygen:

$$3\,O_2 \; \underset{\text{Sparkless electric discharge}}{\rightleftharpoons} \; 2\,O_3 \quad \Delta H = +285.4 \text{ kJ} \tag{6.68}$$

Observe that the reaction is endothermic; that is, ozone is enthalpically uphill relative to ordinary oxygen. A silent discharge produces less heat and thereby discourages the reverse step (Le Chatelier's principle).

Stratospheric ozone is produced from O_2 via a two-step process: photodissociation of O_2 to atomic oxygen by ultraviolet light ($\lambda < 240$ nm), followed by the reaction of atomic oxygen and O_2:

$$O_2 \rightarrow 2\,O \tag{6.69}$$

$$O_2 + O \rightarrow O_3 \tag{6.70}$$

Ozone itself also reacts with atomic oxygen to regenerate O_2:

$$O_3 + O \rightarrow 2\,O_2 \tag{6.71}$$

The reactions are catalyzed by metal ions in cloud droplets and by various free radicals; the details, unfortunately, are outside the scope of our discussion.

The highest concentration of atmospheric ozone (2–8 ppm), that is, the well-known ozone layer, is in the lower part of the stratosphere, 20–30 km above the earth's surface; the exact range varies significantly with the season. Even here, however, the O_3 concentration is only about 2–8 ppm, which is far lower than that of O_2. Despite the low concentrations, the ozone layer absorbs most of the solar UV radiation. Together, atmospheric O_2 and O_3 almost completely screen the lowest wavelength part of the UV range (<200 nm), which is the most harmful. The importance of the ozone layer to life and civilization therefore cannot be exaggerated.

Human-made organohalogens such as chlorofluorocarbons (CFCs; e.g., $CFCl_3$ and CF_2Cl_2) and bromofluorocarbons, which were widely used as refrigerants and propellants until recently, have resulted in significant depletion of the ozone layer. Although otherwise very stable, these compounds decompose photochemically in the atmosphere producing chlorine and bromine atoms, the key culprits in ozone depletion. Thus, a chlorine atom reacts with O_3 to produce hypochlorite radical and O_2:

$$Cl + O_3 \rightarrow ClO + O_2 \qquad (6.72)$$

The hypochlorite radical reacts with a second molecule of O_3 to produce two O_2 molecules and a second chlorine atom, which can then continue to destroy ozone:

$$ClO + O_3 \rightarrow Cl + 2\,O_2 \qquad (6.73)$$

Since the late 1980s, these organohalogens have been gradually phased out by international agreement and largely replaced by products such as R-410A, which is a near-azeotropic (i.e., boiling with little change in composition) mixture of difluoromethane and pentafluoroethane.

Given that ozone oxidizes many, if not most, common substances, space permits only a tiny sampler of its chemical reactions. Thus, it detoxifies cyanide, producing the far less dangerous cyanate anion.

$$CN^- + O_3 \rightarrow NCO^- + O_2 \qquad (6.74)$$

In a simple picture, cyanide may be expected to act as the nucleophile and ozone as the electrophilic oxygen atom donor:

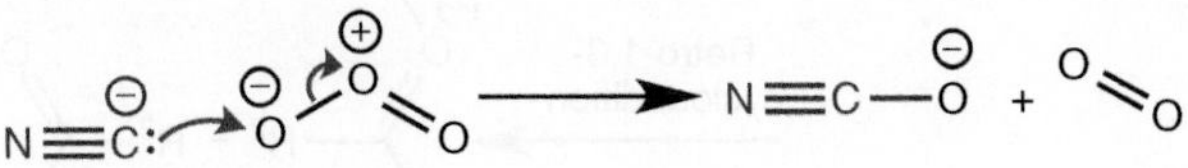

$$(6.75)$$

As in the case of reaction 6.61, don't be overly concerned that an oxygen with a negative formal charge acts as the electrophile. DFT calculations in the authors' laboratory provide support for such a pathway for a variety of OAT reactions involving p-block elements.

Triphenylphosphite forms an ozone adduct at low temperature, which decomposes on warming to triphenylphosphate and singlet oxygen:

$$(PhO)_3P + O_3 \xrightarrow{-78\,°C} (PhO)_3P(O_3)$$

$$\xrightarrow{-15\,°C} (PhO)_3PO + O=O \qquad (6.76)$$

Ozone reacts quantitatively with iodide, so the following reaction is useful for quantitative determination of ozone:

$$O_3 + 2\,KI + H_2O \longrightarrow I_2 + 2\,KOH + O_2 \qquad (6.77)$$

REVIEW PROBLEM 6.10

Suggest polar mechanisms for the above two reactions, that is, equations 6.76 and 6.77.

As far as organic compounds are concerned, the most important reaction of ozone is ozonolysis, where a carbon–carbon double bond is cleaved into two carbonyl fragments, as shown below:

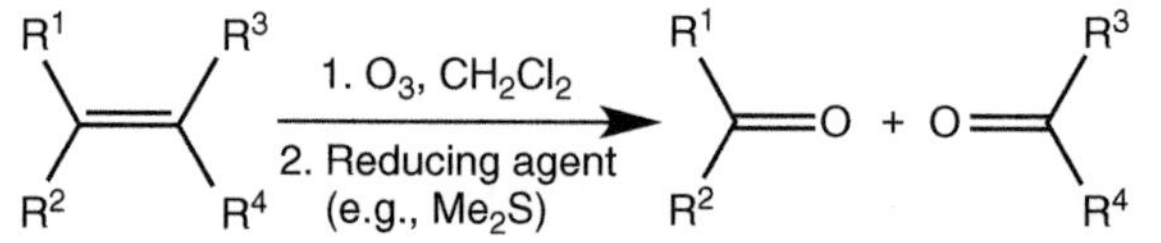

$$
\begin{array}{c}
\text{R}^1 \quad\quad \text{R}^3 \\
\xrightarrow[\substack{\text{2. Reducing agent} \\ \text{(e.g., Me}_2\text{S)}}]{\text{1. O}_3,\ \text{CH}_2\text{Cl}_2} \\
\text{R}^2 \quad\quad \text{R}^4
\end{array}
\qquad
\begin{array}{c}
\text{R}^1 \quad\quad\quad \text{R}^3 \\
\text{O} + \text{O} \\
\text{R}^2 \quad\quad\quad \text{R}^4
\end{array}
\qquad (6.78)
$$

The mechanism, which was worked out by Criegee in 1953, involves somewhat fancy steps. Thus, the first step is a 1,3-dipolar addition of ozone (the 1,3-dipole) to the carbon–carbon double bond (the 1,3-dipolarophile). The product so formed, called a *molozonide*, then undergoes a retro-1,3-cycloaddition to yield a *carbonyl oxide* and a carbonyl compound:

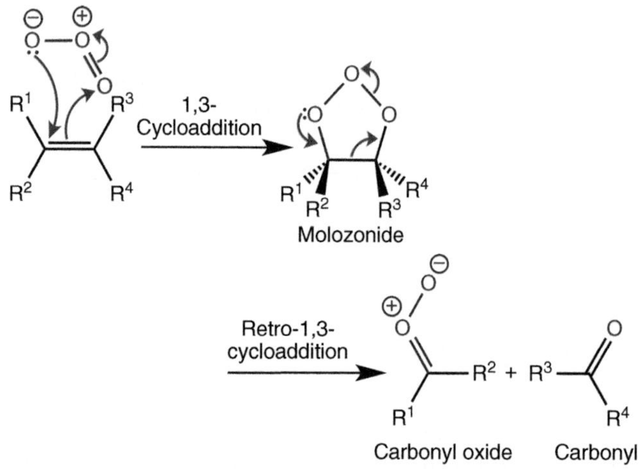

Molozonide

(6.79)

Retro-1,3-cycloaddition

Carbonyl oxide   Carbonyl

The carbonyl compound then flips its position and undergoes a second 1,3-dipolar addition, with the carbonyl oxide as the 1,3-dipole, to form an "ozonide":

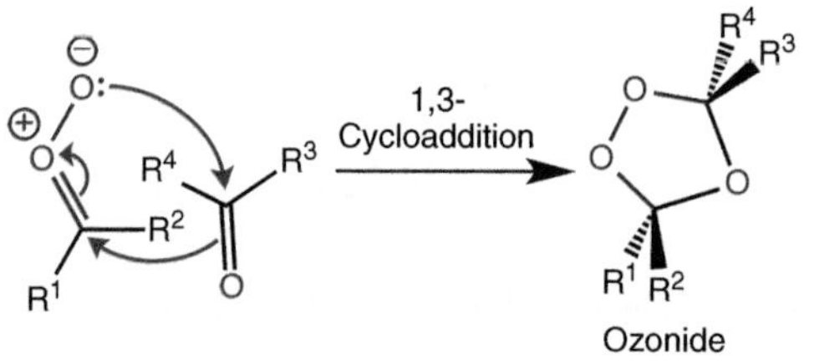

1,3-Cycloaddition

Ozonide

(6.80)

Typical work-up involves treatment of the ozonide with a reducing agent such as zinc and acetic acid or DMS:

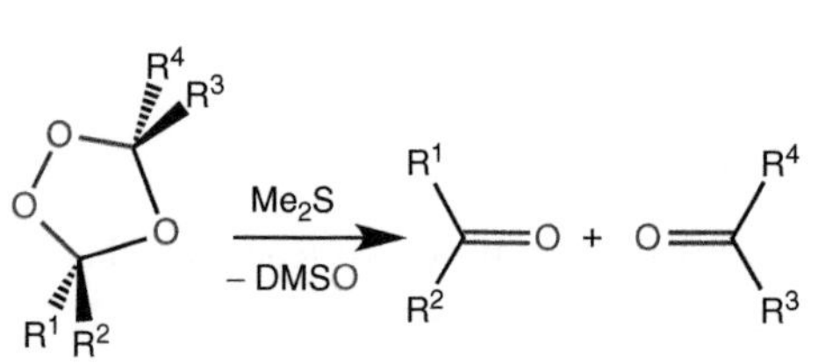

$$\xrightarrow[-\ \text{DMSO}]{\text{Me}_2\text{S}}$$

(6.81)

According to a recent hypothesis by researchers at the Scripps Research Institute, California, antibodies generate ozone by oxidizing water with singlet oxygen, with trioxidane (H_2O_3) as an intermediate:

$$H_2O + O=O \rightarrow H_2O_3 \qquad (6.82)$$

Oxidation of H_2O_3 to ozone would then involve a second molecule of singlet oxygen, a possible mechanistic picture being as follows:

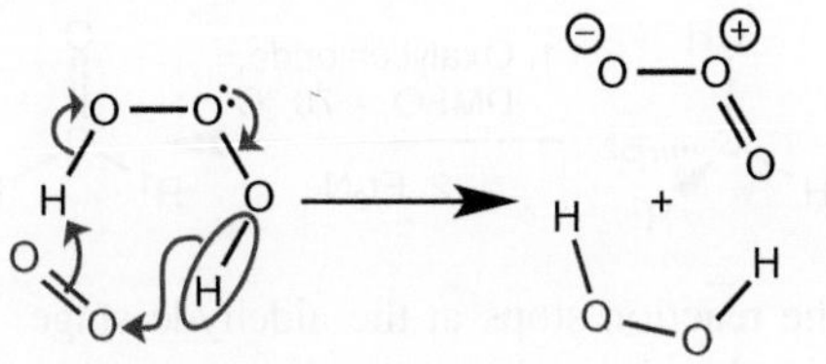

$$(6.83)$$

The researchers also showed that atherosclerotic plaque contains carbonyl compounds, consistent with the ozonolysis of cholesterol (see, e.g., Wentworth, P., *et al. Science* **2003**, *302*, 1053–1056). The carbonyl compounds were derivatized to the arylhydrazones, as shown below, followed by mass spectroscopic characterization:

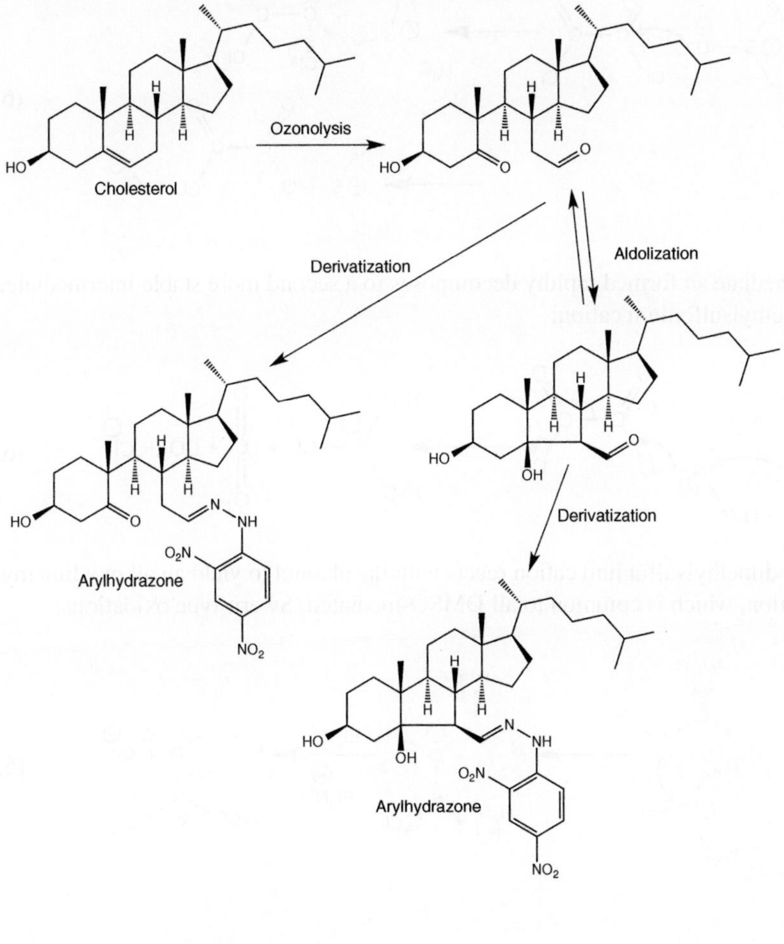

$$(6.84)$$

6.9   SWERN AND RELATED OXIDATIONS

This is a good point to introduce some of the organic applications of higher-valent sulfur compounds. Dimethyl sulfoxide (DMSO) is well known in this regard: it is not only an important polar, aprotic solvent but is also used to oxidize primary and secondary alcohols to the corresponding carbonyl compounds in a number of synthetic reactions, of which the Swern oxidation is the most important. The reagents used in the Swern oxidation are DMSO, oxalyl chloride, and an organic base such as triethylamine:

$$(6.85)$$

For primary alcohols, the reaction stops at the aldehyde stage, with no overoxidation to the carboxylic acid. A broad variety of functional groups are compatible with the reaction conditions.

The oxalyl chloride serves to activate DMSO in the following way:

$$(6.86)$$

The intermediate so formed rapidly decomposes to a second more stable intermediate, the chlorodimethylsulfonium cation:

$$(6.87)$$

The chlorodimethylsulfonium cation reacts with the alcohol to yield an alkoxydimethylsulfonium cation, which is common to all DMSO-mediated, Swern-type oxidations:

$$(6.88)$$

The organic base could then abstract a proton from the alcohol moiety and oxidize it to a carbonyl compound, as follows:

$$\text{(6.89)}$$

This, however, is not what happens. The pK_a of this proton is simply too high (well above 30, depending on R^1 and R^2) for abstraction by triethylamine. Instead, the amine abstracts a proton from one of the methyl groups of the dimethylsulfonium part of the molecule, where the pK_a is about 16–17, to create a sulfur ylide. This is reasonable; a sulfonium cation should stabilize an adjacent negative charge quite effectively:

$$\text{(6.90)}$$

The ylide then rapidly undergoes a 2,3-sigmatropic reaction to yield the carbonyl product and DMS (Me_2S).

$$\text{(6.91)}$$

Of the various by-products formed in this reaction, CO and Me_2S are toxic; as mentioned, Me_2S is also foul-smelling, except in very low concentrations. It's essential, therefore that this reaction is conducted in a well-ventilated fume hood.

There are several variations of the Swern protocol, where the oxalyl chloride is replaced by another activator. Acetic anhydride or trifluoroacetic anhydride is sometimes used. Carbodiimides (R–N=C=N–R) are yet another alternative—the reaction in that case is called a *Pfitzner–Moffatt oxidation*.

REVIEW PROBLEM 6.11

In the Kornblum oxidation, a primary alkyl halide is oxidized to the aldehyde with DMSO and NaHCO$_3$, typically with microwave heating:

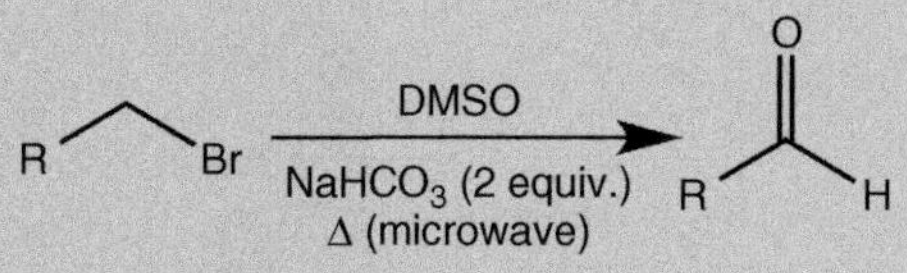

Suggest a mechanism by analogy with the Swern oxidation.

REVIEW PROBLEM 6.12*

A rather remarkable reaction of alkyl sulfoxides is the Pummerer rearrangement, where treatment with acetic anhydride (or another activator) results in rearrangement to an α-acetoxythioether:

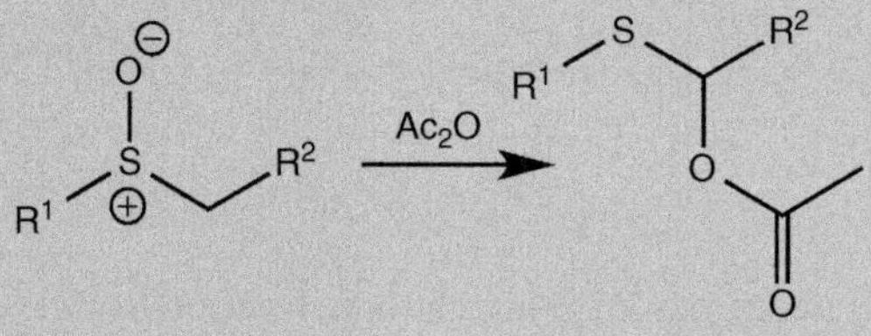

Suggest a mechanism. *Hint*: A pericyclic reaction is involved after the initial acetylation.

6.10 SULFUR YLIDES AND SULFUR-STABILIZED CARBANIONS

Like a phosphonium ion, a sulfonium ion can stabilize an adjacent carbanion center.

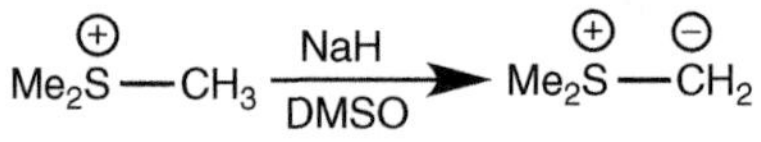

$$\overset{\oplus}{Me_2S}—CH_3 \xrightarrow[\text{DMSO}]{\text{NaH}} \overset{\oplus}{Me_2S}—\overset{\ominus}{CH_2} \tag{6.92}$$

Thus, a pK_a of 16.3 has been estimated for the Ph(Me)(PhCH$_2$)S$^+$ cation for the hydrogens shown in bold, which is about the same as that for the OH group of an alcohol. The anionic carbon in dimethylsulfonium methylide, as the above product is called, reacts with electrophiles such as carbonyl groups and imines via the Johnson–Corey–Chaykovsky (JCC) reaction:

$$(6.93)$$

$$\boxed{Z = O, NR}$$

The reaction is most often used for epoxide synthesis via methylene transfer. An important point concerns the difference in reactivity of sulfonium versus phosphonium ylides. The former gives three-membered rings; the latter gives alkenes via the Wittig reaction. Thermodynamics is believed to account for a good deal of this difference: the P^+–O^- bond in a phosphine oxide (BDE ~544 kJ/mol) is much stronger than the S^+–O^- bond in DMSO (BDE for DMSO → DMS + O: 389 kJ/mol), which would form if the sulfonium ylide reaction resulted in an alkene.

In terms of arrow pushing, the JCC reaction is fairly straightforward. For a carbonyl substrate, the ylidic carbon attacks the carbonyl carbon, the C=O linkage opens up in the usual manner, and the O^- swings back on the now-neutral ylidic carbon to form a three-membered ring:

$$(6.94)$$

The chemistry discussed above is largely that of tetravalent sulfur. Thus, the valence of sulfur is 4 in both sulfonium cations and sulfonium ylides, as well as in DMSO (DMS clearly has divalent sulfur). Another popular sulfur ylide is dimethylsulfoxonium methylide, which is prepared as follows:

$$(6.95)$$

Here, both the starting material and the ylide contain hexavalent sulfur. The pK_a of the starting material, the trimethylsulfoxonium cation (Me_3S^+O), is about 18.2 and is expected to be lower than that of the trimethylsulfonium cation (Me_3S^+). Dimethylsulfoxonium methylide is thus more stable as well as a softer nucleophile than dimethylsulfonium methylide. Qualitatively, however, the two ylides react in a similar way. The most prominent difference between the two ylides is in their reaction with α,β-unsaturated carbonyl compounds: in one case the product is an epoxide, in the other a cyclopropane:

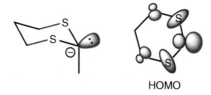

(6.96)

Divalent sulfur can also stabilize adjacent carbanion centers. The best known examples of such carbanions are lithiated 1,3-dithianes, which may be synthesized from aldehydes as shown below:

1,3-Dithiane

(6.97)

The orbital interactions underlying this type of sulfur-mediated carbanion stabilization are somewhat subtle and are best described as a form of hyperconjugation. An examination of the highest occupied molecular orbital (HOMO) shows that the σ-antibonding orbitals for the C–S bonds parallel to the carbanion lone pair (shown in red below) are able to delocalize the negative charge:

HOMO

An equivalent picture in terms of resonance forms is as follows:

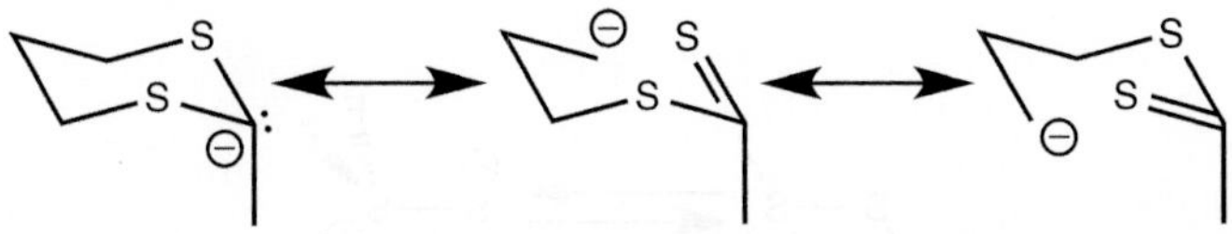

The above two-step reaction sequence is of unusual interest from a synthetic perspective. Note that we have converted a typically electrophilic aldehyde (carbonyl) carbon into a nucleophilic carbanion center in the 1,3-dithiane anion. The overall process thus accomplishes a polarity inversion or, to use the more popular German term, *umpolung*. The 1,3-dithiane anion may now be reacted with a variety of electrophiles, such as alkyl halides, epoxides, and carbonyl compounds, in what is known as the *Corey–Seebach reaction*:

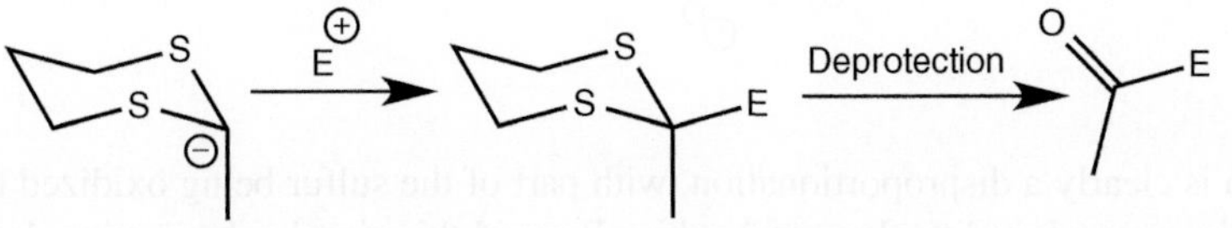

(6.98)

REVIEW PROBLEM 6.14

Aldehydes (like carbonyl compounds in general) react with ethane-1,2-dithiol to yield a 1,3-dithiolane, as shown below:

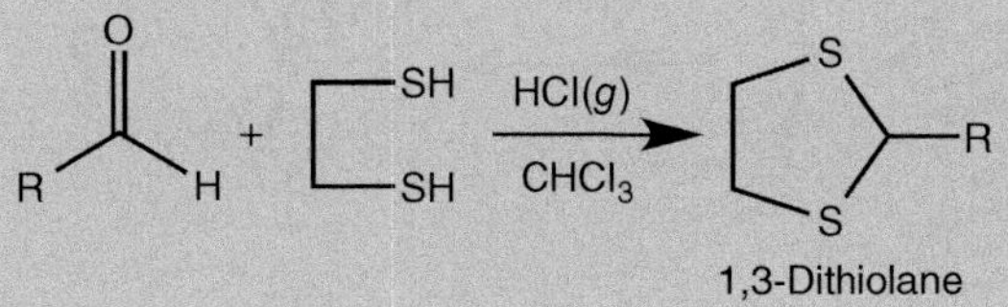

1,3-Dithiolane

1,3-Dithiolanes, however, are rarely used in the Corey–Seebach reaction because anions derived from them undergo fragmentation via a pericyclic pathway, producing ethylene as one of the products. Use arrow pushing to explain this process.

6.11* HYDROLYSIS OF S$_2$F$_2$: A MECHANISTIC PUZZLE

Sulfur forms two stable fluorides—SF$_4$ and SF$_6$. Of the two, SF$_6$ is exceedingly stable, to the point of inertness. By contrast, SF$_4$ is reactive and is used as a fluorinating agent in organic chemistry. In this section, we will focus on a less known sulfur fluoride, S$_2$F$_2$, not because of its inherent importance but because of the complexity of its alkaline hydrolysis, which provides a nice test of our arrow-pushing skills.

Disulfur difluoride, S_2F_2, exists as mixture of two isomers, as shown below. The equilibrium is catalyzed by metal fluorides:

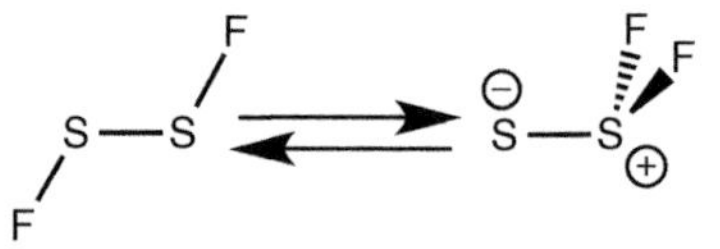

(6.99)

Disulfur difluoride is extremely reactive and is rapidly hydrolyzed by water and alkali:

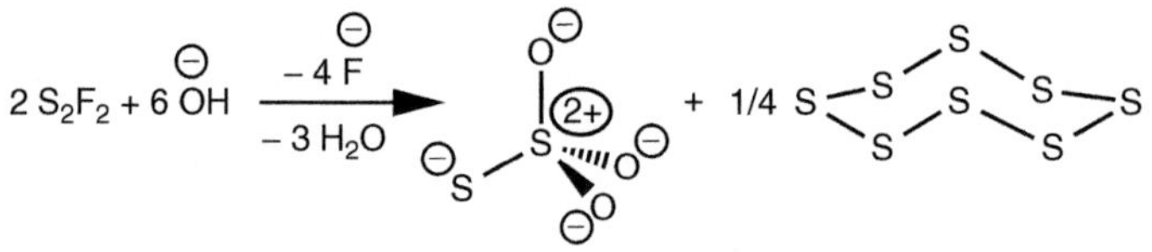

(6.100)

The reaction is clearly a disproportionation, with part of the sulfur being oxidized to thiosulfate and the rest reduced to elemental sulfur. Beyond this simple observation, however, the mechanism might appear utterly impenetrable. To push arrows, where would one even begin?

An interesting clue is provided by the structural similarity of one of the S_2F_2 isomers and thiosulfate, $S_2O_3{}^{2-}$:

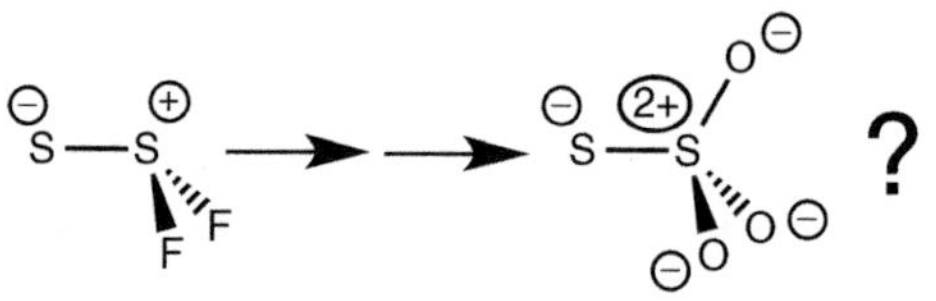

(6.101)

Hydroxide attack on S_2F_2 leads to *thiosulfite*, $S_2O_2{}^{2-}$, as shown below, which is certainly a step closer to thiosulfate:

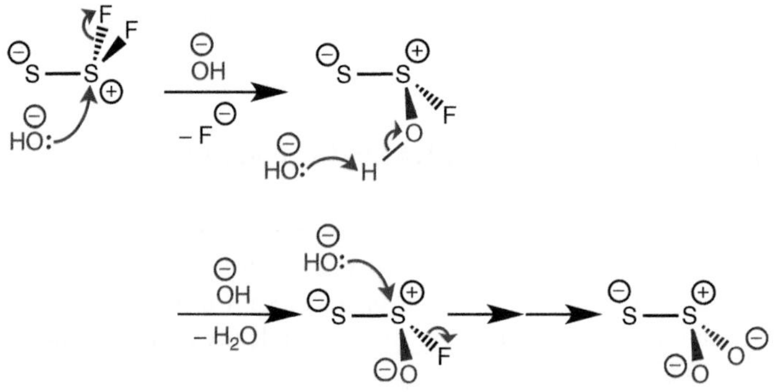

(6.102)

To become *thiosulfate*, the central sulfur in the thiosulfite needs to get oxidized. Electrophilic fluorination, as depicted below, might just provide a way for that to happen:

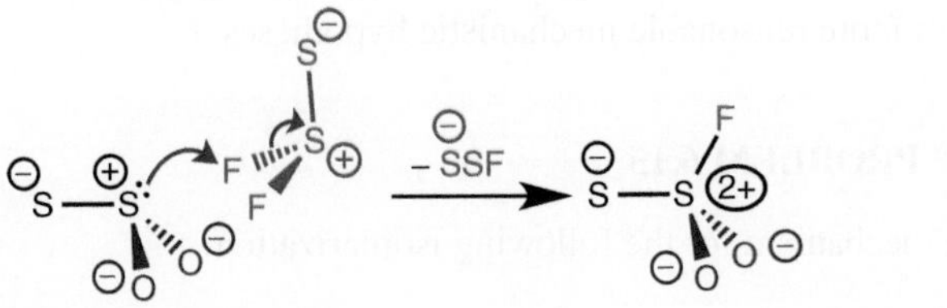

$$(6.103)$$

The product so obtained may now be hydrolyzed to thiosulfate in a straightforward manner:

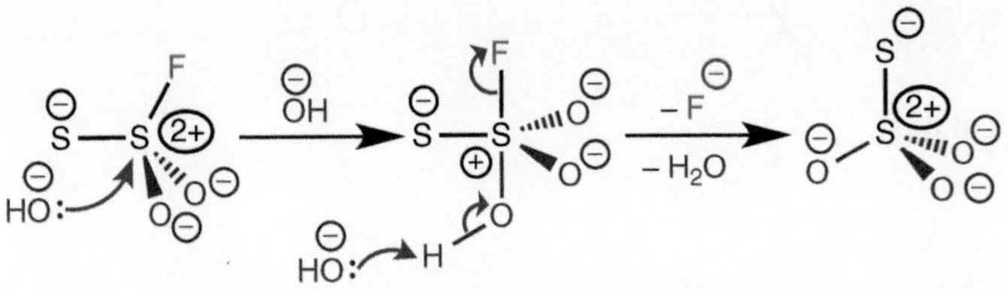

$$(6.104)$$

Now we need to account for the formation of elemental sulfur. The FSS$^-$ anion, the leaving group in the electrophilic fluorination step above, reaction 6.103, appears to be a good candidate for reaction with the linear isomer of S_2F_2 to produce an extended sulfur chain that could eventually lead to S_8, as shown below:

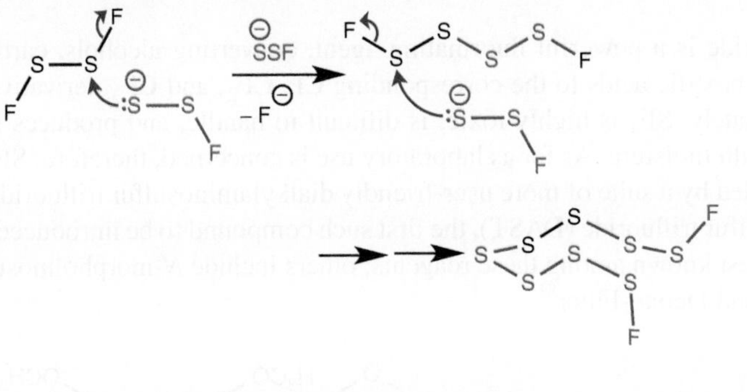

$$(6.105)$$

Like S_2F_2, the S_8F_2 molecule produced above might act as a source of electrophilic fluorine for the thiosulfite anion, while producing elemental S_8:

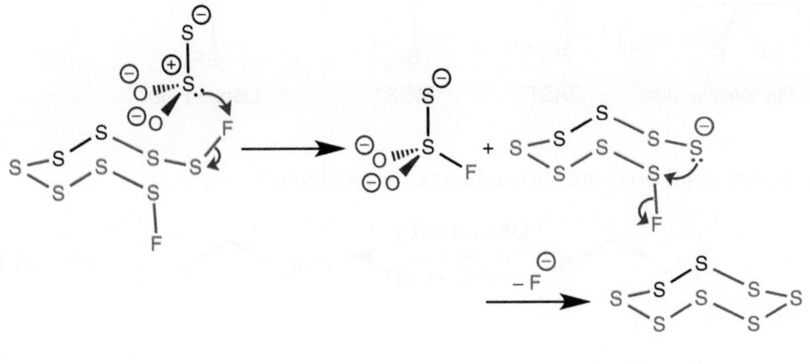

$$(6.106)$$

With that, we have accounted for all the observed products—that was clearly quite a lot of arrow pushing. It's important to remember that the mechanism is speculative, but that is all right. As already emphasized, our goal here is, given a fairly complicated reaction, to come up with one or more reasonable mechanistic hypotheses.

REVIEW PROBLEM 6.15

Suggest a mechanism for the following isomerization:

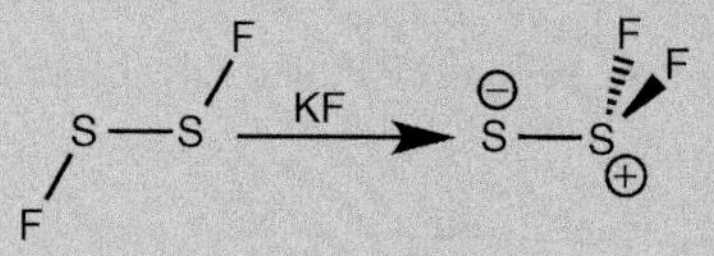

Note: KF is a catalyst.

6.12   HIGHER-VALENT SULFUR FLUORIDES

Sulfur tetrafluoride (SF_4), a corrosive, moisture-sensitive gas, is commercially available. Industrially, it is most conveniently prepared via the following reaction:

$$SCl_2 + Cl_2 + 4\,NaF \rightarrow SF_4 + 4\,NaCl \qquad (6.107)$$

Sulfur tetrafluoride is a powerful fluorinating agent, converting alcohols, carbonyl compounds, and carboxylic acids to the corresponding CF, CF_2, and CF_3 derivatives, respectively. Unfortunately, SF_4 is highly toxic, is difficult to handle, and produces HF (which attacks glass) with moisture. As far as laboratory use is concerned, therefore, SF_4 has been largely superseded by a suite of more user-friendly dialkylaminosulfur trifluoride reagents. Diethylaminosulfur trifluoride (DAST), the first such compound to be introduced (in 1970), is perhaps the best known among these reagents; others include *N*-morpholinosulfur trifluoride (MOST) and Deoxo-Fluor®:

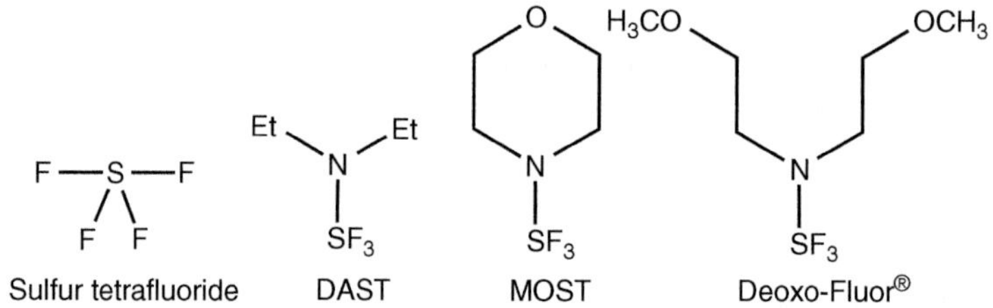

Typical reaction conditions are shown below for an alcohol:

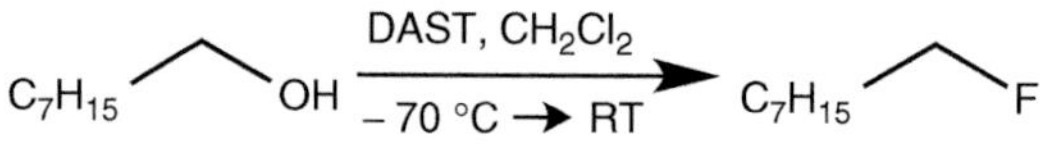

$$(6.108)$$

The mechanism of fluorination effected by SF_4 or any of the dialkylaminosulfur trifluorides is illustrated below for a primary alcohol and DAST. The first step involves coordination of the alcohol to the tetravalent sulfur, followed by the departure of a fluoride leaving group:

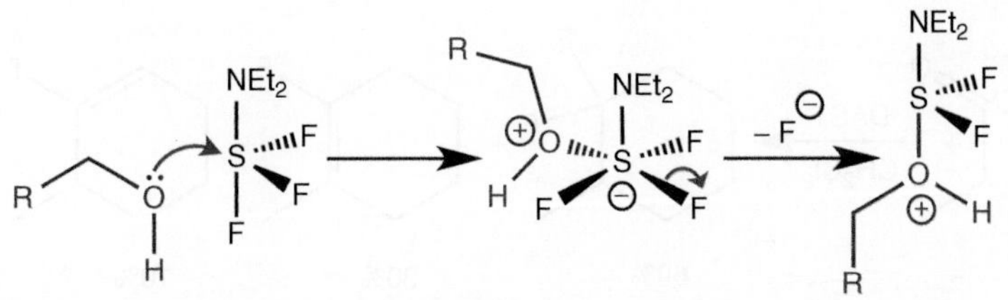

(6.109)

The outgoing fluoride then comes back as a nucleophile, forming the alkyl fluoride product and kicking out an oxygen leaving group, as shown below:

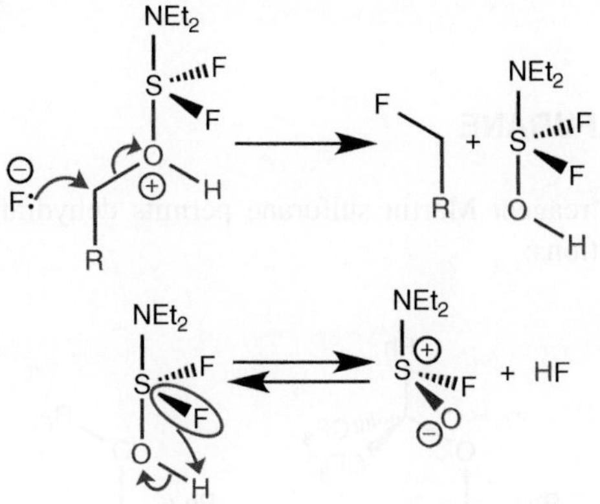

(6.110)

Observe that the sulfur remains tetravalent at the end of the reaction; thus, no redox has taken place.

The "deoxofluorination" step, where oxygen is replaced by fluorine, is shown under reaction 6.110 as an S_N2 displacement. For many alcohols, therefore, deoxofluorination by SF_4 or related reagents involves an inversion of configuration. For other substrates, however, an S_N1 pathway appears to operate, based on product stereochemistry as well as rearranged products.

REVIEW PROBLEM 6.16*

The various aminosulfur trifluorides are prepared via the reaction of SF_4 with a dialkylamino(trialkyl)silane at $\sim-70°$, as shown below for DAST:

$$Et_2NSiMe_3 + SF_4 \rightarrow Et_2NSF_3 + Me_3SiF$$

Suggest a mechanism and comment on the thermodynamic driving forces.

REVIEW PROBLEM 6.17*

Consider the following DAST deoxofluorination of an enolizable ketone:

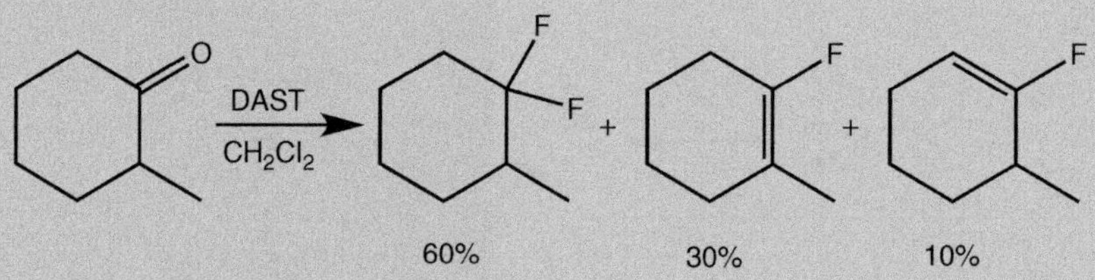

60%   30%   10%

Besides the desired difluoroalkane, two vinyl fluoride by-products are also obtained. Rationalize the various products in terms of a mechanistic picture.

6.13 MARTIN SULFURANE

The tetravalent sulfur reagent Martin sulfurane permits dehydration of alcohols under remarkably mild conditions:

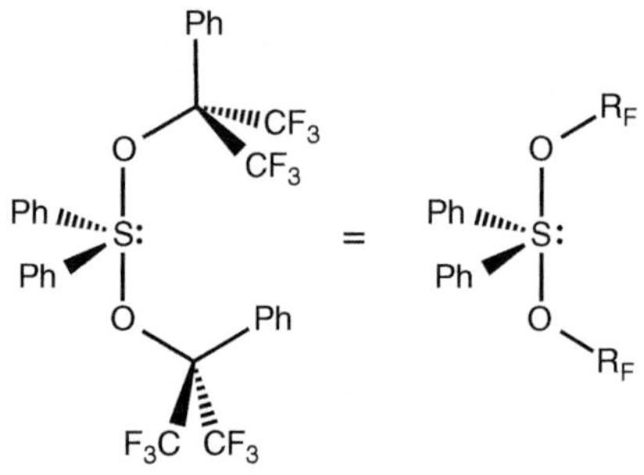

Martin sulfurane

The reagent is expensive, given the fluorinated alkoxy (R_FO) ligands, but is ideally suited for valuable and/or sensitive substrates. It's prepared from diphenyl sulfide and potassium 1,1,1,3,3,3-hexafluoro-2-phenyl-2-propanolate as follows:

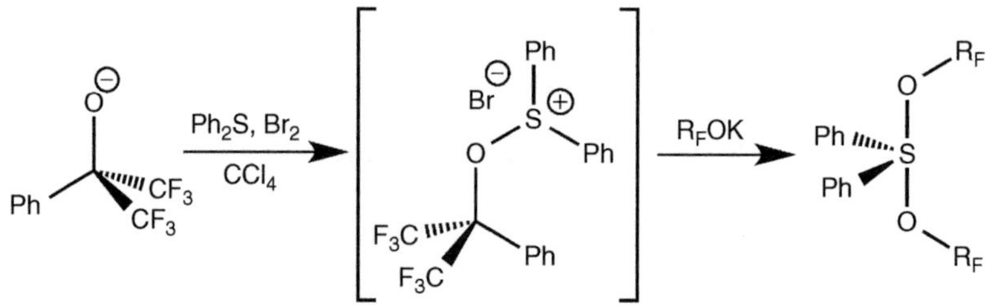

(6.111)

REVIEW PROBLEM 6.18

Write out mechanisms for the reactions involved in the synthesis of Martin sulfurane, reaction 6.111.

The mechanism of action of Martin sulfurane is shown below for the dehydration of cyclohexanol:

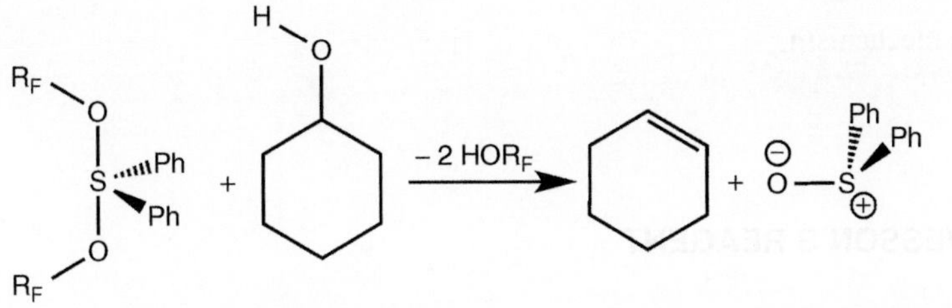

(6.112)

As with SF_4 and related reagents, the alcohol coordinates to the tetravalent sulfur center, followed by stepwise departure of two R_FOH molecules. In the mechanism proposed below, departure of the second R_FOH also results in the alkene product:

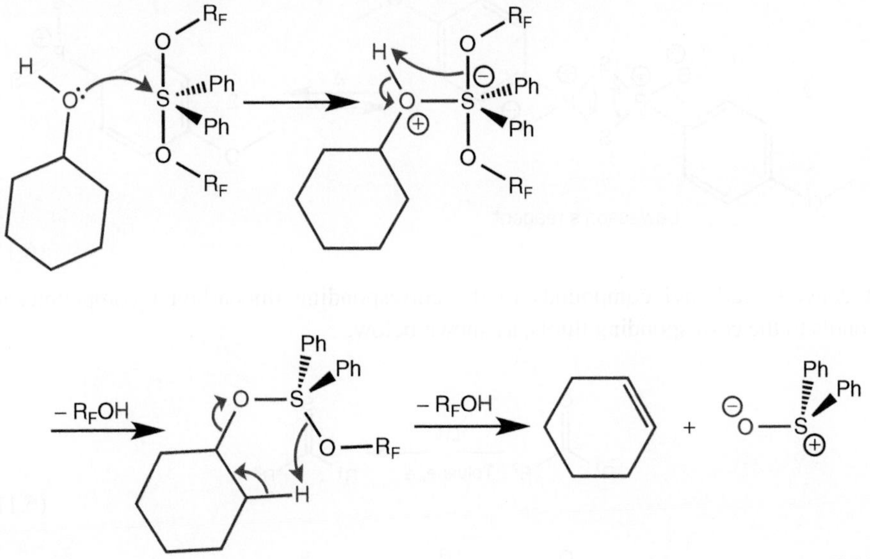

(6.113)

REVIEW PROBLEM 6.19

Martin sulfurane can dehydrate vicinal diols to epoxides, as shown below:

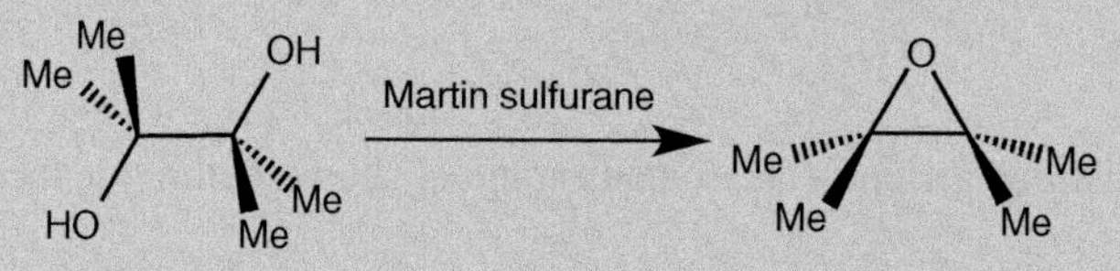

Suggest a mechanism.

6.14 LAWESSON'S REAGENT

Lawesson's reagent (LR) is a phosphorus- and sulfur-containing organic molecule that is used as a thiation agent in organic chemistry. The core of the molecule consists of a four-membered ring of alternating phosphorus and sulfur atoms. The ring opens up on warming, yielding two reactive $ArPS_2$ molecules that act as the actual thiation agents:

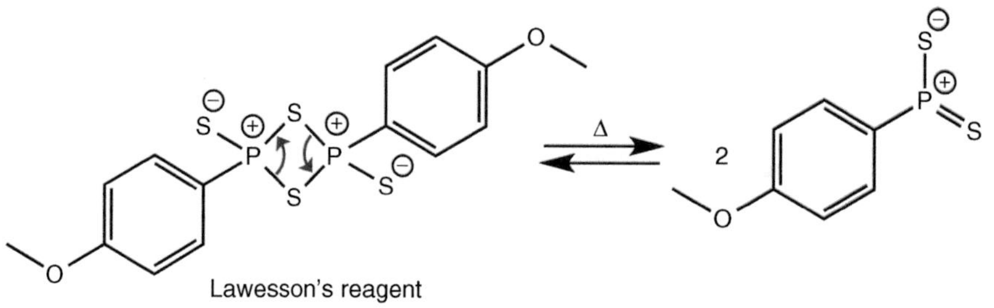

Lawesson's reagent

$$(6.114)$$

LR converts carbonyl compounds to the corresponding thiocarbonyl compounds and alcohols to the corresponding thiols, as shown below:

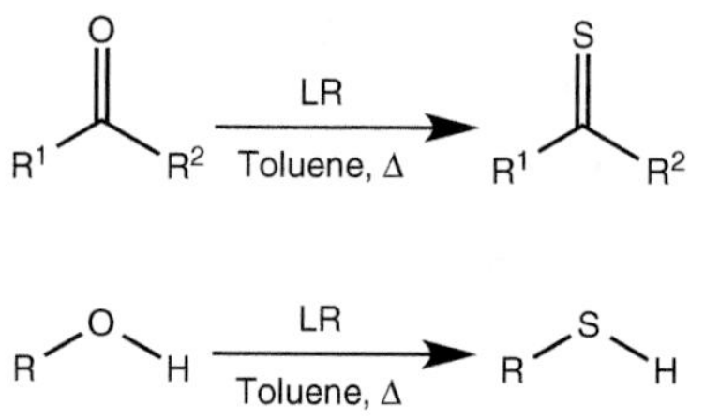

$$(6.115)$$

For carbonyl thiation, the mechanism involves nucleophilic attack by one of the sulfurs of the $ArPS_2$ intermediate on the carbonyl carbon. The "alkoxide" oxygen might then be expected to form a bond with the cationic phosphorus:

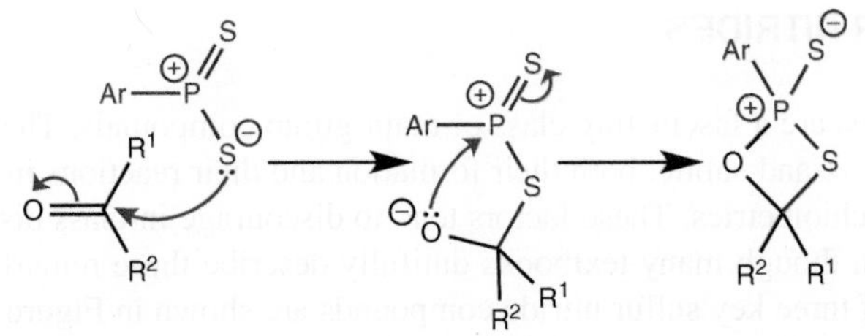

(6.116)

The four-membered ring then falls apart to produce the desired thiocarbonyl compound:

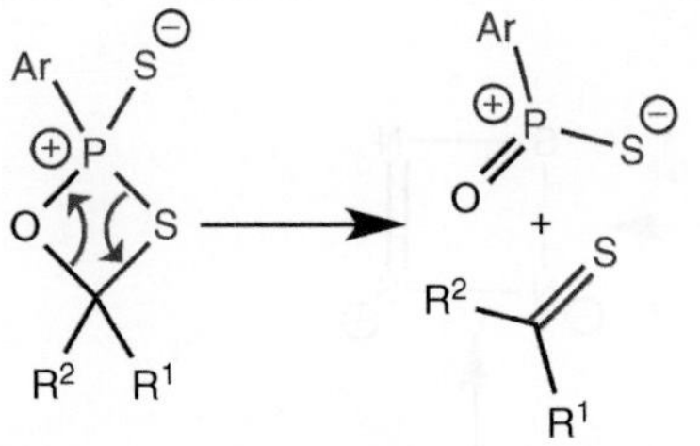

(6.117)

Clearly, the chemistry of LR is a fascinating blend of phosphorus and sulfur chemistry. Its success hinges as much on the nucleophilicity of sulfur as on the oxophilicity of pentavalent phosphorus.

REVIEW PROBLEM 6.20

LR is commercially available but can also be readily synthesized in the laboratory by heating anisole with phosphorus pentasulfide (P_4S_{10}), until the mixture is clear and no more H_2S evolves, followed by recrystallization from toluene or xylene. Suggest a mechanism for the reaction.

Note: LR is foul-smelling, so it must be handled in a well-ventilated fume hood. Glassware that has come in contact with the reagent may be decontaminated with chlorine bleach (NaOCl).

REVIEW PROBLEM 6.21

LR converts "maltol" to "dithiomaltol," as shown below:

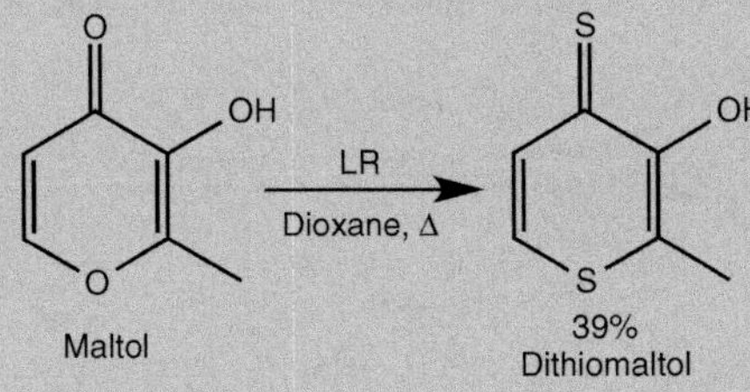

Suggest a mechanism and a rationale for the selective double thiation.

6.15   SULFUR NITRIDES

The sulfur nitrides are a fascinating class of main-group compounds. Their structures and bonding are diverse and subtle; both their formation and their reactions involve stunningly complicated stoichiometries. These factors tend to discourage in-class discussion of these compounds, even though many textbooks dutifully describe these remarkable molecules. The structures of three key sulfur nitride compounds are shown in Figure 6.1. Polythiazyl was the first inorganic conducting polymer to be synthesized. Normally a gold-colored metallic conductor, it becomes superconducting at very low temperatures, below 0.26 K. For reasons of space, our discussion here will focus solely on S_4N_4.

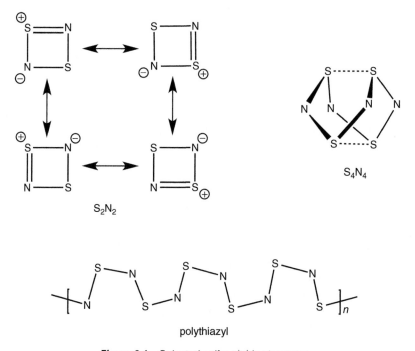

Figure 6.1   *Selected sulfur nitride structures.*

A modern synthesis of S_4N_4 is the following (Maaninen, A.; Shvari, J.; Laitinen, R. S.; Chivers, T. *Inorg. Syn.* **2002**, *33*, 196–199), where we have ignored the transannular interactions shown above (i.e., the dotted lines) and represented S_4N_4 simply as an eight-membered ring:

$$\text{2 Me}_3\text{Si}-\text{N}-\text{S}-\text{N}-\text{SiMe}_3 + 2\,SCl_2 + 2\,SO_2Cl_2$$

$$\longrightarrow 8\,Me_3SiCl + 2\,SO_2 + \quad\quad\quad \tag{6.118}$$

An examination of the starting materials and the products strongly suggests that $[(Me_3Si)_2N]_2S$ provides a key part of the structural framework for S_4N_4, whereas SCl_2 provides the rest of the structural sulfur. Armed with that insight, it's reasonable to deploy the amine nitrogens in nucleophilic attacks on SCl_2:

$$(6.119)$$

The chloride anion produced then carries out a nucleophilic displacement on one of the silyl groups, producing trimethylsilyl chloride in the process:

$$(6.120)$$

An intramolecular nucleophilic displacement, followed by chloride-mediated desilylation, leads to the formation of a bis(trimethylsilyl)-S_2N_2 ring:

$$(6.121)$$

The nitrogens in these rings are nucleophilic, and further N-on-S attacks leads to the creation of a tetrakis(trimethylsilyl)-S_4N_4 ring, which seems clearly indicated as a precursor of the final product, S_4N_4:

$$(6.122)$$

It's worth noting that the above S_4N_4 ring could have arisen in a number of other ways, which do not proceed via an S_2N_2 ring. For example, we could have had acyclic intermediates all through, until the formation of the above S_4N_4 ring. Space doesn't permit a discussion of these many alternative paths, but you are welcome to explore a bit on your own.

The silylated S_4N_4 ring now needs to be oxidized (the trimethylsilyl group may be viewed as a hydrogen surrogate) and that is accomplished with SO_2Cl_2, as follows:

$$(6.123)$$

The cationic nitrogen intermediate thus produced is expected to be stabilized by the adjacent sulfurs. The departing chloride anion now attacks another silyl group, followed by elimination of trimethylsilyl chloride:

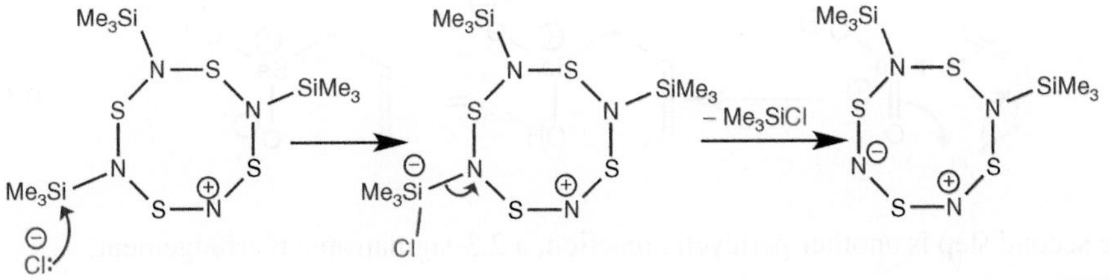

$$(6.124)$$

The process then continues in a similar manner to produce the final product S_4N_4:

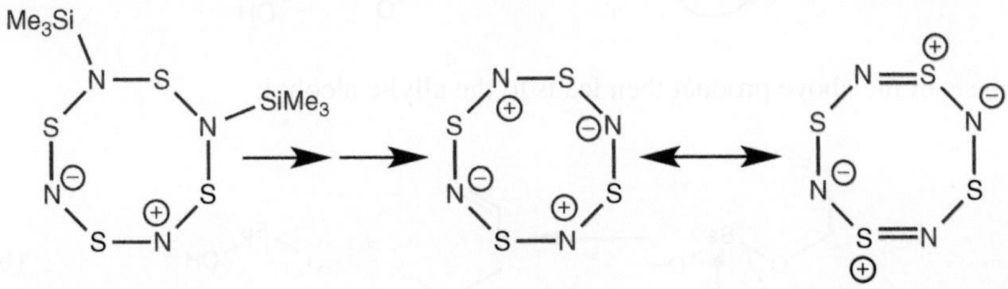

$$(6.125)$$

Despite the complexity of the process, the deliberate choice of sophisticated starting materials leaves little doubt that the chemists who developed the above synthesis foresaw the essentials of the mechanism. That is not only a testament to their insight but also a remarkable demonstration of how mechanistic thinking can guide synthesis design in inorganic chemistry.

6.16* SELENIUM-MEDIATED OXIDATIONS

Despite selenium's toxicity, selenium reagents occupy an important niche in organic chemistry. Selenium dioxide, in particular, allows a number of oxidative transformations that are not as readily accomplished by other methods. One such reaction is allylic hydroxylation, where SeO_2 oxidizes an allylic C–H bond to C–OH without rearrangement of the double bond. Thus, the plant natural product $(-)$-β-pinene is oxidized as follows:

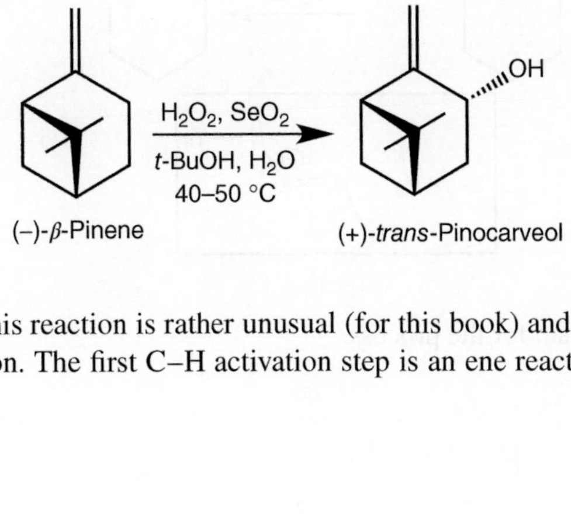

$$(6.126)$$

$(-)$-β-Pinene $(+)$-*trans*-Pinocarveol

The mechanism of this reaction is rather unusual (for this book) and is shown below for a minimal allyl skeleton. The first C–H activation step is an ene reaction (see Section 1.21

for a quick reminder); the product is an allylseleninic acid.

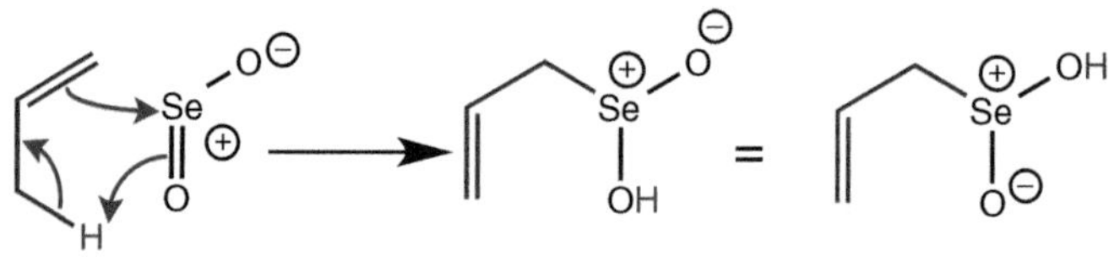

(6.127)

The second step is another pericyclic reaction, a 2,3-sigmatropic rearrangement:

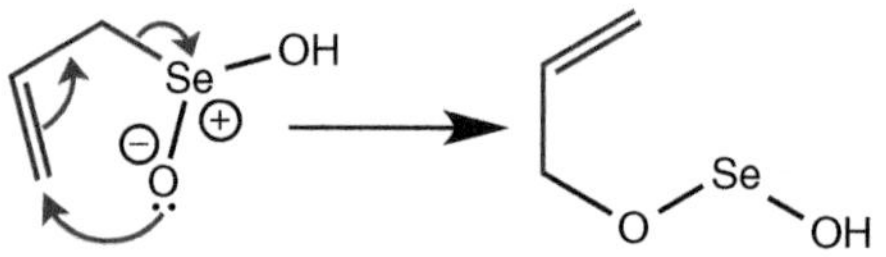

(6.128)

Hydrolysis of the above product then leads to the allylic alcohol:

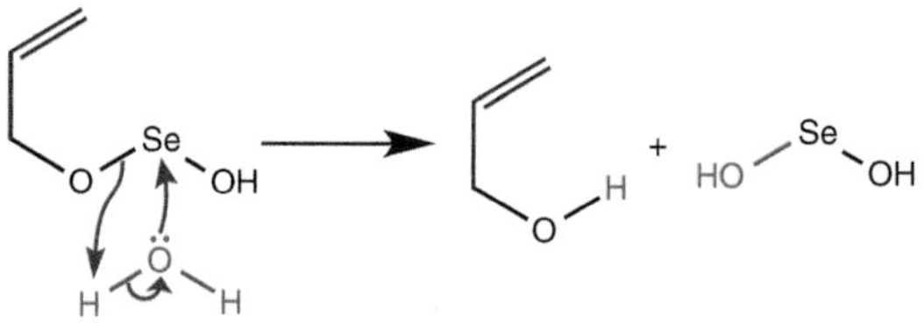

(6.129)

Because of the toxicity of SeO_2, the reaction is often conducted with only a catalytic amount of SeO_2, along with a stoichiometric co-oxidant (e.g., t-BuOOH) which reoxidizes the reduced selenium back to SeO_2.

REVIEW PROBLEM 6.22*

By analogy with SeO_2-mediated allylic hydroxylation, Sharpless and coworkers (*Angew. Chem. Int. Ed.* **1996**, *35*, 454–456) have developed a method for Se-mediated allylic amination:

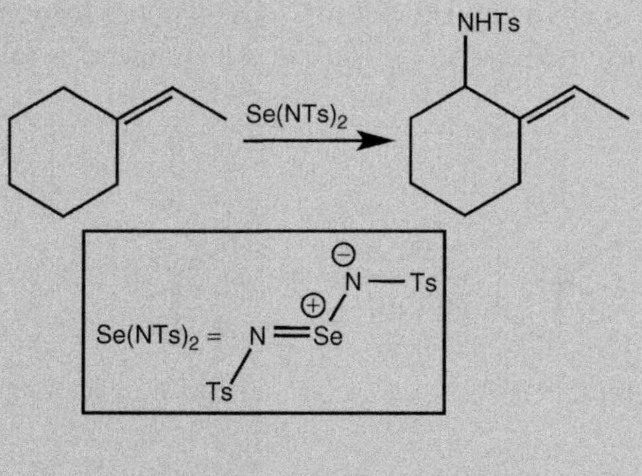

Suggest a mechanism for the process.

Another unique application of SeO_2 is in the α-oxidation of ketones to 1,2-dicarbonyl compounds, as shown below for acetophenone:

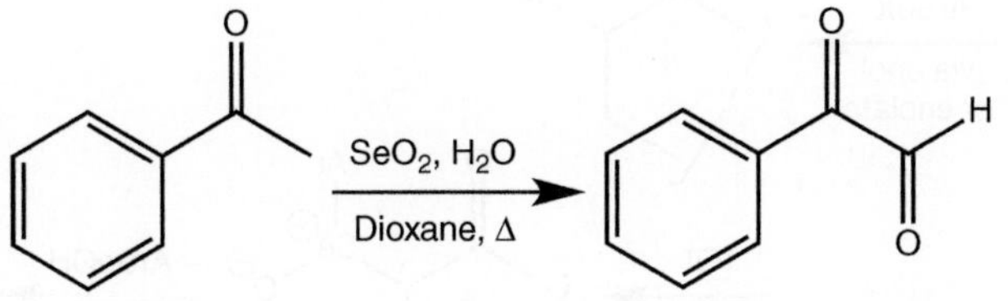

$$(6.130)$$

This reaction is sometimes referred to as the *Riley oxidation*. We may start writing the mechanism in much the same manner as for allylic oxidation, with the carbonyl group taking the place of the carbon–carbon double bond:

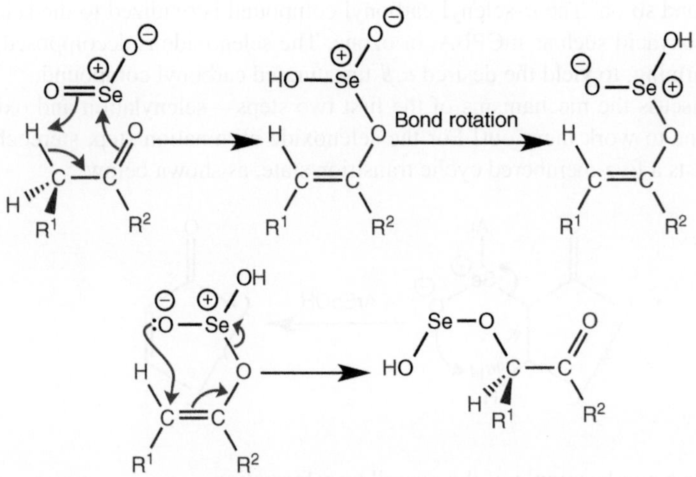

$$(6.131)$$

Abstraction of an α-proton (which is somewhat acidic in a carbonyl compound) then completes the oxidation of the α-carbon to a carbonyl group; the selenium is also detached from the substrate in the same process:

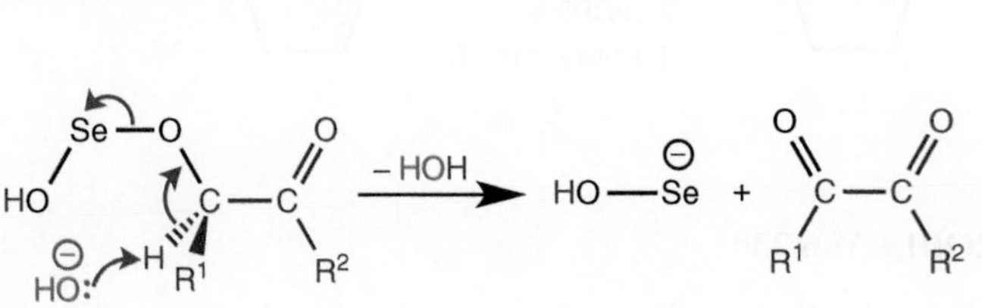

$$(6.132)$$

There are still other selenium-mediated oxidative transformations that do not depend on selenium dioxide. Thus, selenylation, oxidation, and selenoxide elimination constitute a valuable reaction sequence for introducing α,β-unsaturation into a carbonyl compound, as shown below:

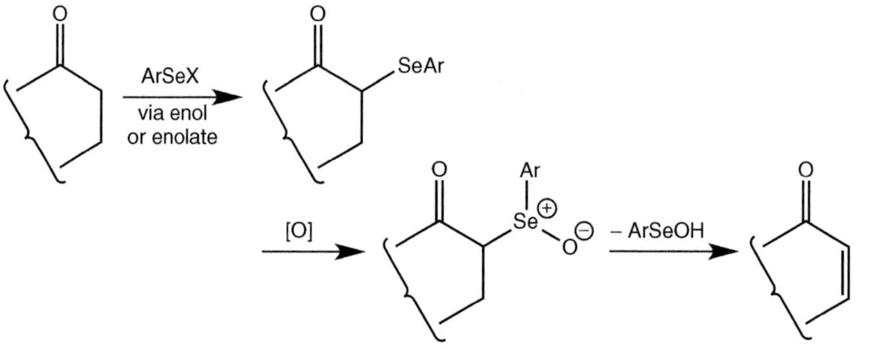

$$(6.133)$$

In a typical procedure, the carbonyl compound is deprotonated to yield the enolate, which is then selenylated by any of a number of selenylating agents ArSeX, where X = Cl, Br, CN, SeAr, and so on. The α-selenyl carbonyl compound is oxidized to the selenoxide by H_2O_2, a peroxyacid such as mCPBA, or ozone. The selenoxide is decomposed *in situ*, typically by warming, to yield the desired α,β-unsaturated carbonyl compound.

We won't discuss the mechanisms of the first two steps—selenylation and oxidation; you are welcome to work them out! For the selenoxide elimination step, stereochemical evidence suggests a five-membered cyclic transition state, as shown below:

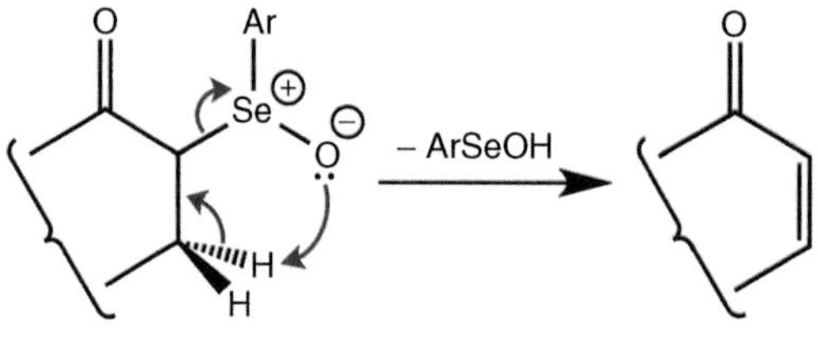

$$(6.134)$$

The following is a good example of the overall transformation:

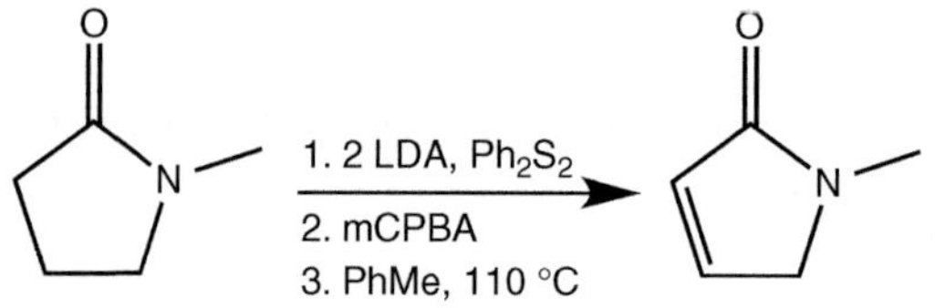

$$(6.135)$$

REVIEW PROBLEM 6.23*

The Grieco–Sharpless elimination provides a mild and powerful method for dehydrating primary alcohols to terminal alkenes, as exemplified by the following reaction sequence:

Suggest mechanistic rationales for the products formed.

6.17 HIGHER-VALENT TELLURIUM: A MECHANISTIC PUZZLE

The tetra- and hexa-valent states are significantly more stable for selenium and tellurium than they are for sulfur. Thus, whereas SCl_4 is exceedingly unstable, decomposing above $-30\,°C$, $SeCl_4$ and $TeCl_4$ are stable under ordinary conditions. With chloride ion acceptors, both tetrachlorides form $[ChCl_3]^+$ (Ch = Se, Te) cations:

$$ChCl_4 + AlCl_3 \rightarrow ChCl_3^+ + AlCl_4^- \tag{6.136}$$

Unlike its Se and S analogs, tellurium tetrachloride adds to carbon–carbon double bonds, giving tetravalent organotelluriums:

$$\tag{6.137}$$

The products thus obtained can then be elaborated to other organotellurium derivatives.

The stability of hexavalent tellurium was clearly demonstrated in a reaction of $TeCl_4$ with an aryllithium, reported by Japanese main-group chemist Kin-ya Akiba and his coworkers (*Tetrahedron* **1997**, *53*, 12195–12202), which led to a disproportionation and produced a highly stable hexaaryltellurium product:

$$2\ TeCl_4 + 8\ ArLi \xrightarrow[\text{Ether}]{-78\ ^\circ C} TeAr_6 + TeAr_2$$

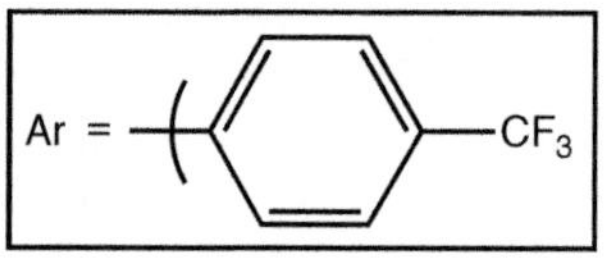

$$Ar = \text{(benzene ring)}-CF_3$$

(6.138)

We chose this reaction for discussion because, like many disproportionation reactions, the mechanism is of considerable interest. An interesting clue to the mechanism was that both $TeAr_3Cl$ and $TeAr_5Cl$ were obtained as side products. Nucleophilic attack by ArLi on $TeCl_4$ produces a whole suite of products, which we may view as intermediates:

$$TeCl_4 + ArLi \longrightarrow ArTeCl_3 \longrightarrow Ar_2TeCl_2 \longrightarrow$$
$$Ar_3TeCl \longrightarrow Ar_4Te \longrightarrow \left[Ar_5Te\right]^{\ominus}$$

(6.139)

Now, for disproportionation to occur, two tetravalent Te species somehow need to hook up into a binuclear species (i.e., a species containing two Te's), which would then rearrange and fall apart into hexa- and di-valent Te compounds. A few different pathways are envisioned below.

Mechanism 1. One possibility involves a chloride-bridged ditellurium intermediate, as shown below:

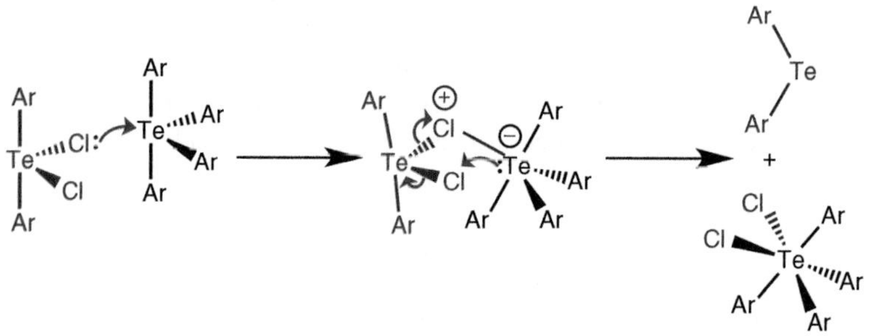

(6.140)

We have thus accounted for both hexa- and di-valent Te products. We only need to arylate $TeAr_4Cl_2$ to produce the $TeAr_6$ final product:

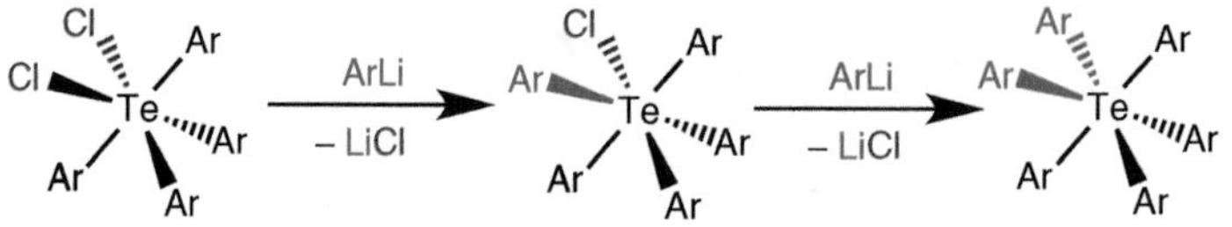

(6.141)

Note that we have also generated the observed by-product $TeAr_5Cl$.

Mechanism 2. Here we consider intermediates with Te–Te bonding. The mixed-valent Te–Te-bonded intermediate below contains both hexa- and tetra-valent Te. Reductive elimination at the tetravalent Te center then produces di- and hexa-valent Te products.

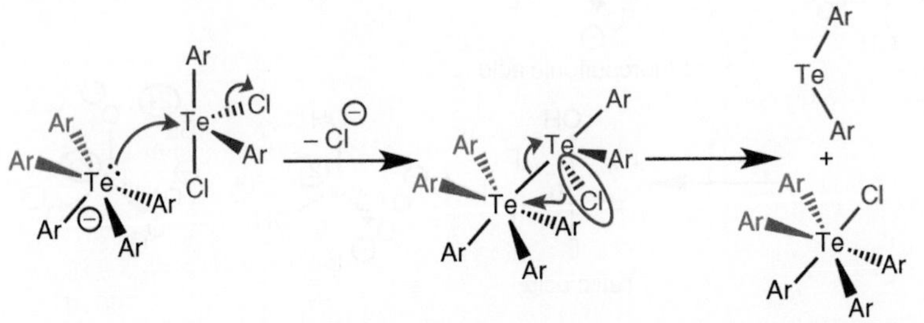

$$(6.142)$$

An additional nucleophilic arylation accounts for the final $TeAr_6$ product:

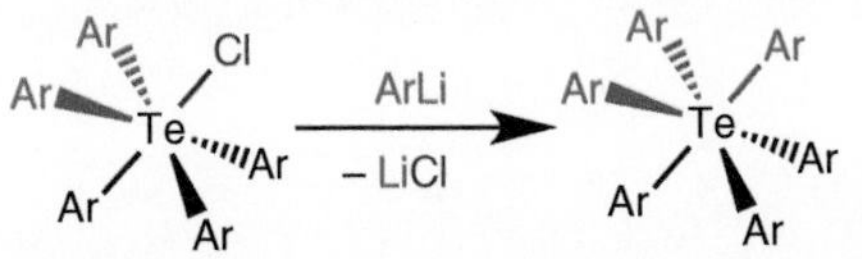

$$(6.143)$$

Mechanism 3. Our last mechanism also involves a Te–Te-bonded intermediate, but it falls apart not via chloride migration but via aryl migration, as shown below:

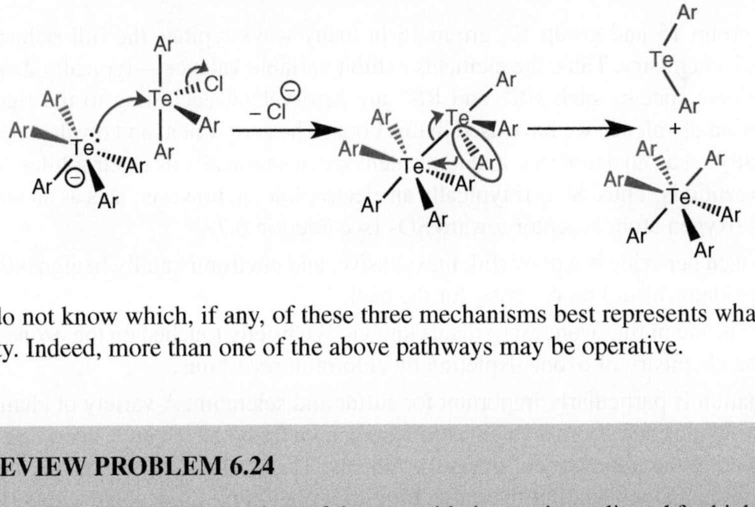

$$(6.144)$$

We do not know which, if any, of these three mechanisms best represents what goes on in reality. Indeed, more than one of the above pathways may be operative.

REVIEW PROBLEM 6.24

The teflate anion (F_5TeO^-) is useful as an oxidation-resistant ligand for higher-valent states of both p-block and d-block elements. Good examples of such teflate-ligated p-block species include $[Pn(OTeF_5)_6]^-$ (Pn = As, Sb and Bi) and $Xe(OTeF_5)^+$. The corresponding acid, teflic acid ($HOTeF_5$), a fluorinated analog of orthotelluric acid [$Te(OH)_6$], can be obtained by treatment of barium tellurate with fluorosulfonic acid, as follows:

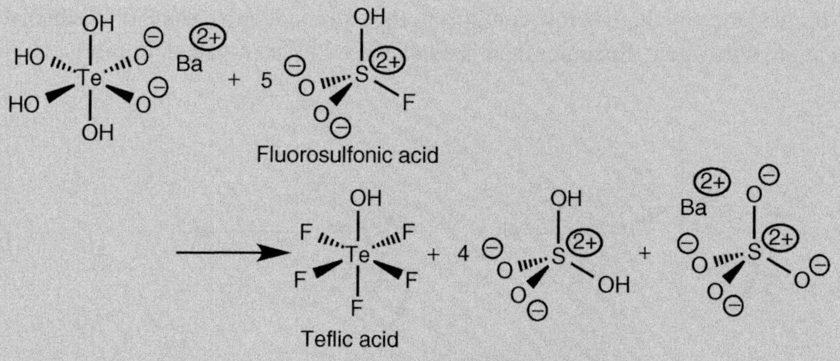

Suggest a mechanism for the above reaction. (*Hint*: Note that the reaction essentially involves ligand exchange between the Te and S centers.)

6.18 SUMMARY

Some highlights of what we discussed in this chapter are as follows:

1. Like group 15 and group 17, group 16 in many ways typifies the full richness of p-block chemistry. Thus, the elements exhibit variable valence—typically 2, 4, and 6. Anionic species such HS^- and RS^- are zero-valent, according to the rigorous definition of valence we have used in this book. The zero-valent and divalent species typically act as nucleophiles, and tetravalent species typically as electrophiles. There are exceptions. Thus, SCl_2 is typically an electrophile; it, however, acts as an nucleophilic oxygen atom acceptor toward SO_3 (see Section 6.7).

2. Hydrogen peroxide is a powerful, inexpensive, and environmentally benign oxidant. Few oxidants afford more "bang for the buck."

3. Ozone is one of the strongest oxidants known. We briefly touched on the ozone layer and the chemistry of ozone depletion by chlorofluorocarbons.

4. Catenation is particularly important for sulfur and selenium. A variety of chain and ring structures built from divalent atoms are known for both elements, including most of the common allotropes of the two elements. These structures are readily broken down by nucleophiles, but they also form quite readily. Sulfides and thiols (RSH) are readily oxidized, forming S–S linkages.

5. Sulfur and selenium form typical acidic oxides, ChO_2 and ChO_3, which form oxoacids with water. The oxides are also redox-active. Sulfur dioxide acts as both an oxidant and a reductant; selenium dioxide is an important oxidant in organic chemistry.

6. Tetravalent selenium-oxo units (including SeO_2) interact with many organic compounds, particularly carbon–carbon double bonds and carbonyl groups, via pericyclic reactions.

7. Many higher-valent p-block oxides, broadly defined, effect oxygen atom transfers to lower-valent p-block compounds. DFT calculations in the authors' laboratory suggest that the mechanisms are best viewed as direct S_N2-like displacements and do not involve oxo-bridged intermediates.

8. Sulfur tetrafluoride and aminosulfur trifluorides are important fluorinating agents that typically operate via a non-redox mechanism. Similarly, Martin sulfurane is a tetravalent sulfur-based dehydrating agent for alcohols, which also operates via a non-redox pathway.

9. Sulfonium ions and cationic tetravalent sulfur centers stabilize adjacent anions. Sulfoxides, in the form we have depicted them in this chapter, are good examples of such stabilization, as are sulfur nitrides such as S_2N_2 and S_4N_4. Sulfonium ylides provide yet another example of the phenomenon.

10. Although sulfur and phosphorus ylides attack carbonyl compounds in the same way initially, the reaction pathways subsequently diverge. The tetravalent sulfur center typically undergoes reductive elimination, producing DMS and an epoxide (the JCC reaction, Section 6.10). In the case of phosphorus ylides, the pentavalent phosphorus center remains pentavalent and ends up as a highly stable phosphine oxide (the Wittig reaction, Section 5.9b), while the carbonyl group ends up as an alkene.

FURTHER READING

1. Devillanova, F. A., ed. *Handbook Of Chalcogen Chemistry: New Perspectives in Sulfur, Selenium and Tellurium*; Royal Society, 2006. 740 pp. *An excellent source for newer aspects of chalcogen chemistry that are not yet adequately covered in standard textbooks.*

2. Chivers, T. *A Guide To Chalcogen-nitrogen Chemistry*; World Scientific: Singapore, 2004. 318 pp. *A full treatment of a fascinating class of compounds.*

3. Furukawa, N.; Sato, S. "New Aspects of Hypervalent Organosulfur Compounds," *Topp. Curr. Chem.* **1999**, *205*, 89–129.

7

The Halogens

Balard did not discover bromine, rather bromine discovered Balard.

Comment by Justus von Liebig about Antonie Jerome Balard

Halogens (often given the generic symbol X) are ubiquitous in all areas of chemistry, so you have already encountered a good deal of halogen chemistry by now. They occur most commonly as monovalent groups or substituents and as mononegative halide anions; the latter act as both nucleophiles and leaving groups. Beyond these stereotypes, however, halogen chemistry is considerably more diverse. A few general remarks should help set the stage for our in-depth survey.

- With the exception of fluorine, which is essentially always monovalent, with an oxidation state of −1, the other halogens exhibit a variety of valences, particularly 3, 5, and 7. The higher-valent states typically occur in the company of strongly electronegative atoms, notably oxygen and fluorine. The various relevant species will be discussed in due course, as we work through the chapter; it's useful, however, to introduce the trivial names of the various oxoacids and oxoanions (for X = Cl, Br, I) right at the outset. These traditional names are widely used, and it's important that you become familiar with them as quickly as possible.

X Oxidation State, Valence		Oxoacids		Oxoanions
+I, 1	HOX	Hypohalous acid	XO^-	Hypohalite
+III, 3	HXO_2	Halous acid	XO_2^-	Halite
+V, 5	HXO_3	Halic acid	XO_3^-	Halate
+VII, 7	HXO_4	Perhalic acid	XO_4^-	Perhalate

- Halide ions may be oxidized to the molecular halogens, with the larger anions more easily oxidized:

$$I^- > Br^- > Cl^- > F^-$$

Arrow Pushing in Inorganic Chemistry: A Logical Approach to the Chemistry of the Main-Group Elements, First Edition. Abhik Ghosh and Steffen Berg.
© 2014 John Wiley & Sons, Inc. Published 2014 by John Wiley & Sons, Inc.

Stated differently, the larger the anion, the stronger it is as a reducing agent. Thus, iodide is quite a strong reducing agent, while fluoride is nonreducing. By the same token, molecular halogens are oxidants, with oxidizing power decreasing down the group:

$$F_2 > Cl_2 > Br_2 > I_2$$

Feel free to have a quick look at Table 1.5 for a sense of the relative X_2/X^- reduction potentials. The data will clearly show F_2 to be one of the strongest oxidants. Chlorine, much more moderate by comparison, arguably offers "more bang for the buck" in the sense that it's cheap, much easier to handle, and remarkably versatile.

- Halides are nucleophilic. The Swain–Scott nucleophilicity increases down the group (see, e.g., Table 1.1):

$$I^- > Br^- > Cl^- > F^-$$

As emphasized in Section 1.2, however, the relative nucleophilicities of the halides is a somewhat tricky issue, and in polar, aprotic solvents the order of nucleophilicities is actually the reverse of the above.

- The higher-valent states of chlorine, bromine, and iodine are powerful electrophiles and/or oxidants. Again, have a look at Table 1.5 and note the very high reduction potentials of hypochlorous acid (HOCl), chlorous acid (HClO$_2$), and chlorine dioxide (ClO$_2$). A number of higher-valent halogen compounds interact with organic compounds, forming high-valent species that subsequently fall apart via reductive elimination. Thus, higher-valent organoiodine compounds have long been an important class of reagents in organic chemistry. Recently, even trivalent bromine compounds are finding unique applications in organic synthesis, a topic we will touch on toward the end of this chapter.

REVIEW PROBLEM 7.1

Bromine water (Br_2/H_2O) added to an aqueous solution of sodium iodide liberates molecular iodine. Suggest a mechanistic explanation.

REVIEW PROBLEM 7.2

If you examine Table 1.1, you are likely to arrive at the conclusion that fluoride is not a particularly good nucleophile. Recently, Stephen DiMagno and coworkers at the University of Nebraska synthesized anhydrous tetrabutylammonium fluoride (TBAF$_{anh}$) via the following reaction, and found it to be a highly active form of nucleophilic fluoride in dry DMSO and other polar aprotic solvents (Sun, H.; Dimagno, S. *J. Am. Chem. Soc.* **2005**, *127*, 2050–2051):

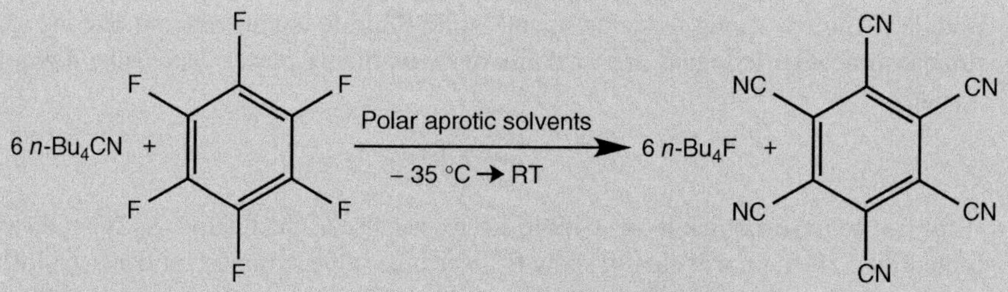

The following two reactions illustrate well the exceptional reactivity of TBAF$_{anh}$:

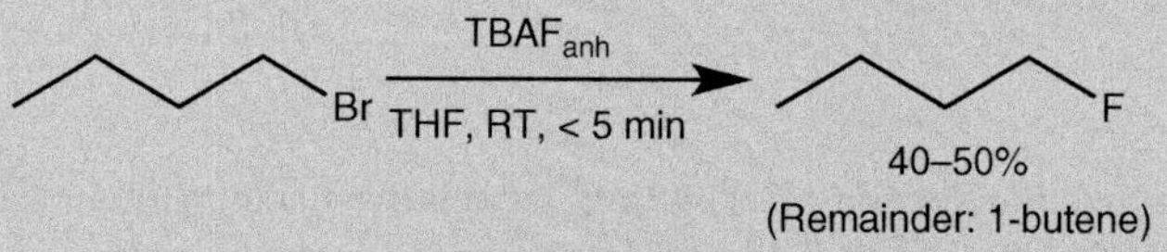

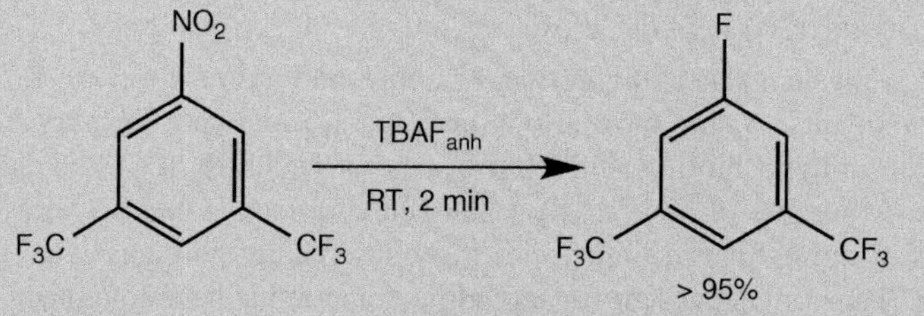

What might account for the remarkable reactivity of the TBAF$_{anh}$ reagent? What might be a reason for the researchers' choice of the highly toxic tetrabutylammonium cyanide as a starting material for synthesizing TBAF$_{anh}$?

As a postscript to this story, we may add that, as a nucleophilic fluorinating agent, [^{18}F]-TBAF$_{anh}$ has been found to be significantly superior to cryptand-activated K^{18}F, which is the standard ^{18}F source in positron-emission tomography (PET).

7.1   SOME NOTES ON ELEMENTAL HALOGENS

As reactive electronegative elements, the halogens do not occur in their elemental form in Nature. They occur most commonly as halide anions, and the molecular elements are synthesized—and indeed were discovered—via the oxidation of the corresponding halide ions. As mentioned above, the ease of oxidation of the halide anions increases down the group, and the oxidizing power of the molecular elements decreases down the group. Thus, as an oxidant, fluorine is the strongest, followed by chlorine, and so on. The story of the discovery of the halogens provides for excellent illustrations of these trends. Here we'll briefly recount the discovery of bromine by the French chemist Antonie Jérôme Balard (1802–1876); for the other elements, we'll refer you to *Further reading*.

Born of poor parents in Montpellier, France, Balard was adopted and educated by his godmother. At age 17, he became a laboratory assistant ("préparateur") at the local *École de Pharmacie*. While studying salt marsh flora, he concentrated sea water, crystallized out sodium sulfate, which precipitated first, and attempted to find uses for the remaining mother liquor. With certain reagents, he found that the brine turned brown. Saturating the liquors with chlorine and distilling, he isolated a pungent, dark-red liquid, with a boiling point of 47 °C and a density three times that of water. He attempted to decompose it into simpler substances, but soon recognized that he had discovered an element similar to chlorine and iodine. The results were communicated to the Académie des Sciences and eventually published under the title "Sur une substance particulière contenue dans l'eau de la mer (About a particular substance present in sea water)" in the *Annales de Chimie et de Physique*. Balard's findings caused a sensation, not least because of his tender age; he was 23 years old at the time of his discovery. Life thereafter was kind to Balard. Despite the trauma of childhood poverty, which affected him deeply, he became a highly successful chemist, making many significant discoveries and ultimately becoming a professor at the prestigious Collège de France.

Others also played a role in the discovery of bromine. A young German student named Carl Löwig discovered bromine a year before Balard, but did not publish his findings after he came across Balard's paper. The great German chemist Justus von Liebig also prepared bromine but apparently did not recognize its elementary nature, viewing it instead as a chloride of iodine. Did Liebig's comment referring to bromine's powerful stench, cited at the beginning of this chapter, reflect a tinge of envy or bitterness? We don't know; it may well have been pure, good-natured humor.

In general, the X–X bonds in the elements are weak (bond dissociation energies (BDEs): F–F 159; Cl–Cl 243; Br–Br 192, I–I 151, all in kJ/mol) and easily broken. Both homolysis and nucleophile-induced heterolysis are important reaction pathways for the diatomic elements:

$$X-X \begin{cases} \xrightarrow{\text{Nu}^{\ominus}} \text{Nu}-X + X^{\ominus} \\ \xrightarrow[\text{or } h\nu]{\Delta} X\cdot + X\cdot \end{cases} \qquad (7.1)$$

A classic reaction for preparing a molecular halogen (other than fluorine)—indeed, the reaction that led to the discovery of chlorine as an element—involves oxidation of a hydrohalic acid with manganese dioxide:

$$\text{MnO}_2 + 4\text{HX} \rightarrow \text{MnX}_2 + X_2 + 2\text{H}_2\text{O} \qquad (7.2)$$

Although transition metals are outside the scope of this book, some indication of the mechanism can be given. In essence, the acid protonates some of the oxide ligands, thereby increasing the electrophilicity of the Mn(IV) centers, which are then reduced by electron transfer from Cl$^-$. The chlorine atoms so formed couple to form Cl$_2$ molecules.

$$\begin{aligned} X{:}^{\ominus} \quad \text{Mn(IV)} &\longrightarrow X\cdot + \cdot\text{Mn(III)} \\ X{:}^{\ominus} \quad \cdot\text{Mn(III)} &\longrightarrow X\cdot + \cdot\text{Mn(II)} \\ X\cdot \quad \cdot X &\longrightarrow X-X \end{aligned} \qquad (7.3)$$

Note that in the "mechanism" above we have only shown electrons that are involved in the electron transfer. The +IV, +III, and +II states of manganese are typically associated with 3, 4, and 5 unpaired d electrons, respectively.

REVIEW PROBLEM 7.3

Explain Balard's discovery with the help of equations/mechanisms. Would it have been reasonable to formulate the red liquid as iodine chloride?

The elemental halogens are important electrophiles. An important and familiar example is the electrophilic addition of bromine to carbon–carbon double bonds. As mentioned in Section 1.12, the first product of electrophilic addition is a cyclic bromonium ion.

Molecular iodine is less reactive and does not add to a carbon–carbon double bond in the absence of a catalyst. It does, however, react with somewhat stronger nucleophiles such as pyridine:

$$2\,I_2 + 2\,py \rightarrow [I(py)_2]^+[I_3]^- \quad (7.4)$$

The first step of the reaction is probably a straightforward nucleophilic displacement:

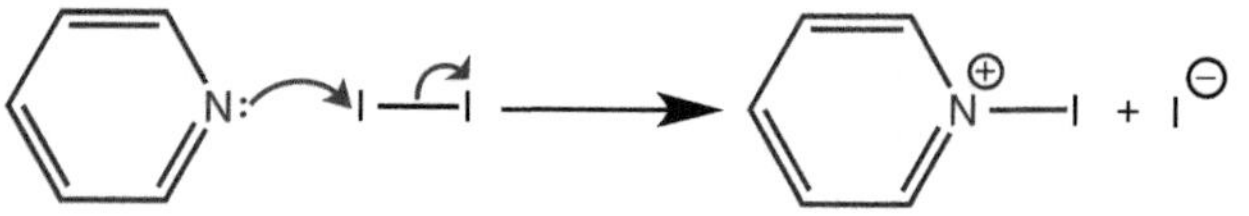

(7.5)

Both the ions produced react further with nucleophiles, as shown below:

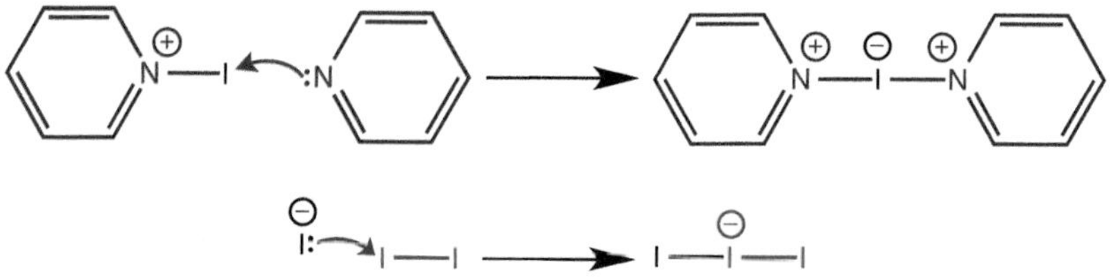

(7.6)

The triiodide anion, $[I_3]^-$, is a well-known species, yellow in dilute aqueous solution and brown when concentrated. Indeed, the formation of the triiodide anion underlies the solubility of elemental iodine in aqueous potassium iodide but not in pure water.

REVIEW PROBLEM 7.4

Note that the central iodine atom in the above final products, $[I(py)_2]^+$ and $[I_3]^-$, are hypervalent. Use valence shell electron pair repulsion (VSEPR) theory to predict the structures (linear/bent) of these two ions.

As with pyridine, the interaction of cryptand and I_2 leads to a cationic I^+ complex (depicted below), with I_3^- as the counterion:

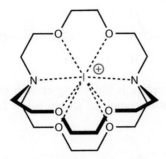

Silver ions interact strongly with halide ions, and the stability of silver halides may drive otherwise unlikely reactions such as the following:

$$I_2 + AgClO_4 \rightarrow AgI + IClO_4 \tag{7.7}$$

As an extremely weak base and a poor nucleophile, perchlorate (ClO_4^-) would not normally attack I_2, but silver ion coordination drives the process, as shown below:

$$\tag{7.8}$$

Molecular iodine exhibits contrasting "Janus-like" behavior in its interactions with transition metals (Rogachev, A. Yu.; Hoffmann, R. *J. Am. Chem. Soc.* **2013**, *135*, 3262–3275). With some metals, I_2 acts as a normal donor ligand. In certain platinum(II) compounds, however, I_2 acts as an acceptor, with the Pt feeding its d_{z2} electron pair into the I–I σ^* orbital. The two bonding scenarios appear to lead to distinct geometries: when I_2 binds as a normal ligand, the result is a bent M–I–I angle; on the other hand, when Pt(II) acts as the "ligand" and I_2 as the acceptor, the result is a linear coordination geometry, as depicted below:

A quick word about astatine may be of interest here. All isotopes of astatine are intensely radioactive; even the most stable, ^{210}At, has a half-life of only 8.1 h. Astatine chemistry is therefore understandably relatively little explored. It is utterly fascinating, however. Thus, condensed, bulk astatine (which is yet to be prepared) has been predicted to be metallic. Experimentally, there is significant evidence for At^+ cations in aqueous solution. Two possible closed-shell structures for such species are the following:

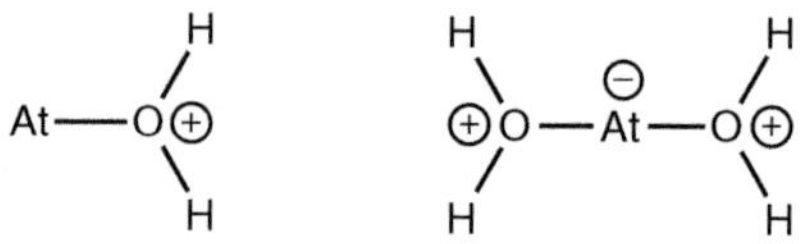

Sometimes, a given nucleophile is not strong enough to cleave a halogen molecule but does form a "charge-transfer complex," shown below for acetonitrile (CH_3CN) and bromine:

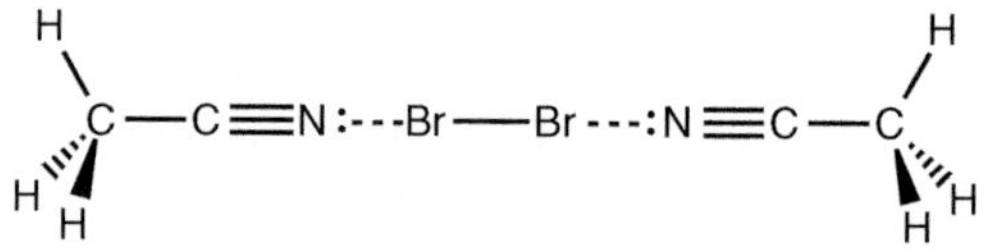

The dotted lines indicate a weak interaction, that is, possibly not full-fledged bonds. The term "charge transfer" implies that, although electron transfer or bond formation does not happen in the ground state, an electron does jump from one component to the other in an excited state, giving rise to characteristic features in the optical spectrum. At low temperatures, charge-transfer complexes with even stronger nucleophiles such as phosphines may be isolated:

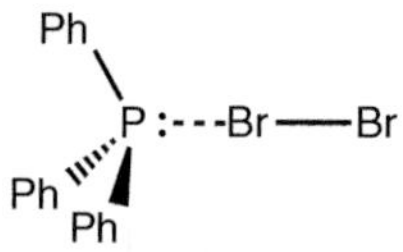

7.2 ALKALI-INDUCED DISPROPORTIONATION OF MOLECULAR HALOGENS

The reaction of halogen molecules (X = Cl, Br, I) with *cold* aqueous alkali to yield hypohalite (OX^-) and halide (X^-) may be viewed as a classic inorganic S_N2 displacement, rather similar to the reaction with pyridine mentioned above:

$$X_2 + 2\,NaOH \rightarrow NaX + NaOX + H_2O \qquad (7.9)$$

The halogen–halogen bond breaks easily because of its weakness:

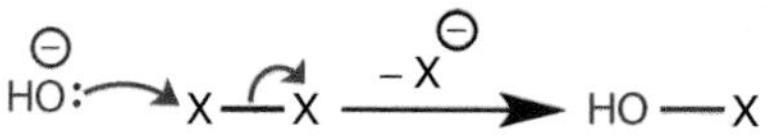

$$(7.10)$$

Hydroxide then deprotonates the hypohalous acid to produce a hypohalate anion:

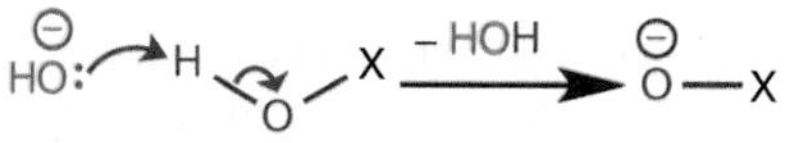

$$(7.11)$$

With hot alkali, however, the same reactants give different products. The hypohalite anions, the initial products of the reaction, disproportionate on warming to halate (XO_3^-) and halide (X^-, X = Cl, Br, I):

$$3 \, XO^- \rightarrow XO_3^- + 2 \, X^- \tag{7.12}$$

This is not an easy mechanism to envision. Multiple negatively charged hypohalite anions must somehow come together to generate the products. The simplest mechanism involves an S_N2-type attack on a hypohalite oxygen, with halide as the leaving group:

$$\tag{7.13}$$

A negatively charged oxygen atom as an electrophile is admittedly a rather awkward proposition. Certainly, alternative pathways are conceivable. To make a long story short, however, DFT calculations appear to favor the above path: hydrogen bonding by water and the Na^+ counterions provide the electrostatic stabilization necessary for various anionic reactants to come together.

A milder version of the above disproportionation can lead to dichlorine monoxide (Cl_2O). The reaction involves the interaction of Cl_2 with moist Na_2CO_3:

$$2 \, Cl_2 + 2 \, Na_2CO_3 + H_2O \rightarrow 2 \, NaHCO_3 + 2 \, NaCl + Cl_2O \tag{7.14}$$

As usual, it's useful to consider the product structures. The carbonate ion (CO_3^{2-}) ends up as bicarbonate (HCO_3^-). That strongly suggests that the oxygen in Cl_2O derives from water, there being no other oxygen source. We therefore begin by using carbonate to strip a proton off water:

$$\tag{7.15}$$

The OH^- so produced reacts with Cl_2 exactly as described above to produce HOCl. A second carbonate ion may then deprotonate the HOCl, producing ClO^-.

$$\tag{7.16}$$

The hypochlorite ion, ClO^-, is the key nucleophile that reacts with Cl_2 to produce Cl_2O.

$$\text{Cl—Cl} \longleftarrow :\overset{\ominus}{O}\text{—Cl} \xrightarrow{-\overset{\ominus}{Cl}} \text{Cl}\diagup^{O}\diagdown\text{Cl} \tag{7.17}$$

REVIEW PROBLEM 7.5

Bleaching powder, written simplistically as $Ca(OCl)Cl$, is manufactured by passing chlorine over CaO:

$$CaO + Cl_2 \rightarrow Ca(OCl)Cl$$

Suggest a mechanism.

7.3   ACID-INDUCED COMPROPORTIONATION OF HALATE AND HALIDE

We discussed the disproportionation of molecular halogens to halate and halide in hot alkali (Section 7.2). The reverse process, called *comproportionation*, where two different oxidation states end up as one at the end of a reaction, happens in acidic solution. Thus, halate (XO_3^-), halide (X^-, $X = Cl$, Br, I), and protons react together to yield halogen molecules and water:

$$XO_3^- + 5\,X^- + 6\,H^+ \rightarrow 3\,X_2 + 3\,H_2O \tag{7.18}$$

Given that the process is mediated by acid, a proton-coupled attack by X^- on XO_3^- seems like a reasonable first step:

$$\tag{7.19}$$

We use the term "proton-coupled attack" as somewhat of a hedge to indicate that we do not have a good sense of the exact timing of the protonation, i.e., whether it occurs before, after, or synchronously with attack by X^-. A second proton-coupled attack by X^- on the above intermediate would then produce the first X_2 molecule and halite (XO_2^-), the first reduction product of halate (XO_3^-).

$$\xrightarrow{-\,HOH} \quad + \quad X\text{—}X \tag{7.20}$$

A very similar process then results in the reduction of halite (XO_2^-) to hypohalous acid (HOX), while producing the second molecule of X_2:

$$(7.21)$$

Proton-coupled attack by halide on the hypohalous acid then produces the third and final molecule of X_2:

$$(7.22)$$

7.4 HYPOFLUOROUS ACID, HOF

Hypofluorous acid, HOF, is prepared by passing F_2 over ice at $-43\,°C$:

$$F_2 + H_2O \rightarrow HOF + HF \qquad (7.23)$$

A heterolytic mechanism, analogous to the reaction of chlorine or bromine with water, may be envisioned as follows:

$$(7.24)$$

The F–F bond is so weak, however, that a homolytic (radical) path mediated by fluorine atoms is also a possibility:

$$(7.25)$$

The process may be viewed as a chain reaction, where the first step is the radical initiation, the second and the third steps are propagation, and the last step is radical termination.

Despite its name, hypofluorous acid is not significantly acidic and does not ionize in water. It is, however, highly reactive and even oxidizes water to H_2O_2:

$$HOF + H_2O \rightarrow H_2O_2 + HF \tag{7.26}$$

The fact that an O–O bond is formed provides a key clue to the mechanism. The electronegativity of fluorine (3.98) is considerably higher than that of oxygen (3.44), so, as far as the O–F bond in HOF is concerned, fluorine is always the negative end of the dipole. Assuming polar mechanisms, water is thus the likely nucleophile, the oxygen atom in HOF the electrophile, and F^- the leaving group. The weakness of the O–F bond (BDE ~190 kJ/mol) also makes this a reasonable proposition:

$$\tag{7.27}$$

Alternatively, a cyclic mechanism might be envisioned:

$$\tag{7.28}$$

At room temperature, HOF decomposes explosively:

$$2\,HOF \rightarrow O_2 + 2\,HF \tag{7.29}$$

Again, the fact that an O–O bond forms provides the key clue. Now, however, HOF is both the nucleophile and the electrophile:

$$\tag{7.30}$$

Once again, an alternative cyclic mechanism may be operative:

$$\tag{7.31}$$

With three highly electronegative atoms strung together, the intermediate HOOF is not expected to be a stable molecule and it readily falls apart to HF and O_2. The following water-mediated cyclic mechanism may provide a low energy pathway:

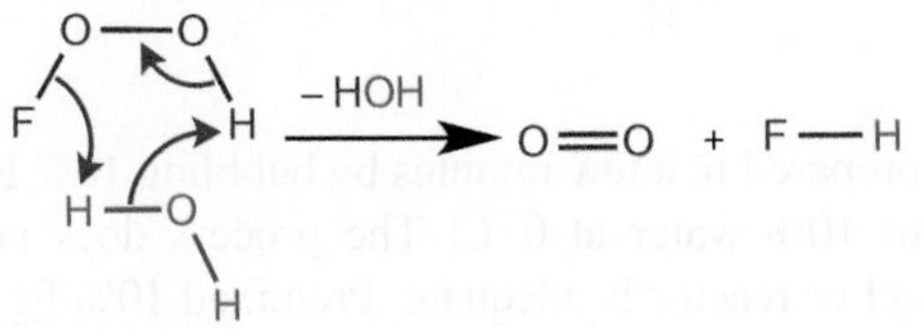

$$(7.32)$$

Alternatively, we could envision an HF-catalyzed cyclic mechanism:

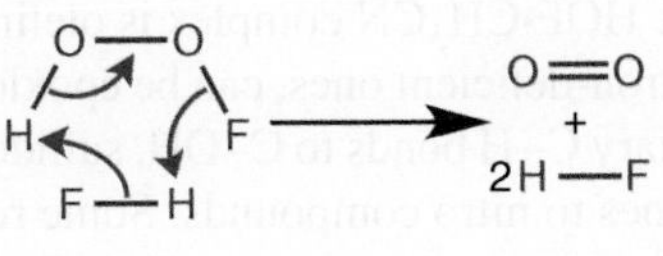

$$(7.33)$$

Given its instability, it's impressive that solid HOF has been characterized by single-crystal X-ray crystallography.

The mechanistic discussion above was clearly quite speculative! We believe, however, that such hypothesizing is useful. It's preferable, in our view, to be able to come up with proposals about how a reaction *might* happen than to have no idea whatsoever.

With organic compounds, HOF typically reacts via polar mechanisms, with the oxygen atom as the electrophile and F^- as the leaving group. Some representative transformations effected by HOF are as follows:

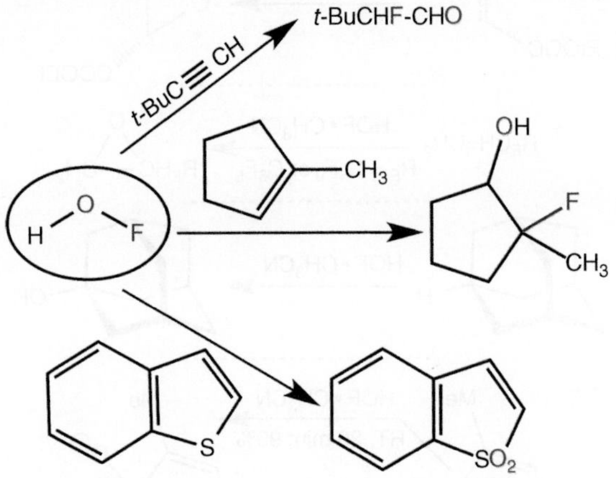

$$(7.34)$$

REVIEW PROBLEM 7.6

Suggest mechanisms for the set of reactions in (7.34).

Recently, the HOF·CH$_3$CN complex (shown below) has been found to be an excellent oxygen atom transfer agent toward a variety of functional groups:

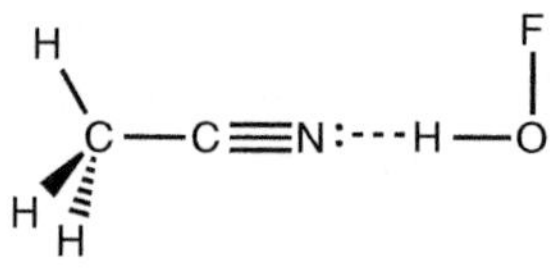

The complex is easily prepared in a few minutes by bubbling 10% F_2 in N_2 through aceto-nitrile containing about 10% water at 0 °C. The process does not require any special apparatus. A glass vessel or reactor is adequate. Premixed 10% F_2 in N_2 is commercially available, which, contrary to popular belief, can be handled with no more trouble than other corrosive gases such as Cl_2.

A major application of the HOF·CH_3CN complex is olefin epoxidation. Practically any type of olefin, including electron-deficient ones, can be epoxidized by this reagent. In addition, the reagent oxidizes tertiary C–H bonds to C–OH, sulfides (including unreactive ones) to sulfones, and primary amines to nitro compounds. Some representative applications are depicted below:

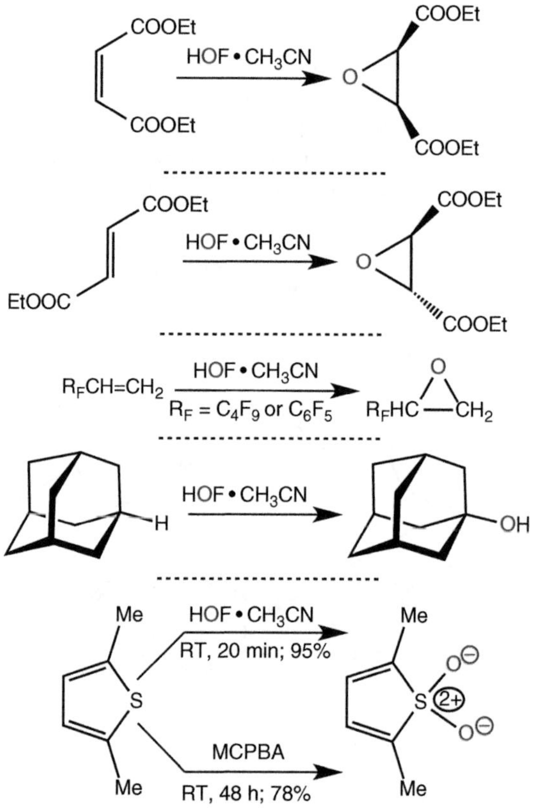

(7.35)

7.5 ELECTROPHILIC FLUORINATING AGENTS: *N*-FLUORO COMPOUNDS

The 1980s saw the emergence of a suite of *N*-fluoro compounds, which have since become the reagents of choice for electrophilic fluorination of organic compounds, that is, as "F^+" donors. Some popular *N*-fluoro reagents are depicted in Figure 7.1

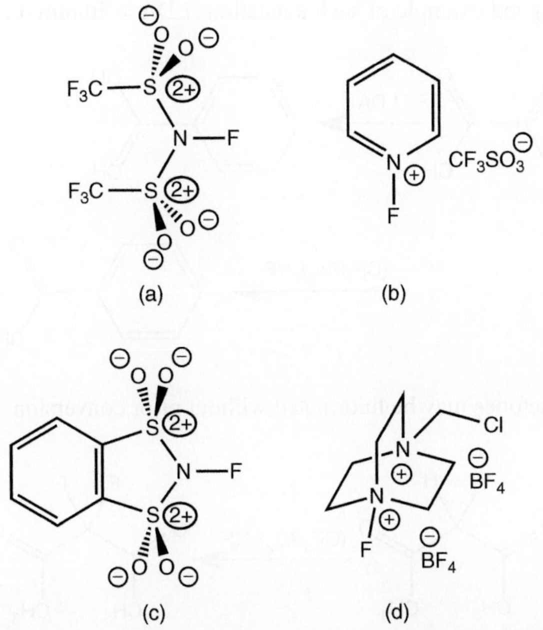

Figure 7.1 *Representative N-fluoro reagents. (a) N-fluorotrifluoromethanesulfonimide. (b) N-fluoropyridinium triflate. (c) N-fluoro-o-benzenedisulfonimide (NFOBS). (d) Selectfluor™-(BF₄)₂.*

The reagents are all commercially synthesized by fluorination of the corresponding N-H compounds with N_2-diluted elemental fluorine, as shown below for *N*-fluorotrifluoromethanesulfonimide:

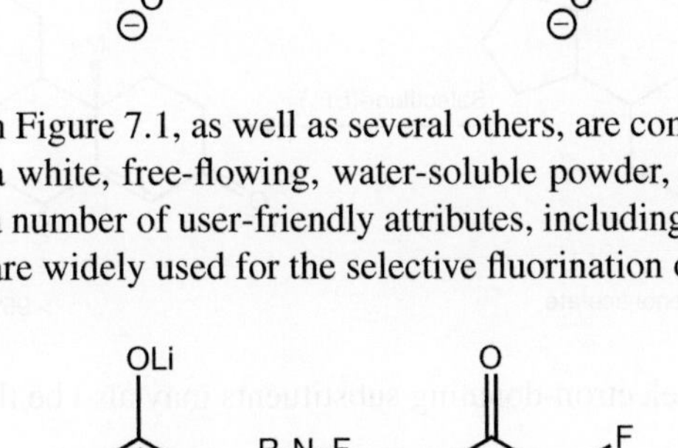

$$(7.36)$$

The reagents shown in Figure 7.1, as well as several others, are commercially available; of these, Selectfluor™, a white, free-flowing, water-soluble powder, has proved particularly popular, by virtue of a number of user-friendly attributes, including relatively low toxicity.

N-fluoro reagents are widely used for the selective fluorination of enolates:

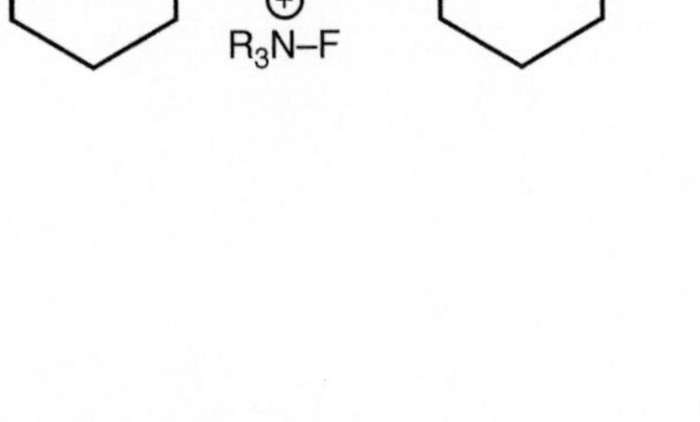

$$(7.37)$$

The following is a good example of such a reaction (LDA = lithium diisopropylamide):

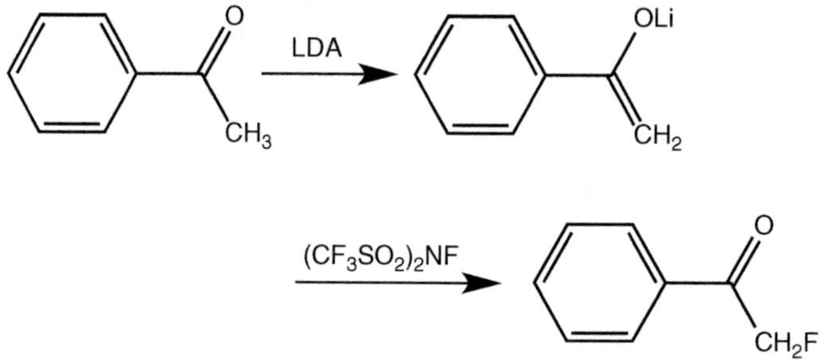

$$(7.38)$$

Interestingly, β-diketones may be fluorinated without prior conversion to enolates:

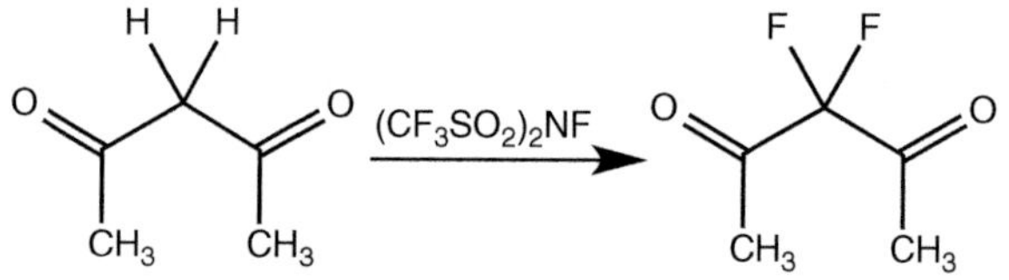

$$(7.39)$$

In addition, a number of steroidal silyl enol ethers and enol esters have been selectively fluorinated by Selectfluor, as shown below:

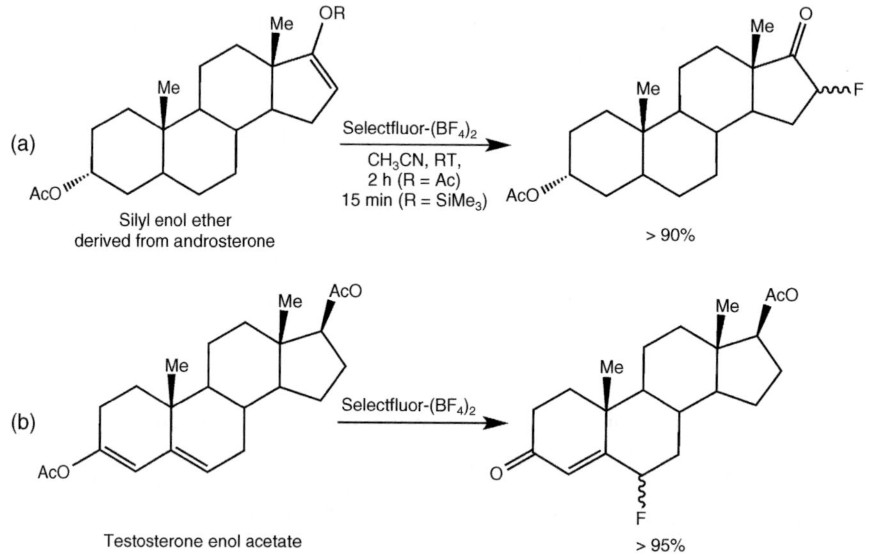

(a)

Selectfluor-$(BF_4)_2$

CH$_3$CN, RT,
2 h (R = Ac)
15 min (R = SiMe$_3$)

Silyl enol ether
derived from androsterone

> 90%

(b)

Testosterone enol acetate

Selectfluor-$(BF_4)_2$

> 95%

$$(7.40)$$

Finally, aromatics with electron-donating substituents may also be fluorinated by *N*-fluoro reagents.

REVIEW PROBLEM 7.7

(a) Draw a mechanism for reaction 7.37

(b) Why does reaction 7.39 proceed without the need for prior enolate formation?

(c) Suggest a radical mechanism for enolate fluorination based on single-electron transfers.

Given the importance of fluorination in drug design, chiral *N*-fluoro reagents were also sought after and developed. The first such reagents to be developed were the *N*-fluorocamphorsultams (a sultam is a cyclic sulfonamide):

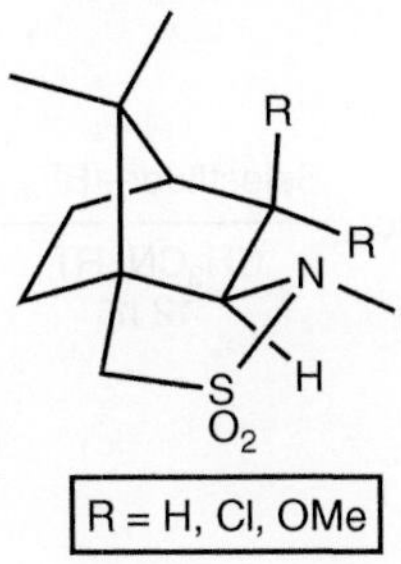

R = H, Cl, OMe

These, however, led to products with modest enantiomeric excesses (% major enantiomer − % minor enantiomer). Subsequently, a variety of chiral *N*-fluoro reagents were obtained via the interaction of cinchona alkaloid derivatives (the substrates in the reaction below) and Selectfluor, as shown below:

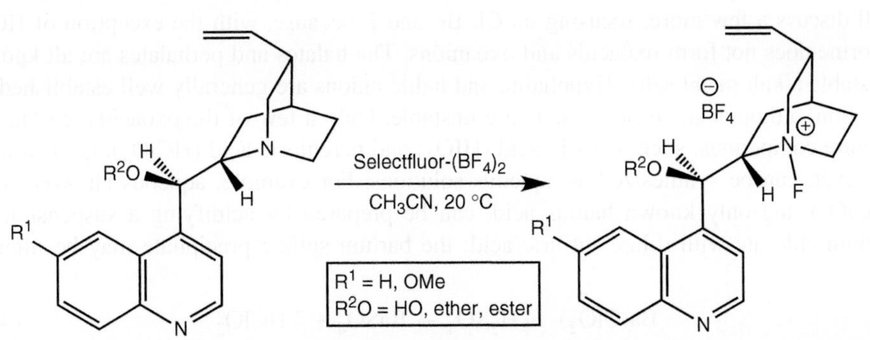

R^1 = H, OMe
R^2O = HO, ether, ester

(7.41)

These reagents generally lead to good to excellent enantiomeric excesses. Their downside, however, is that they are too expensive for general use. Most current research on enantioselective, electrophilic fluorination relies on the less expensive achiral reagents shown in Figure 7.1, in conjunction with standard organic or metal-based, chiral catalysts.

REVIEW PROBLEM 7.8

Carbon nucleophiles other than enolates, such as organosilanes and organostannanes, may also be fluorinated with N-fluoro reagents. Suggest mechanisms for the following two transformations:

(a)

(b)

7.6   OXOACIDS AND OXOANIONS

We have already discussed a number of oxo species in our discussion so far. Here we will discuss a few more, focusing on Cl, Br, and I, because, with the exception of HOF, fluorine does not form oxoacids and oxoanions. The halates and perhalates are all known as stable alkali metal salts. Hypohalite and halite anions are generally well established in solution, although some of the salts are unstable. Only a few of the oxoacids are known as pure compounds, such as iodic acid (HIO_3) and perchloric acid ($HClO_4$). Quite a few, however, can be synthesized as aqueous solutions. For example, aqueous chlorous acid ($HClO_2$), the only known halous acid, can be prepared by acidifying a suspension of barium chlorate with dilute sulfuric acid; the barium sulfate precipitate may be filtered off:

$$Ba(ClO_2)_2 + H_2SO_4 \rightarrow BaSO_4 + 2\,HClO_2 \qquad (7.42)$$

Chloric and bromic acids can be prepared similarly by the action of H_2SO_4 on the appropriate barium salts. Iodic acid, by contrast, is conveniently prepared by hydrolysis of readily available I_2O_5:

$$I_2O_5 + H_2O \rightarrow 2\,HIO_3 \qquad (7.43)$$

The mechanism is expected to be rather simple; water attacks one of the pentavalent iodine centers, with iodate ($IO_3{}^-$) acting as the leaving group:

$$(7.44)$$

REVIEW PROBLEM 7.9

As mentioned above, iodic acid can be isolated as a pure solid. It consists of discrete HIO_3 molecules. What would you predict for the molecular structure based on VSEPR considerations?

In the rest of this section, we will focus on sodium chlorite ($NaClO_2$), particularly on its role as a source of chlorine dioxide (ClO_2), which is a stable radical (like NO). The chemistry is both important and, from an arrow-pushing perspective, instructive. Chlorine dioxide is widely used as a bleaching agent for paper pulp and also as a disinfectant for municipal water. In regard to the latter, ClO_2 has an advantage over Cl_2 in that it does not produce toxic trihalomethanes from organic contaminants. Sodium chlorite is also used as a mild disinfectant in mouthwash, toothpaste, eye drops, and contact lens cleaning solution, among other things.

Acidification of a $NaClO_2$ solution with strong acid results in the formation of ClO_2:

$$5\ NaClO_2 + 4\ HCl \rightarrow 4\ ClO_2 + 5\ NaCl + 2\ H_2O \qquad (7.45)$$

The ClO_2 arises via rapid disproportionation of chlorous acid ($HClO_2$), which forms initially, under strongly acidic conditions. A more insightful way to write the reaction is:

$$5\ HClO_2 \rightarrow 4\ ClO_2 + H^+ + Cl^- + 2\ H_2O \qquad (7.46)$$

Given that the reaction involves the generation of Cl^- (where both chlorite oxygens have been stripped off), the mechanism has to involve several steps. Let's generate the first ClO_2 by the most direct route imaginable, viz. an electron transfer from ClO_2^- to $HClO_2$, and see where that takes us:

$$(7.47)$$

A similar electron transfer from a second ClO_2^- anion may then reduce the hypochlorite radical ($ClO^\bullet$) formed above:

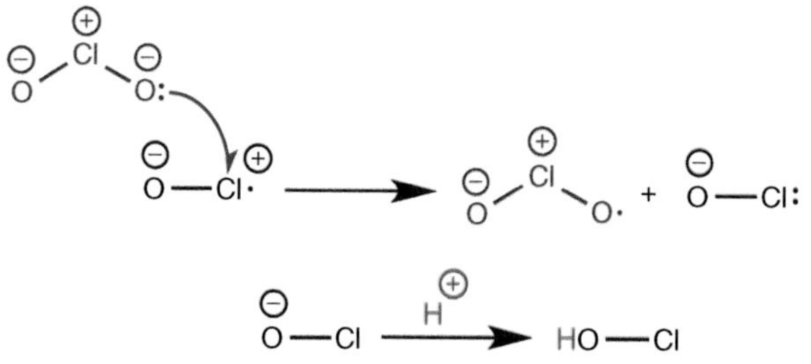

$$(7.48)$$

Chlorite (ClO_2^-) may now be envisioned as doing a one-electron reduction of HOCl, or its protonated form (as shown below), producing a chlorine atom:

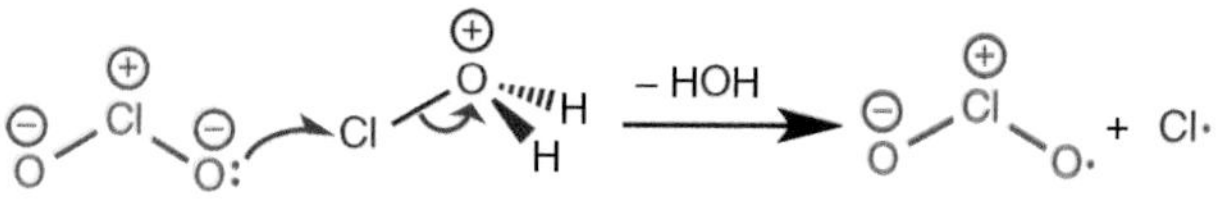

$$(7.49)$$

A final electron transfer reduces atomic chlorine to chloride, as required by the overall stoichiometry of the reaction:

$$(7.50)$$

REVIEW PROBLEM 7.10

Chlorine dioxide is conveniently prepared in the laboratory by oxidizing chlorite with, for example, molecular chlorine:

$$2\,NaClO_2 + Cl_2 \rightarrow 2\,ClO_2 + 2\,NaCl$$

Suggest a mechanism.

REVIEW PROBLEM 7.11

Chlorine dioxide can be converted back to $NaClO_2$ by reaction with Na_2O_2. Write a balanced equation and suggest a mechanism.

Like the molecular halogens, ClO_2 undergoes disproportionation in alkaline solution; the products are chlorite (ClO_2^-) and chlorate (ClO_3^-):

$$2\,ClO_2 + 2\,OH^- \rightarrow ClO_2^- + ClO_3^- + H_2O \qquad (7.51)$$

The stoichiometry of the reaction further indicates that one OH⁻ must be adding to ClO_2. A potential stumbling block in writing this out, however, is that the Lewis structure used for chlorine dioxide until now is not particularly representative of its real electronic structure. It's better to view ClO_2 as a resonance hybrid:

Hydroxide can now be added to the Cl of the middle resonance structure:

$$(7.52)$$

The product $ClO_3{}^{•2-}$, a one-electron-reduced chlorate anion, could then reduce ClO_2 to $ClO_2{}^-$:

$$(7.53)$$

7.7 HEPTAVALENT CHLORINE

The last three sections provided an introduction to some of the higher oxidation states of halogens: +I, +III, and +V. In general, the valence was also the same as the oxidation state, which, as we have emphasized, is not always the case (see Section 1.24). The heptavalent state is another important one for chlorine and iodine; we'll focus on chlorine in this section. The most familiar representatives of heptavalent chlorine are perchloric acid and perchlorate salts. Anhydrous perchloric acid ($HClO_4$) is one of the strongest common inorganic acids, and the perchlorate anion ($ClO_4{}^-$) is widely used as a weakly coordinating anion. *Caution*: Many perchlorate salts are explosive, so they should be handled with care and in small quantities.

A rather remarkable reaction involves the interaction of P_4O_{10} and $HClO_4$. When discussing the chemistry of P_4O_{10}, we emphasized its dehydrating properties; with perchloric acid, it extracts the elements of water, producing dichlorine heptoxide (Cl_2O_7), which may thus be viewed as the anhydride of perchloric acid (2 $HClO_4$ – H_2O):

$$2\ HClO_4 + P_4O_{10} \rightarrow Cl_2O_7 + H_2P_4O_{11} \tag{7.54}$$

This remarkable dehydrating action of P_4O_{10} reflects both the ability of the oxo groups to act as proton acceptors and the oxophilicity of the pentavalent phosphorus centers. Protonation of one of the terminal oxygens of P_4O_{10} is a logical first step of the mechanism. The nearest phosphorus is expected to become even more oxophilic as a result and therefore more susceptible to nucleophilic attack by perchlorate, ClO_4^-:

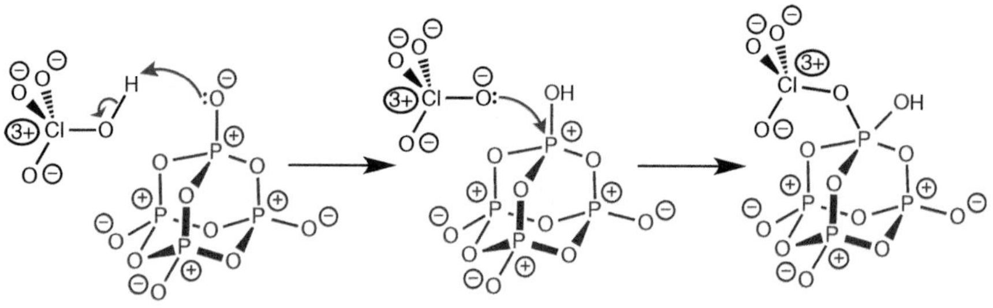

(7.55)

Additional protonation of a nearby oxygen of the P_4O_{10} skeleton is expected to prime the coordinated perchlorate to nucleophilic attack by a second perchlorate, resulting in Cl_2O_7.

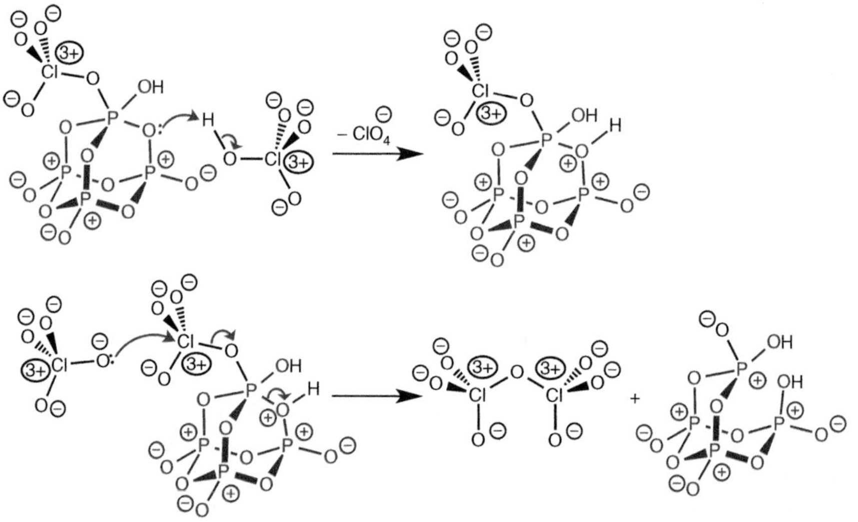

(7.56)

REVIEW PROBLEM 7.12

The unusual oxide Cl_2O_6 may be prepared via the interaction of ClO_2F and $HClO_4$. Suggest a structure for Cl_2O_6 and a mechanism underlying the preparative route.

REVIEW PROBLEM 7.13

Cl_2O_7, a classic acidic oxide, reacts with water to give perchloric acid. Draw a mechanism.

REVIEW PROBLEM 7.14

Cl_2O_6 is hydrolyzed by aqueous alkali to chlorate and perchlorate. Draw a mechanism.

We won't discuss standard preparative routes for perchloric acid or perchlorates (primarily for reasons of space), except for the following. Thermal decomposition of potassium chlorate ($KClO_3$) results in potassium perchlorate ($KClO_4$):

$$4 \text{ KClO}_3 \rightarrow 3 \text{ KClO}_4 + \text{KCl} \qquad (7.57)$$

Or, in ionic form:

$$4 \text{ ClO}_3^- \rightarrow 3 \text{ ClO}_4^- + \text{KCl} \qquad (7.58)$$

As in another example (disproportionation of hypochlorite) earlier in the chapter (Section 7.2), the mechanism of this reaction is somewhat hard to come to terms with because multiple negatively charged ClO_3^- ions must come together to generate the products. Clearly, the K^+ counterions play a major role in facilitating the interaction, as does the high temperature under which the reaction occurs.

By analogy with the earlier example, the mechanism may then be envisioned as simply a series of oxygen atom transfers:

$$(7.59)$$

An awkward aspect of each of the above steps is that the nucleophile is a chlorine with a positive formal charge. Once again, when dealing with higher-valent compounds, this need not be particularly upsetting. Formal charges are simply an electron-bookkeeping device and do not provide a realistic indication of the electrostatic character of atoms.

In our final example in this section, we will consider the synthesis of perchloryl fluoride, $FClO_3$, another heptavalent chlorine compound:

$$KClO_4 + 3\ HF + 2\ SbF_5 \rightarrow FClO_3 + KSbF_6 + H_3O^+SbF_6^- \qquad (7.60)$$

Observe that this reaction employs fluoroantimonic acid, one of the most potent superacids. The perchlorate anion is thus readily protonated, even though perchloric acid is one of the strongest common acids.

$$(7.61)$$

A fluoride anion, derived from HF, can then attack the electrophilic Cl in protonated perchloric acid, followed by departure of water:

$$(7.62)$$

Perchloryl fluoride ($FClO_3$) is a gas with a characteristic sweet smell. Although a high energy species from a thermodynamic point of view, $FClO_3$ is kinetically stable, decomposing thermally only at 400 °C! These properties make it of interest as an oxidizer for rocket fuels. In addition, unlike ClF_5 and BrF_5, it does not corrode the fuel tanks.

The chlorine atom of $FClO_3$ is significantly electrophilic, reacting with a variety of anionic nucleophiles. As a result, it's useful for introducing perchloryl (ClO_3) groups into organic compounds. Thus, RO^- gives $ROClO_3$ and PhLi gives $PhClO_3$.

$$(7.63)$$

REVIEW PROBLEM 7.15

Suggest mechanisms for the two $FClO_3$ reactions in (7.63) above.

7.8 INTERHALOGEN COMPOUNDS

A fair number of interhalogen compounds with the general formula XY_n where $n = 1, 3, 5$, or 7, are known, the more well-established ones being those listed in Table 7.1.

A couple of quick observations: For a given halogen X as the central atom, observe that the choice of Y = F generally leads to the highest valence for X. Second, note that ICl_3 exists as a dimer, with two of chlorines acting as bridging ligands.

REVIEW PROBLEM 7.16

Draw a Lewis structure for $(ICl_3)_2$.

A common reaction of the uncharged interhalogens is with powerful halide acceptors, resulting in interhalogen cations:

$$ClF_3 + AsF_5 \rightarrow [ClF_2^+][AsF_6^-] \tag{7.64}$$

$$BrF_5 + 2SbF_5 \rightarrow [BrF_4^+][Sb_2F_{11}^-] \tag{7.65}$$

$$IF_7 + BF_3 \rightarrow [IF_6^+][BF_4^-] \tag{7.66}$$

Certain halogen fluorides also act as fluoride anion acceptors:

$$ClF + CsF \rightarrow Cs^+[ClF_2^-] \tag{7.67}$$

$$ClF_3 + CsF \rightarrow Cs^+[ClF_4^-] \tag{7.68}$$

$$IF_7 + CsF \rightarrow Cs^+[IF_8^-] \tag{7.69}$$

TABLE 7.1   A List of Relatively Stable Interhalogens Containing No More than Two Different Halogens

	F	Cl	Br	I
F	F_2			
Cl	ClF, ClF_3, ClF_5	Cl_2		
Br	BrF, BrF_3, BrF_5	$BrCl$	Br_2	
I	IF, IF_3, IF_5, IF_7	$ICl, (ICl_3)_2$	IBr	I_2

In addition, mention should also be made of the following ligand exchange reaction of IF_7:

$$IF_7 + POF_3 \rightarrow IOF_5 + PF_5 \tag{7.70}$$

We already discussed possible mechanisms of this reaction in Section 1.19. We pointed out that such reactions typically involve bridged intermediates. Thus, for this reaction, an oxo-bridged intermediate was considered plausible.

The bromine fluorides BrF_3 and BrF_5 are important compounds. Bromine trifluoride and trivalent organobromine compounds are rapidly gaining acceptance as exciting new reagents in organic chemistry; we will have much more to say about them in Sections 7.14 and 7.15. BrF_5 is of interest as an oxidizer in liquid rocket propellants and as a fluorinating agent for the processing of uranium. Both compounds can be prepared by direct reaction of the elements under different conditions.

The uncharged heptavalent fluorides ClF_7 and BrF_7 do not exist, presumably for steric reasons. The unique heptavalent cations $[ClF_6]^+$ and $[BrF_6]^+$, however, may be prepared from the corresponding pentafluorides, XF_5 ($X = Cl$, Br), with such strong oxidants as PtF_6 or KrF^+:

$$2\,ClF_5 + PtF_6 \rightarrow ClF_6{}^+PtF_6{}^- + ClF_4{}^+PtF_6{}^- \tag{7.71}$$

$$XF_5 + KrF^+SbF_6{}^- \rightarrow XF_6{}^+SbF_6{}^- + Kr \quad (X = Cl,\ \ Br) \tag{7.72}$$

REVIEW PROBLEM 7.17

Use arrow pushing to rationalize reactions 7.64 through 7.69.

REVIEW PROBLEM 7.18

BrF_5 reacts violently with water to yield bromic acid and HF:

$$BrF_5 + 3\,H_2O \rightarrow HBrO_3 + 5\,HF$$

Write out a detailed mechanism.

7.9*   HALOGENS IN ORGANIC SYNTHESIS: SOME CLASSICAL REACTIONS

We are now in a good position to appreciate a variety of halogen-based reagents that are part of the modern organic chemist's repertoire. Before we do so, however, it may be useful to briefly recapitulate some of the classical applications of halogens in organic chemistry.

As briefly discussed in Chapter 1 (Section 1.11), molecular bromine adds to carbon–carbon double bonds by way of an ionic mechanism. The first step of the addition results in a cyclic bromonium ion, which is then opened up by a bromide ion. Such a mechanism accounts for the stereospecificity of the process, which may be described as an *anti* addition; that is, the two bromine atoms add to opposite faces of the double bond. It's worth noting that bromonium ions are not just a mechanistic hypothesis; indeed, with the highly sterically hindered alkene biadamantylidene, a bromonium ion has actually been isolated and characterized:

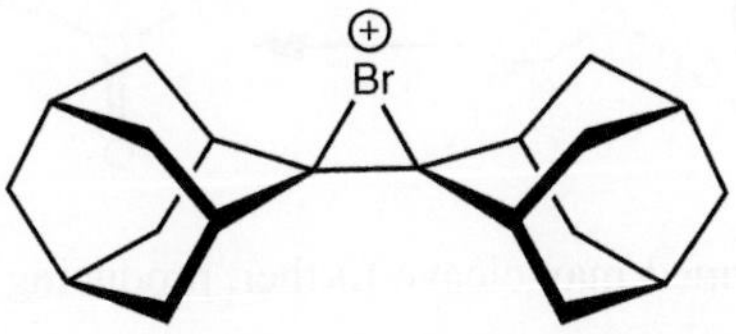

REVIEW PROBLEM 7.19*

The following substitution reaction, which occurs on exposure of the substrate to sodium formate in formic acid, proceeds with full retention of configuration:

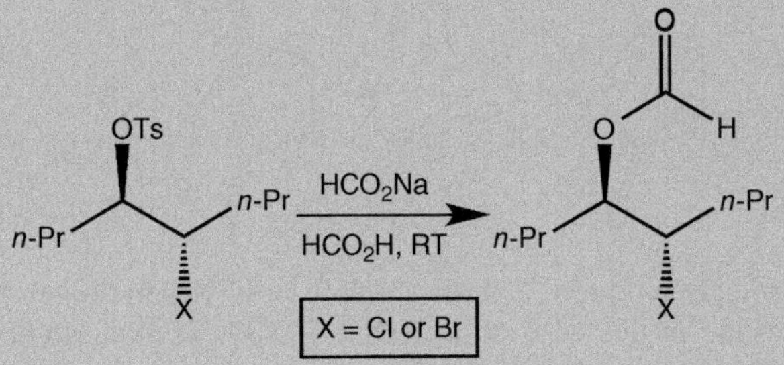

Suggest a mechanistic rationale. (*Hint*: Think of halonium ion intermediates)

Radical reagents, on the other hand, are perfectly suited for allylic bromination. *N*-bromosuccinimide (NBS) suspended in carbon tetrachloride, along with a radical initiator such as dibenzoyl peroxide or light, is the classic reagent for this purpose:

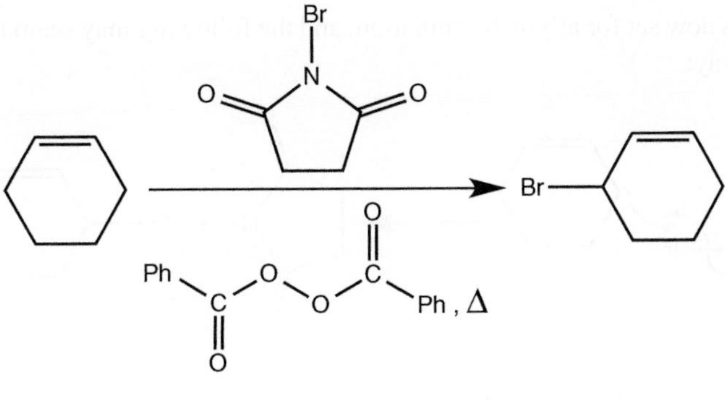

(7.73)

As you'll see, the mechanism is not particularly complex, but there are a few subtle points that are worth appreciating. The role of the radical initiator provides a natural starting point for our discussion. The very weak O–O bond (BDE ~138 kJ/mol) cleaves homolytically under the influence of heat or light:

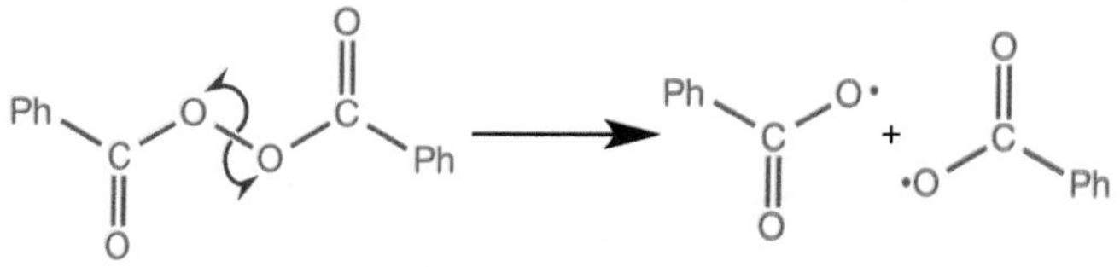

$$(7.74)$$

The benzoyloxy radical formed may cleave further, producing phenyl radicals and highly stable CO_2:

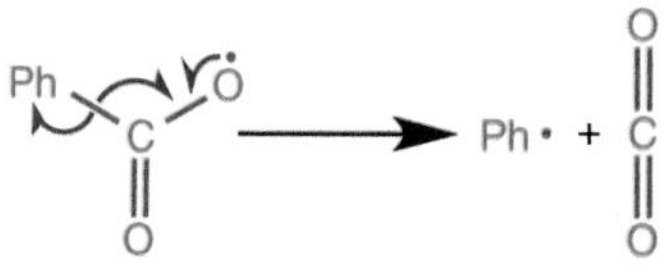

$$(7.75)$$

The weak N–Br bond in NBS (BDE ~243 kJ/mol) is also susceptible to homolytic cleavage:

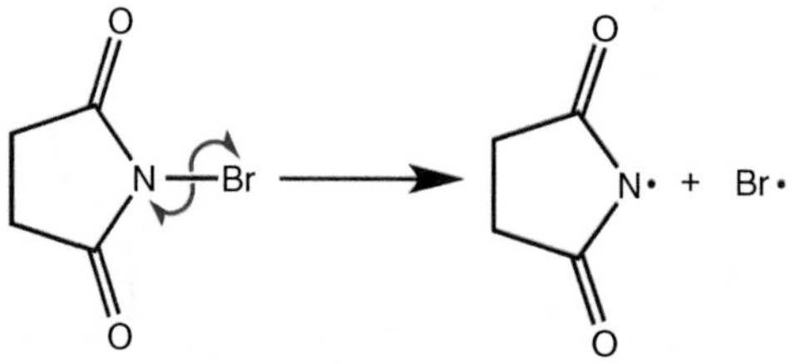

$$(7.76)$$

The various radicals produced can all abstract an allylic hydrogen from the organic substrate, cyclohexene in this case. With BDEs of ~372 kJ/mol, allylic C–H bonds are considerably weaker than regular alkyl-H bonds, whose dissociation energies vary from about 404 kJ/mol for a tertiary C–H bond to as high as 439 kJ/mol for methane. Allylic hydrogen abstraction is shown below for the benzoyloxy radical, a major radical species in the system:

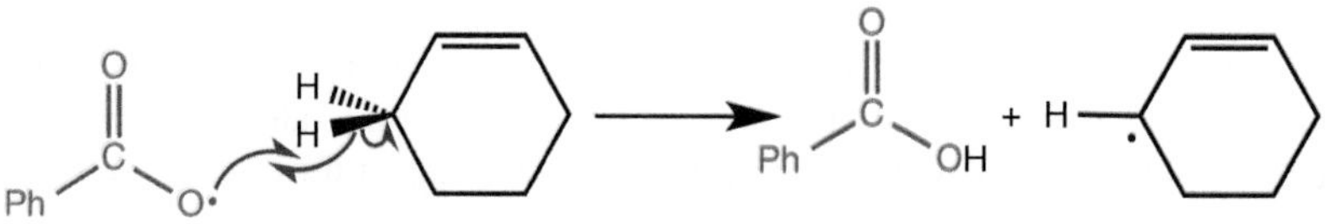

$$(7.77)$$

The stage is now set for allylic bromination, and the following may seem like a plausible reaction pathway:

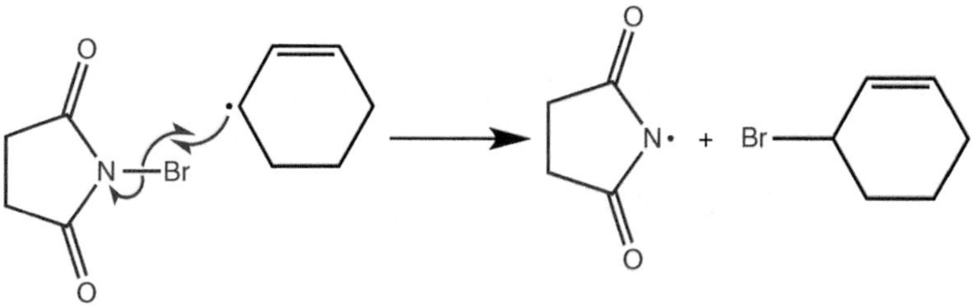

$$(7.78)$$

In fact, however, the above is *not* believed to be a significant reaction pathway. NBS is essentially insoluble in carbon tetrachloride so it's generally used only as a suspension. The concentration of available NBS is expected to be too low for the above reaction to proceed to a significant extent. Instead, the actual brominating agent is believed to be Br_2:

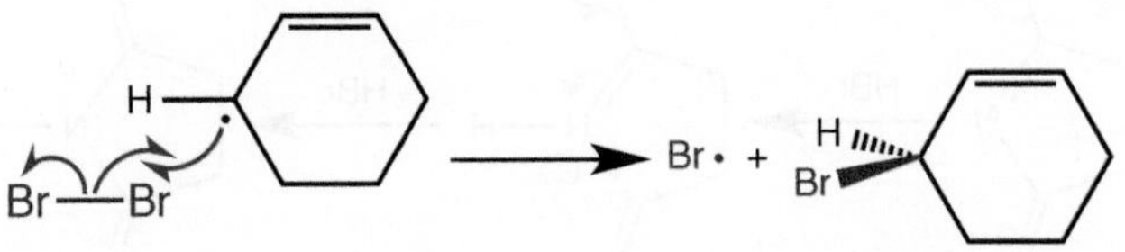

$$\text{(7.79)}$$

The bromine atoms produced also engage in allylic C–H abstraction, producing HBr in the process.

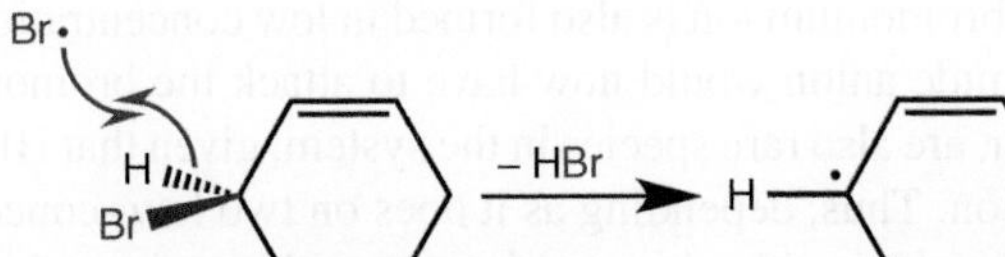

$$\text{(7.80)}$$

The bromine atoms may also undergo radical recombination with the allylic radical:

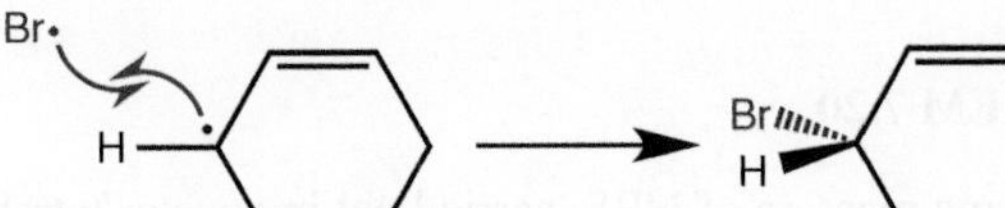

$$\text{(7.81)}$$

Two questions arise at this point: (1) How does molecular bromine arise in this system? (2) If molecular bromine is indeed formed, why does it not add to the double bond via a polar mechanism (involving a bromonium ion)?

1. The answer to the first question is that the small quantities of HBr present in the system react with NBS via a polar mechanism to produce small quantities of Br_2. In essence, a bromide anion is thought to attack the electrophilic bromine of a protonated NBS molecule:

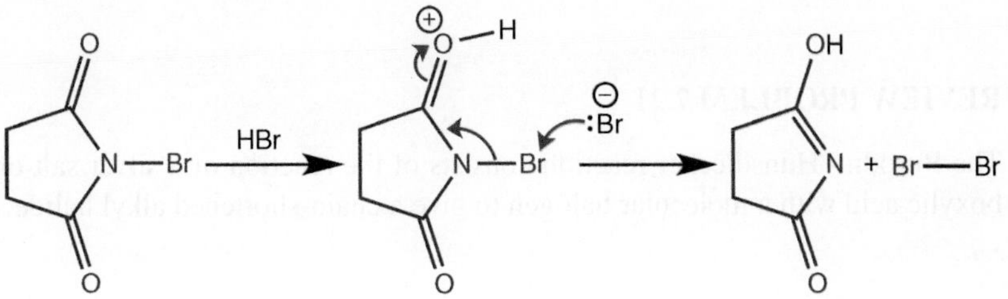

$$\text{(7.82)}$$

The "enolic" form of succinimide so produced then tautomerizes to the normal imide structure, under HBr catalysis.

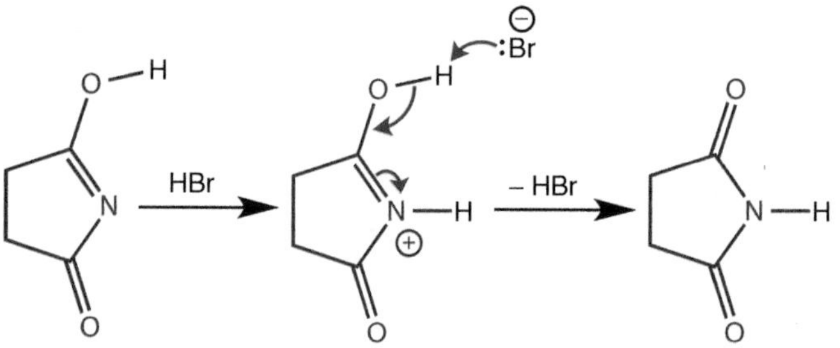

(7.83)

2. Let us now suppose that molecular bromine, formed in the above manner, reacts with the alkene to generate a bromonium ion. Because of the low–concentration of Br_2 in the system, the bromonium ion is also formed in low concentration. For Br_2 addition to occur, a bromide anion would now have to attack the bromonium ion. Bromide anions, however, are also rare species in the system, given that HBr is formed only in low concentration. Thus, depending as it does on two low–concentration species, a bromonium ion and bromide, this reaction channel is an improbable one and cannot compete with the radical pathway, which explains the absence of Br_2 addition under these conditions. This picture is backed by quantitative kinetic evidence, which we won't go into.

REVIEW PROBLEM 7.20

Consider the following reaction of NBS, carried out in a water/tetrahydrofuran mixture, a "polar opposite" of carbon tetrachloride, which is the standard solvent for allylic brominations:

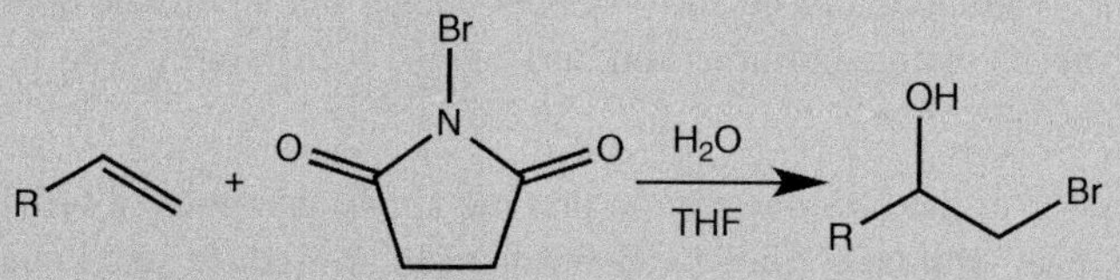

Suggest a mechanism for the reaction.

REVIEW PROBLEM 7.21

The Borodin–Hunsdiecker reaction consists of the reaction of a silver salt of a carboxylic acid with a molecular halogen to give a chain-shortened alkyl halide.

Suggest a mechanism. *Hint*: The reaction depends critically on the high affinity of Ag^+ for halide ions. Also consider the possibility of radical intermediates.

The majority of classical applications of halogens in organic chemistry, however, involve the use of X_2 or X^+ equivalents as electrophiles. One such classic reaction is the haloform reaction, which involves reaction of a methyl ketone with a mixture of X_2 (X = Cl, Br, I) and NaOH/H_2O, effectively NaOX:

$$(7.84)$$

In the old days, before spectroscopic analysis became routine, this reaction, when carried out with iodine (X = I), was used as a qualitative visual test for methyl ketones because the iodoform (CHI$_3$) produced is easily recognizable as a pale yellow solid. Nowadays, the reaction is used mainly as a means of shortening the carbon chain of a methyl ketone. The mechanism involves repeated generation of a terminal enolate from the methyl ketone and its reaction with HOX, a weak acid and the putative halogenating agent under the reaction conditions:

$$(7.85)$$

The trihalomethylketone formed is cleaved by hydroxide, generating a chain-shortened carboxylate and a trihalomethide anion. The latter picks up a proton from the protic reaction medium, producing the haloform (CHX$_3$):

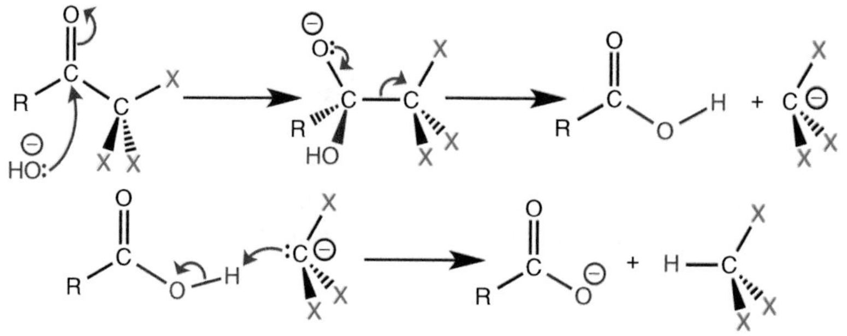

(7.86)

Another classic use of halogens (bromine, in particular) in organic chemistry is the Hofmann rearrangement or degradation of a primary amide (RCONH$_2$) to an isocyanate (RNCO). The reagent is bromine and sodium hydroxide, effectively NaOBr, essentially the same as that in the haloform reaction. "Rearrangement" refers to the fact that the carbon connectivity is changed from R–C–N to R–N–C; in other words, the carbon chain is shortened, which is why the reaction is also referred to as Hofmann degradation.

The reaction begins with deprotonation of the amide to an enolate-like intermediate, as shown below:

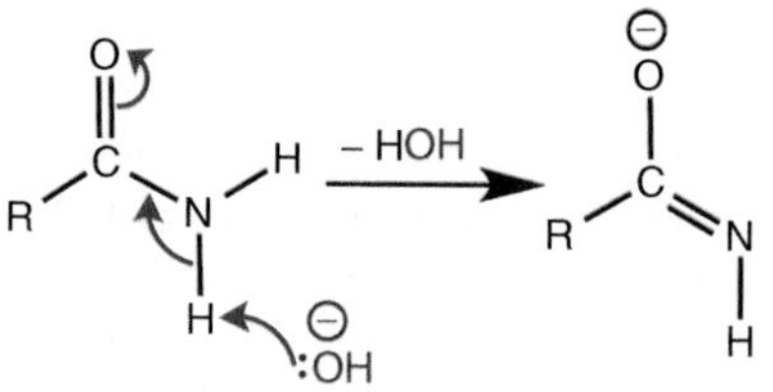

(7.87)

Hypobromous acid (HOBr) then presumably acts as the brominating agent:

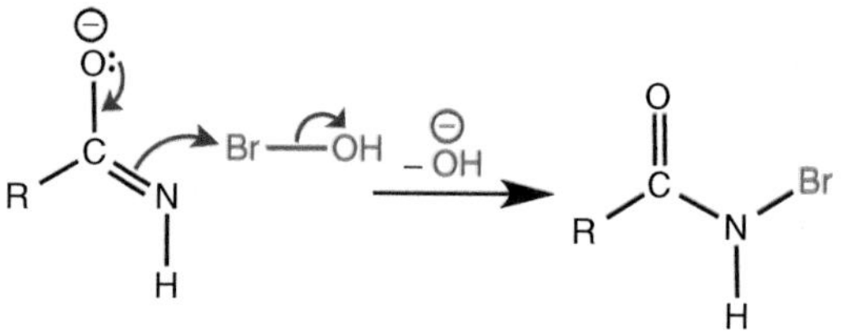

(7.88)

Deprotonation of the *N*-bromoamide so formed then triggers the rearrangement to isocyanate:

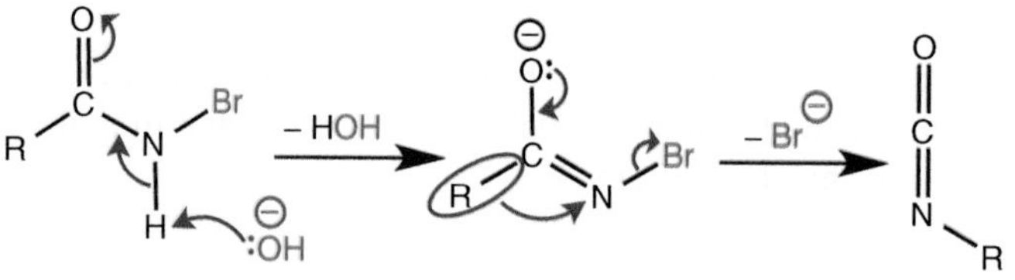

(7.89)

Very often, the isocyanate is not isolated but is allowed to hydrolyze to the amine under the reaction conditions:

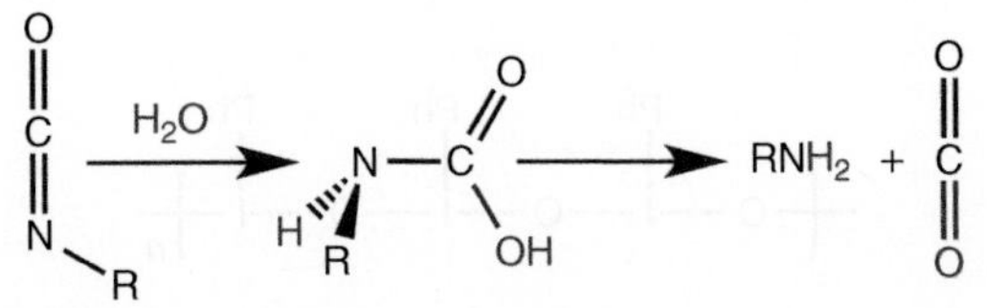

(7.90)

7.10 AN INTRODUCTION TO HIGHER-VALENT ORGANOIODINE COMPOUNDS

Higher-valent organoiodine compounds represent a major contribution of modern halogen chemistry to organic synthesis. The first such compound, dichloroiodobenzene ($PhICl_2$), however, was prepared well over a century ago:

$$PhI + Cl_2 \rightarrow PhICl_2 \qquad (7.91)$$

Given the variety of iodine valence states in organoiodine compounds, the λ nomenclature (see Section 1.26) will prove useful for our discussion. Dichloroiodobenzene is thus a λ^3-iodane, the superscript being essentially the valence of the atom in question. As shown below, a number of other λ^3-iodanes such as iodobenzene diacetate are also readily accessible from iodobenzene; the pentavalent iodine compound iodoxybenzene is shown below in a different color to distinguish it from the other trivalent compounds.

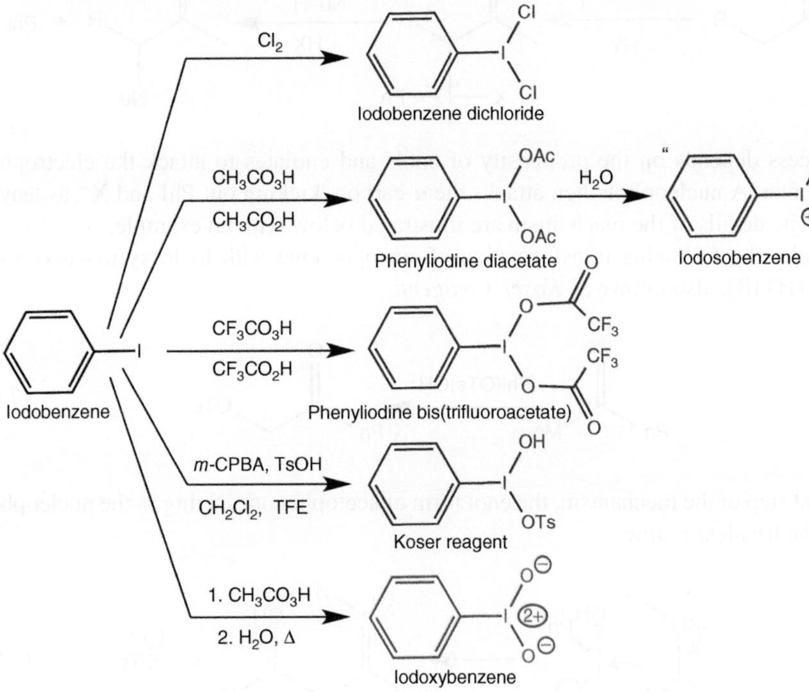

(7.92)

Iodobenzene diacetate can be hydrolyzed to iodosobenzene, which is conventionally written as Ph–I=O or Ph–I$^+$–O$^-$; this, however, is only an empirical formula, the actual structure being a polymer:

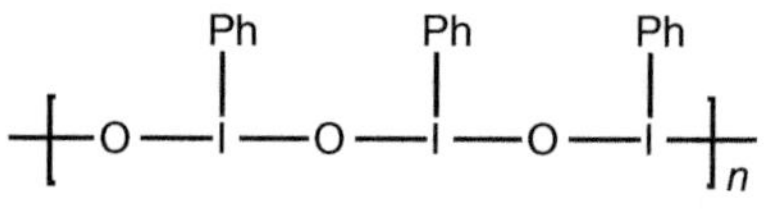

Peroxyacid treatment, followed by heating in water, converts iodobenzene to iodoxybenzene, a pentavalent iodine compound, as shown above.

To varying degrees, all these compounds have found use as reagents in organic synthesis. In addition, substituted iodobenzenes have led to additional sophisticated reagents. Below we examine the mechanisms of action of a few of these.

7.11 λ^3-IODANES

Aryl-λ^3-iodanes have gained popularity as oxidants on account of their stability, ease of handling, and selectivity. One of their chief applications is functionalization of enolizable carbonyl compounds:

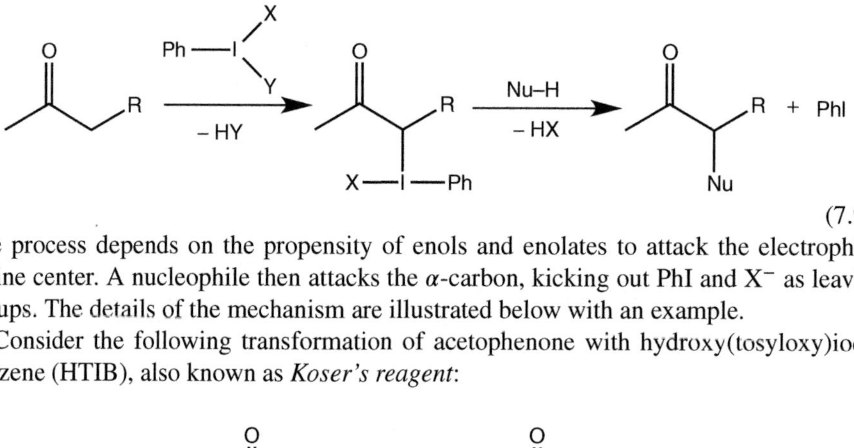

$$(7.93)$$

The process depends on the propensity of enols and enolates to attack the electrophilic iodine center. A nucleophile then attacks the α-carbon, kicking out PhI and X$^-$ as leaving groups. The details of the mechanism are illustrated below with an example.

Consider the following transformation of acetophenone with hydroxy(tosyloxy)iodobenzene (HTIB), also known as *Koser's reagent*:

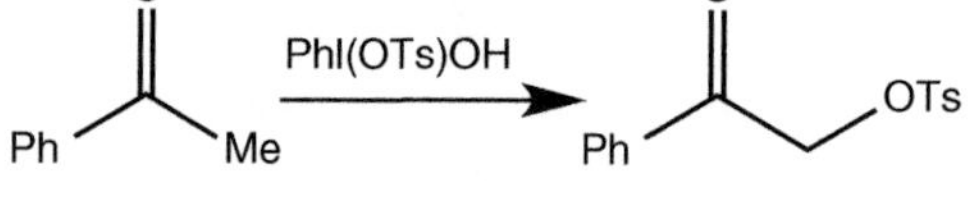

$$(7.94)$$

In the first step of the mechanism, the enol form of acetophenone, acting as the nucleophile, attacks the trivalent iodine:

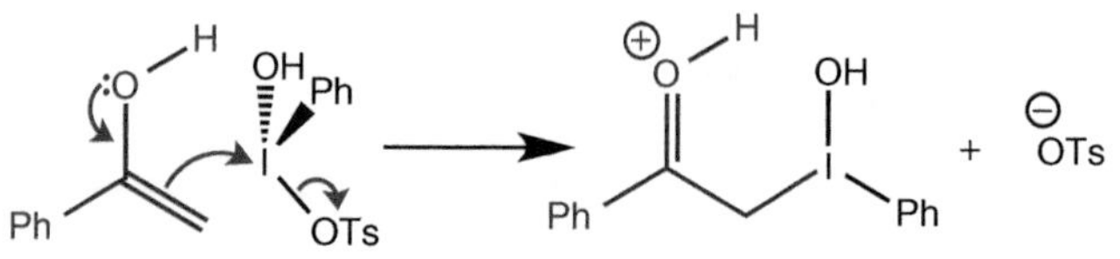

$$(7.95)$$

Tosylate ($^-$OTs) is then a more than adequate nucleophile for kicking out PhI, a superb leaving group:

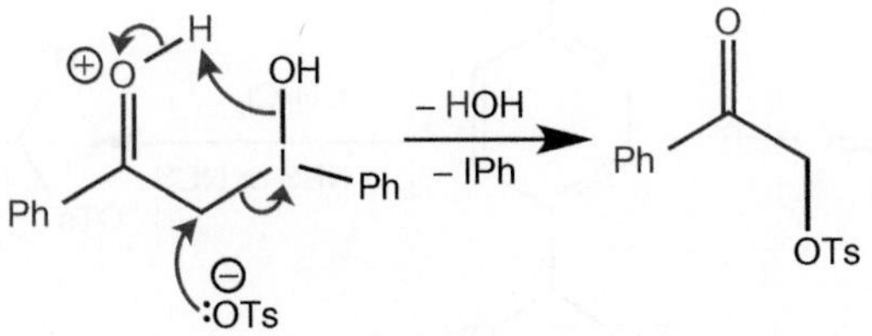

(7.96)

The product of the reaction, an α-tosyloxyketone, is a valuable intermediate which can be functionalized to a variety of products via nucleophilic displacement of the tosylate.

Ligand exchange of monoaryl-λ^3-iodanes with an arylsilane or arylstannane leads to analogous diaryl reagents, which have unique uses. For example, diaryl-λ^3-iodanes bring about α-arylation of enolates, as shown below:

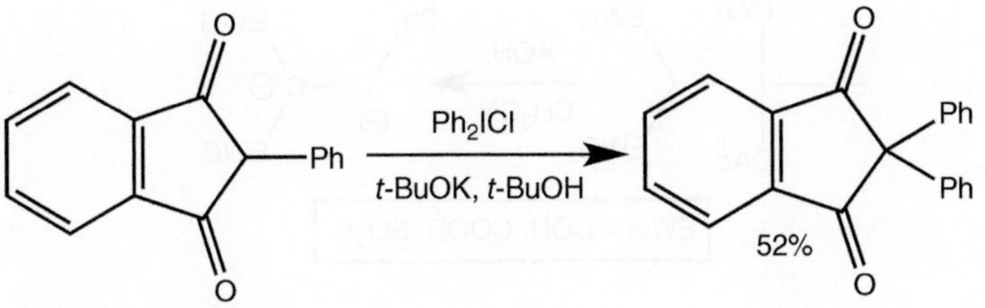

(7.97)

The mechanism is likely to involve nucleophilic attack by the enolate on the trivalent iodine center. Subsequent reductive elimination at iodine results in 2,2-diphenylindane-1,3-dione, the observed product, as shown below:

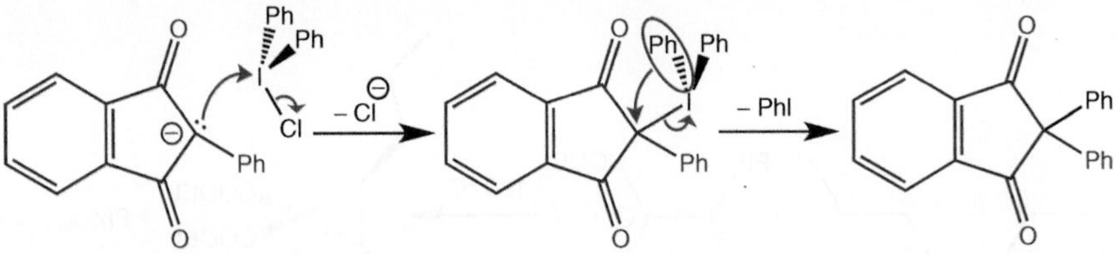

(7.98)

REVIEW PROBLEM 7.22

Poly{[4-(hydroxy)(tosyloxy)iodo]styrene}, a polymeric version of Koser's reagent, has been reported to effect "halotosyloxylation" of alkynes with iodine or NBS or NCS, as shown below. The advantage of the polymeric reagent is that it can be readily separated from the product, regenerated, and reused.

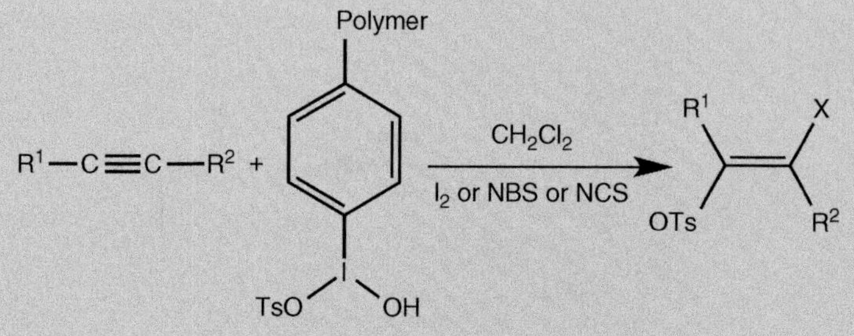

Suggest a mechanism.

An interesting application of phenyliodine diacetate is the reaction with carbon acids such as β-diketones under mild, basic conditions to yield iodonium ylides (EWG = electron-withdrawing group):

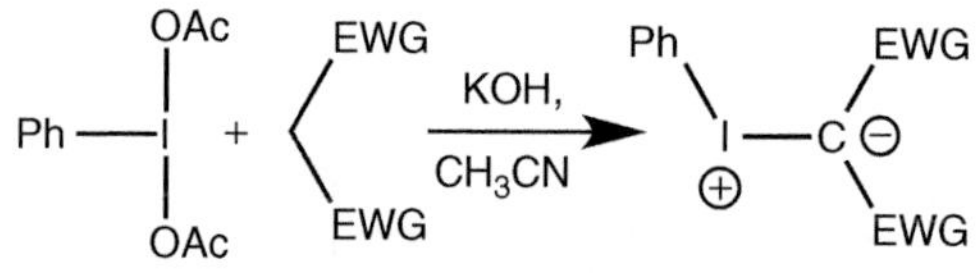

$$\text{EWG = COR, COOR, SO}_2\text{Ar}$$

(7.99)

Under the same conditions, arylsulfonamides are converted to phenyliodonium imides or iminoaryl-λ^3-iodanes.

These iodonium ylides function as carbene and nitrene transfer agents. Carbene transfer typically requires a transition-metal catalyst. A few uncatalyzed reactions, however, have been documented, such as the following remarkably stereospecific transformations of (Z)- and (E)-hept-3-ene:

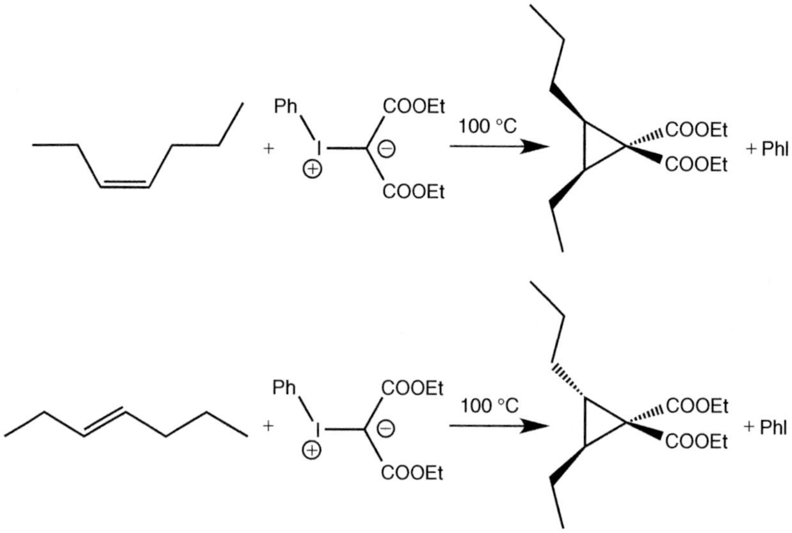

(7.100)

Somewhat regrettably, we will skip a proper mechanistic discussion of iodonium ylide-mediated group transfer chemistry. As mentioned, the majority of such processes require transition-metal catalysis, a topic outside the scope of this book.

REVIEW PROBLEM 7.23

Antonio Togni and coworkers have recently described aryl(trifluoromethyl)-λ^3-iodanes that function as *electrophilic* trifluoromethylating agents (Eisenberger, P.; Gischig, S.; Togni, A. *Chem. Eur. J.* **2006**, *12*, 2579–2586); these reagents nicely complement the *nucleophilic* Ruppert–Prakash reagent (Me$_3$SiCF$_3$), which we mentioned in Review problem 4.5.

Togni reagent 1   Togni reagent 2

The Togni reagents, as these compounds have come to be known, trifluoromethylate a variety of functional groups, including alcohols, phenols, thiols, β-ketoesters and α-nitroesters:

(c), (d), 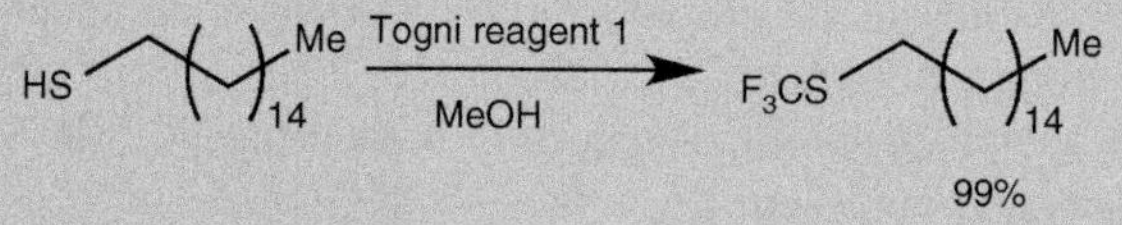 (e) reaction schemes with Togni reagent 1

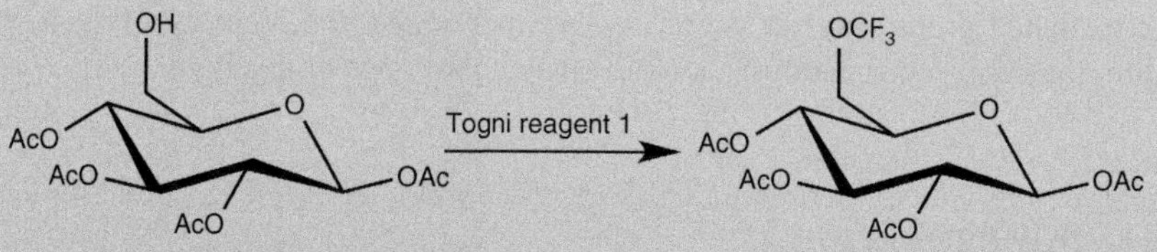

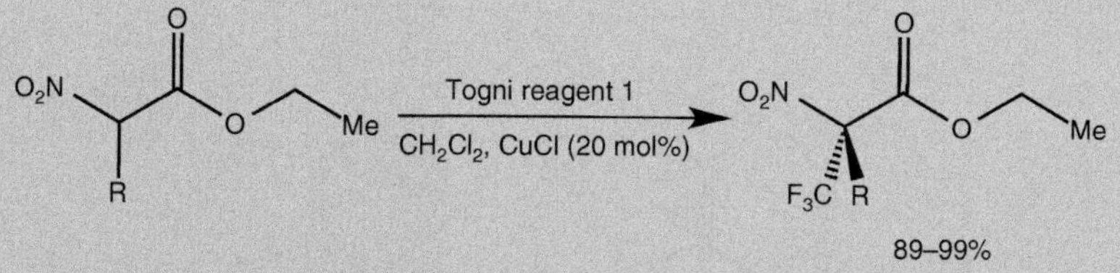

Propose reasonable mechanisms for the above transformations.

Note: The last reaction, which involves an α-nitroester, warrants a couple of comments. Note that a copper catalyst has been added; this is sometimes done for Togni reagents. Unfortunately, a discussion of the role of the copper is outside the scope of this book. Second, the α-nitro-α-trifluoromethyl product may be transformed to the corresponding α-amino acids, which in turn may be applied to the synthesis of fluorinated pharmaceuticals.

REVIEW PROBLEM 7.24

This is a bit of a tough puzzle, but the reactants are rather simple and there really aren't that many choices regarding what must be happening: Ozone reacts with iodine dissolved in cold, anhydrous perchloric acid to yield iodine(III) perchlorate:

$$I_2 + 6\,HClO_4 + O_3 \rightarrow 2\,I(ClO_4)_3 + 3\,H_2O$$

Suggest a mechanism.

7.12 λ^5-IODANES: IBX AND DESS–MARTIN PERIODINANE

Two of the most important λ^5-iodane reagents are 2-iodoxybenzoic acid (IBX) and Dess–Martin periodinane (DMP). The best current syntheses of the two reagents are as follows, where $KHSO_5$ is potassium hydrogen peroxysulfate:

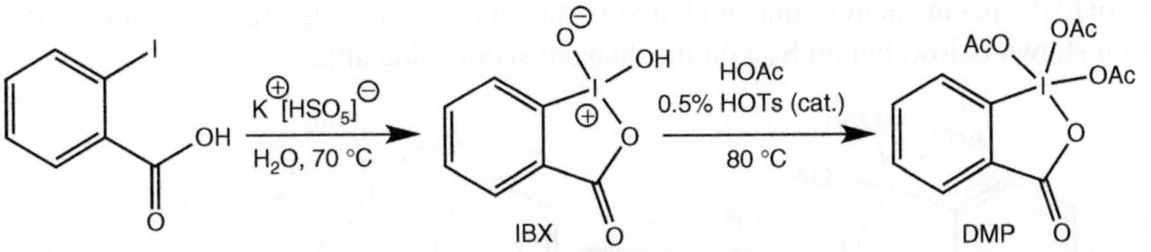

(7.101)

Acetylation of the O$^-$/OH group results in DMP being much more soluble in organic solvents than IBX. Both are excellent and selective reagents for oxidation of primary alcohols to aldehydes and secondary alcohols to ketones. Both have a number of advantages over DMSO-based (see Section 6.9) and transition-metal-based protocols.

Let us first consider the mechanism of oxidation of a primary alcohol by IBX:

$$RCH_2OH \xrightarrow{\text{IBX}} RCHO \qquad (7.102)$$

Coordination of the alcohol to the pentavalent iodine of IBX is a reasonable starting point:

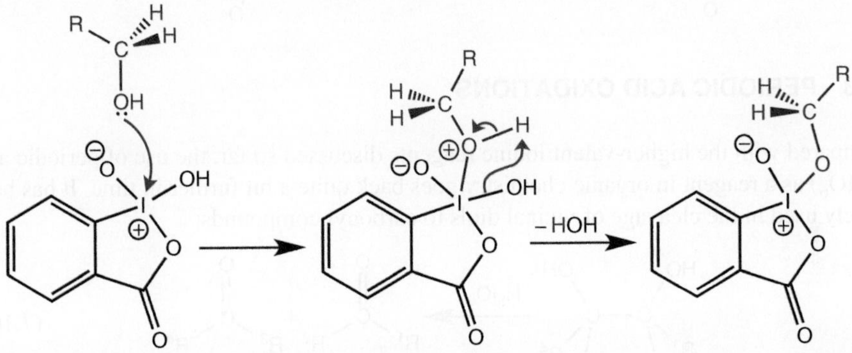

(7.103)

The alkoxy-λ^5-iodane intermediate so produced might then be expected to fall apart via a cyclic transition state to a carbonyl compound (an aldehyde in our example) and an aryl-λ^3-iodane as the final products:

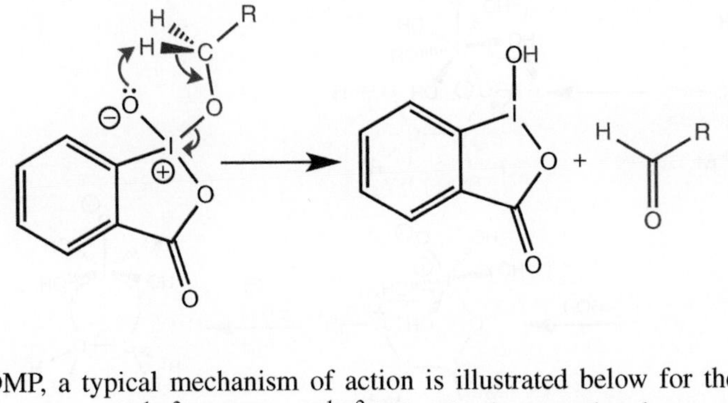

(7.104)

For DMP, a typical mechanism of action is illustrated below for the oxidation of a secondary alcohol (R^{1}R^{2}CHOH → R^{1}R^{2}CO). The first step involves coordination of the alcohol to the pentavalent iodine, with loss of one of the acetate ligands. Mechanistic details are not shown below, but an S_N2-Si mechanism seems plausible:

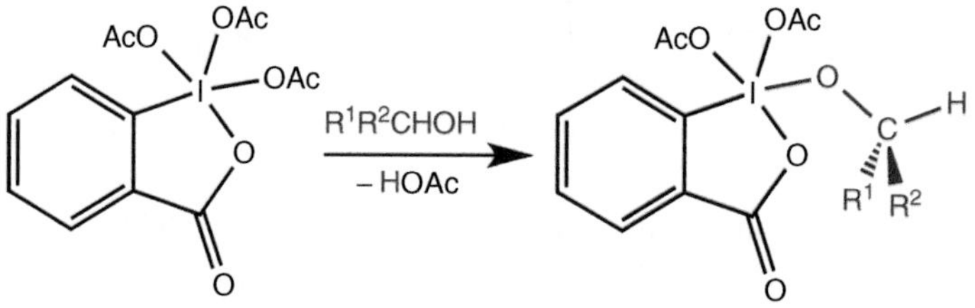

$$(7.105)$$

Abstraction of an α-proton from the alkoxide then results in a ketone and an aryl-λ^3-iodane, as shown below.

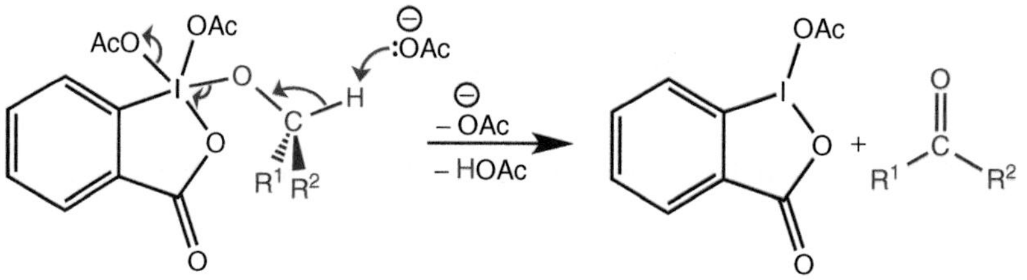

$$(7.106)$$

7.13 PERIODIC ACID OXIDATIONS

Compared with the higher-valent iodine reagents discussed so far, the use of periodic acid (H_5IO_6) as a reagent in organic chemistry goes back quite a bit further in time. It has been widely used in the cleavage of vicinal diols to carbonyl compounds:

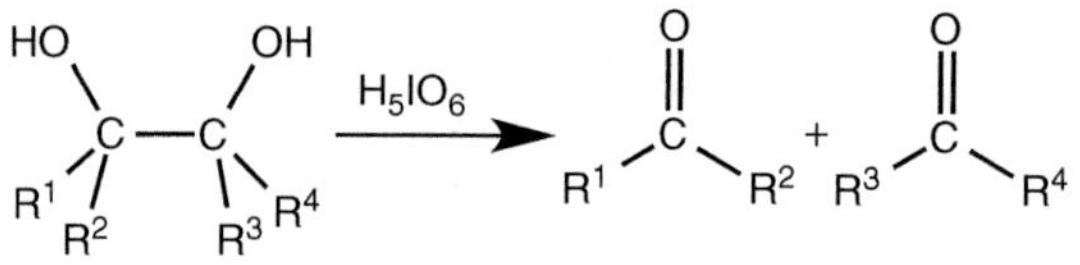

$$(7.107)$$

As with lead tetraacetate, the first steps of diol cleavage involve formation of a cyclic ester:

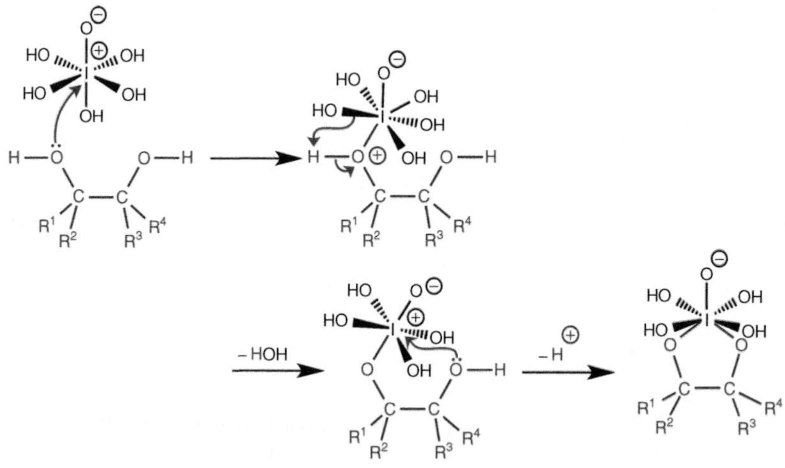

$$(7.108)$$

The cyclic periodate ester breaks down spontaneously, as shown below, creating a pair of carbonyl groups and a pentavalent, pentacoordinate iodine intermediate:

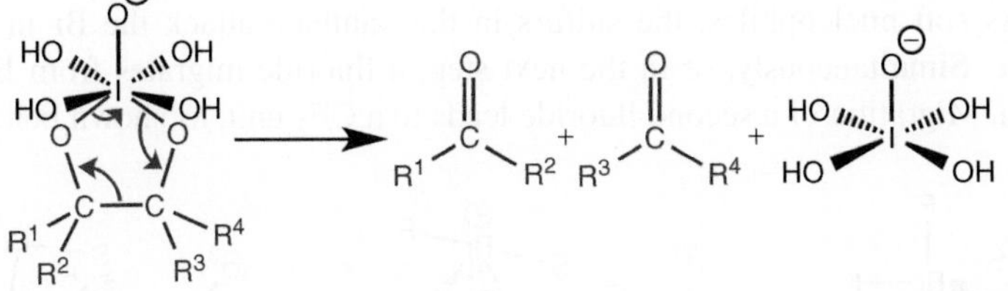

$$(7.109)$$

The iodine intermediate, which may be viewed as a hydrated form of iodic acid, should then eliminate water to yield iodate (IO_3^-). A water-catalyzed pathway is shown below:

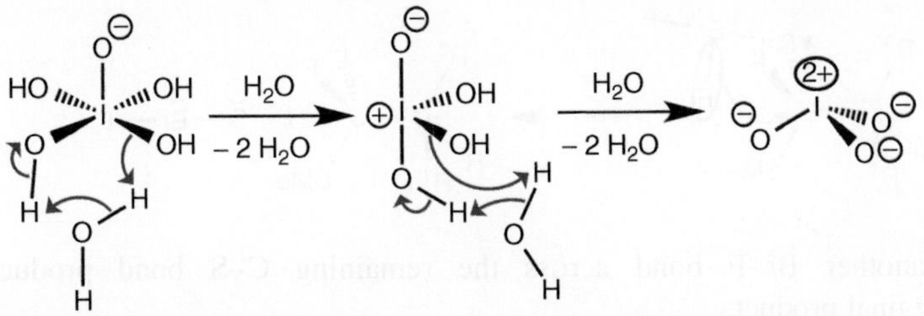

$$(7.110)$$

Diol cleavage played an important role in natural product structure analysis in the days before instrumental analysis. Understandably, the importance of periodic acid oxidations has diminished somewhat in modern organic chemistry.

7.14 BROMINE TRIFLUORIDE

Compared to higher-valent iodine reagents, aryl-λ^3-bromanes have only attracted significant attention in the last few years. They are synthesized from bromine trifluoride (BrF_3), an increasingly important reagent in its own right and our topic for this section; aryl-λ^3-bromanes are discussed in the next section (Section 7.15). Bromine trifluoride is commercially available and, with care and common sense, can be handled safely in a well-equipped organic/inorganic chemistry laboratory. That said, there's no denying that it's a dangerously corrosive substance, which etches glass and reacts violently with water and oxygen-containing solvents.

Bromine trifluoride is very useful for creating CF_2 and CF_3 groups, which are present in a variety of drugs, anesthetics, and other biomedical products. Thus, carboxylic acids can be converted to CF_3 compounds with BrF_3, albeit after initial conversion to the xanthate esters, as shown below:

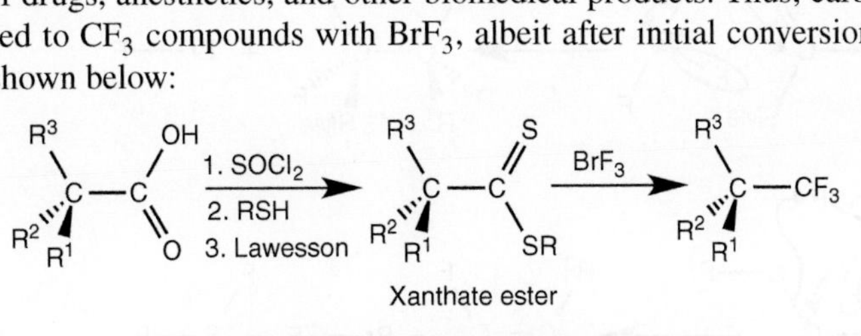

$$(7.111)$$

Yields: 80% (R^1, R^2, R^3 = alkyl/aryl)
40% (R^1, R^2, = alkyl/aryl; R^3 = H)
10% (R^1 = alkyl; R^2, R^3 = H)

A few quick comments on the first three steps of the transformation may be useful: $SOCl_2$ converts the COOH to COCl, RSH then converts COCl to C(O)SR (thioester), and Lawesson's reagent (which we discussed in Section 6.14) then converts the thioester to the xanthate. As soft nucleophiles, the sulfurs in the xanthate attack the Br in BrF_3, a soft electrophile. Simultaneously, or in the next step, a fluoride migrates from Br to the xanthate carbon. Migration of a second fluoride leads to a CF_2 unit, as shown below:

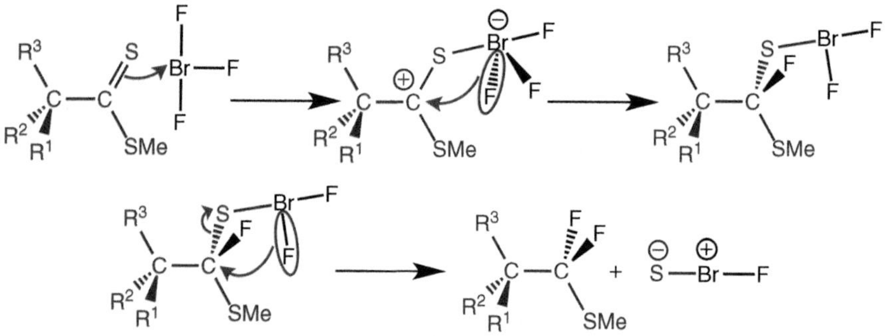

$$(7.112)$$

Addition of another Br–F bond across the remaining C–S bond produces the CF_3-containing final product.

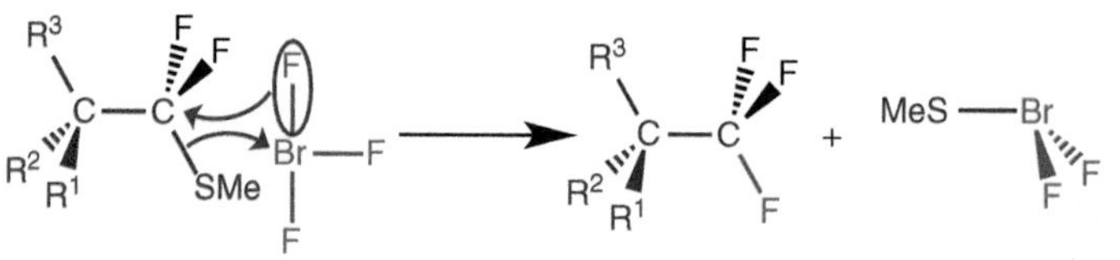

$$(7.113)$$

The fate of the sulfur-containing wastes has not been addressed in detail in the literature, but they can be destroyed with a solution of NaOCl (bleach).

An interesting aspect of the above mechanism is that it has not involved any net oxidation or reduction. Thus, the bromine has remained trivalent throughout. In that sense, the process may remind you of fluorination by SF_4 or DAST (Section 6.12). An alternative mechanism, where fluorine is the site of electrophilic attack and BrF_3 undergoes reductive elimination, is also conceivable.

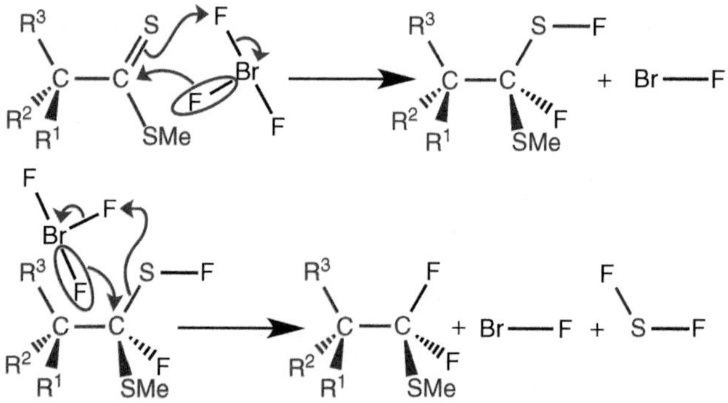

$$(7.114)$$

Although detailed mechanistic studies are lacking, HSAB considerations appear to favor the first non-redox mechanism, where sulfur, a soft nucleophile, attacks bromine, a soft electrophile.

REVIEW PROBLEM 7.25*

BrF_3 is also useful for transforming carbonyl groups to CF_2 units. First, however, the carbonyl groups need to be derivatized to hydrazones or oxime ethers:

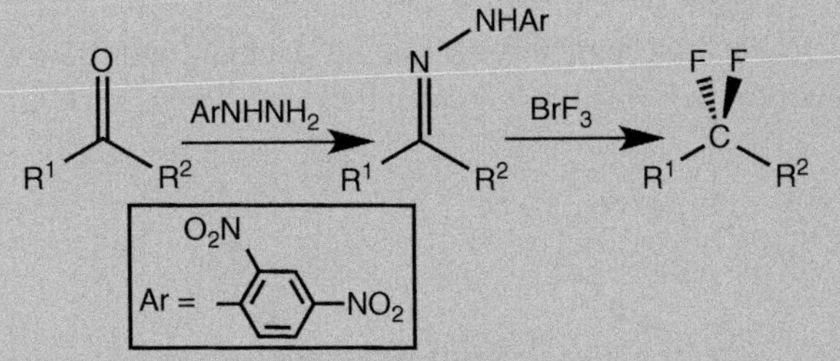

Suggest a mechanism for the BrF_3-mediated transformation.

Pyridine (py) reacts with BrF_3 to form a complex, which may be written as $BrF_3 \cdot py$ (Hagooly, Y.; Rozen, S. *Org. Lett.* **2012**, *14*, 1114–1117). It is obtained very simply as a white precipitate by mixing equimolar quantities of BrF_3 and pyridine in $CHCl_3$ or $CFCl_3$ without any additional purification. It is a user-friendly, solid form of BrF_3 that is also somewhat milder in its reactivity. Typically, it is fully functional in fluorination reactions where BrF_3 would otherwise have been used.

REVIEW PROBLEM 7.26*

Suggest a mechanism for the following reaction:

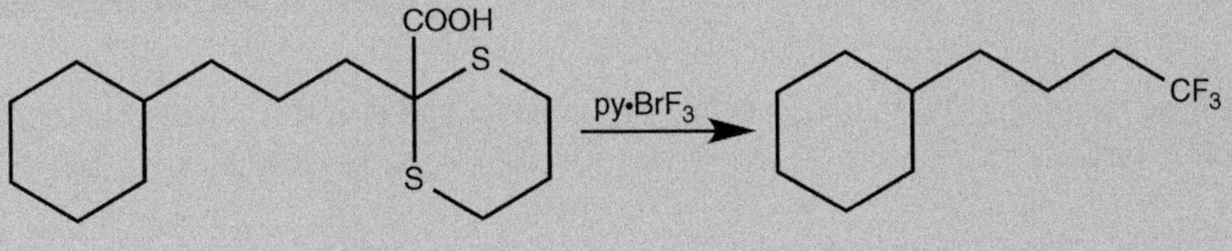

BrF_3 is sufficiently reactive to fluorinate certain tertiary C–H bonds, as shown below, where R is a carbalkoxy (CO_2R') or fluorinated alkyl group:

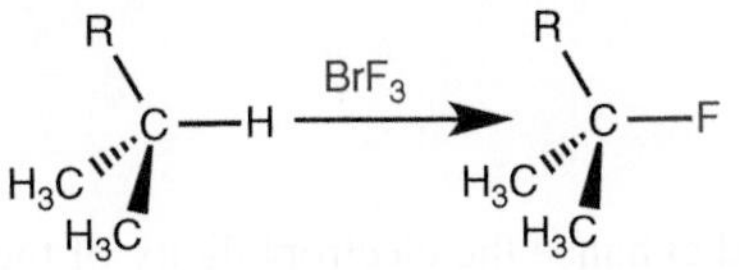

(7.115)

A cyclic transition state, with BrF_3 undergoing reductive elimination, appears plausible:

$$(7.116)$$

7.15* ARYL-λ^3-BROMANES

In a significant recent development, Masahito Ochiai and coworkers synthesized aryl-λ^3-bromanes from BrF_3 and a nucleophilic aryl donor such as an arylsilane or arylstannane:

$$(7.117)$$

As expected from the higher electronegativity of bromine relative to iodine, aryl-λ^3-bromanes are both more electrophilic and more susceptible to reductive elimination than analogous aryl-λ^3-iodanes. Thus, the aryl-λ^3-bromanes have allowed both faster and more convenient versions of known reactions, as well as a number of new reactions.

REVIEW PROBLEM 7.27*

Suggest a mechanism for reaction 7.117.

Alkyne couplings are a good example of the novel and useful reactivity exhibited by aryl-λ^3-bromanes.

$$(7.118)$$

Coordination of BF_3 should enhance the electrophilicity of the λ^3-bromane, priming it as an acceptor of a migrating alkynyl group:

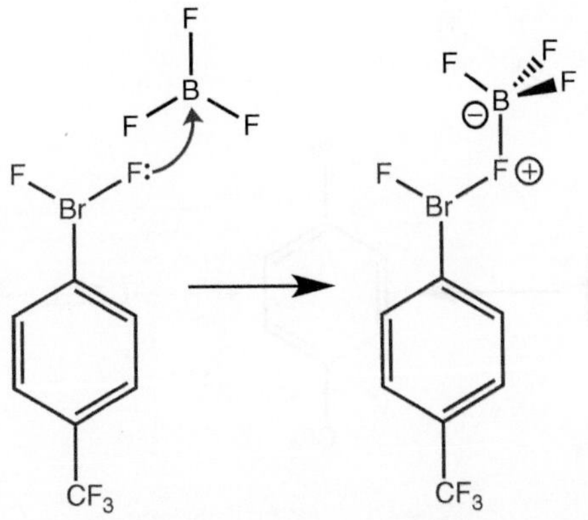

(7.119)

The coordinated BF_3 might then latch on to the tin of the alkynylstannane, creating an anionic tin center and thereby setting the stage for alkynyl migration:

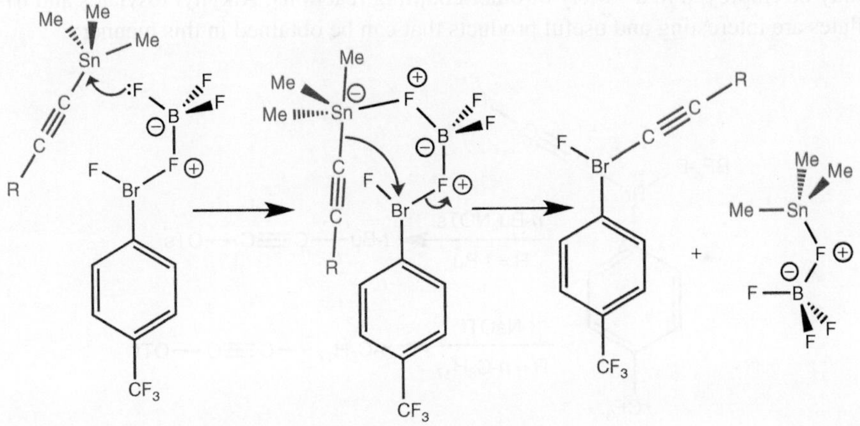

(7.120)

The (alkynyl)aryl-λ^3-bromane so formed can undergo a second round of the same chemistry to yield a (dialkynyl)aryl-λ^3-bromane intermediate, which undergoes spontaneous reductive elimination to generate the diyne final product:

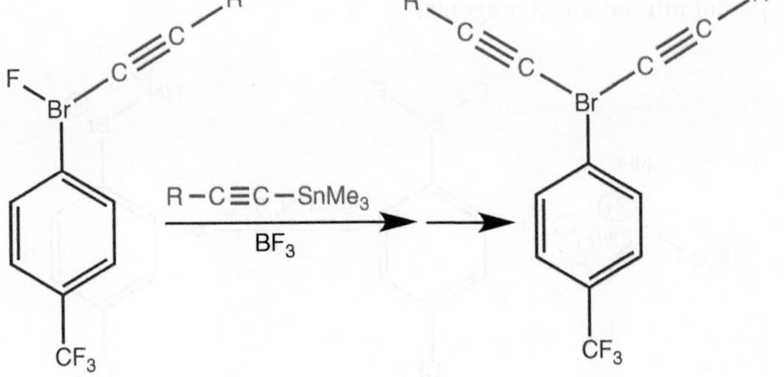

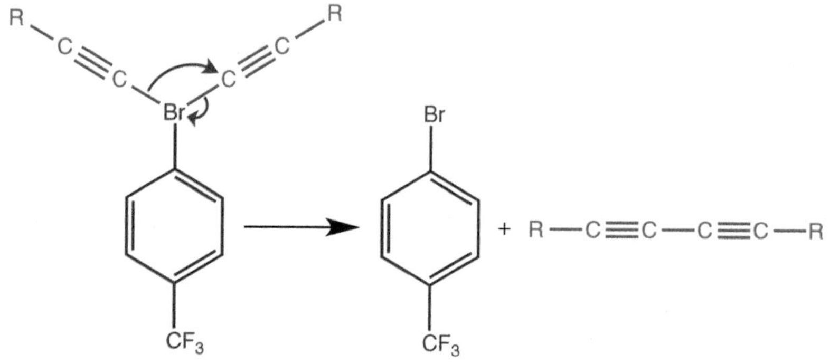

(7.121)

REVIEW PROBLEM 7.28*

The (alkynyl)aryl-λ^3-bromanes mentioned above are stable enough to be isolated and may be employed in a variety of other coupling reactions. Alkynyl tosylates and triflates are interesting and useful products that can be obtained in this manner:

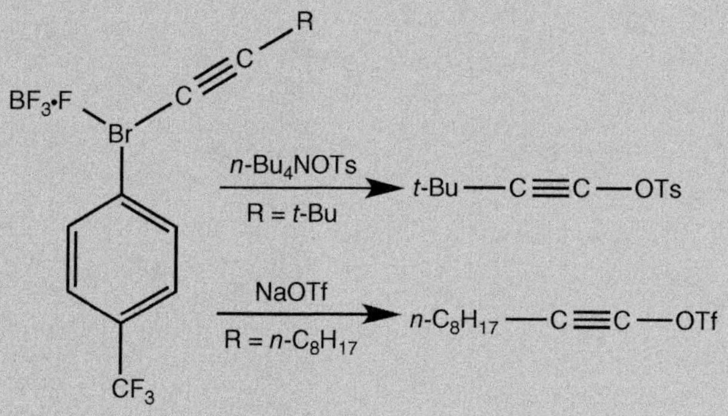

Suggest mechanisms for the above two reactions.

Interaction of trifluoromethanesulfonamide (also known as *triflamide* or *triflylamide*, TfNH$_2$) with aryl(difluoro)-λ^3-bromanes leads to triflylimido-λ^3-bromanes, ArBrNTf, which act as potent nitrene transfer agents.

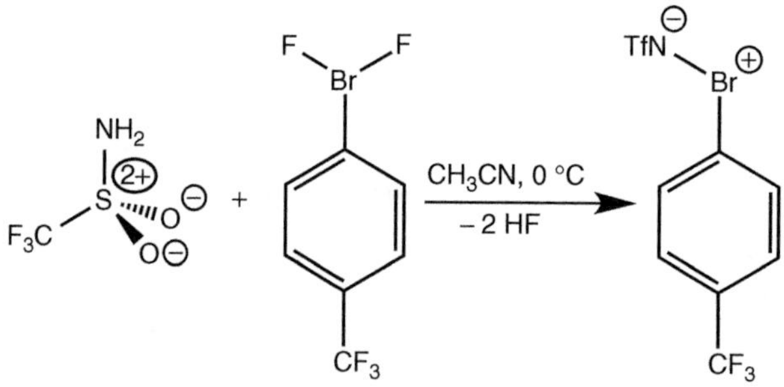

(7.122)

Interestingly, an analogous synthesis cannot be accomplished with *p*-toluenesulfonamide (tosylamide, TsNH$_2$), indicating the need for a group that is more electron withdrawing than Ts.

Formation of the Br–N bond may be expected to involve nucleophilic attack by the triflamide nitrogen on the trivalent bromine; this would be followed by elimination of HF:

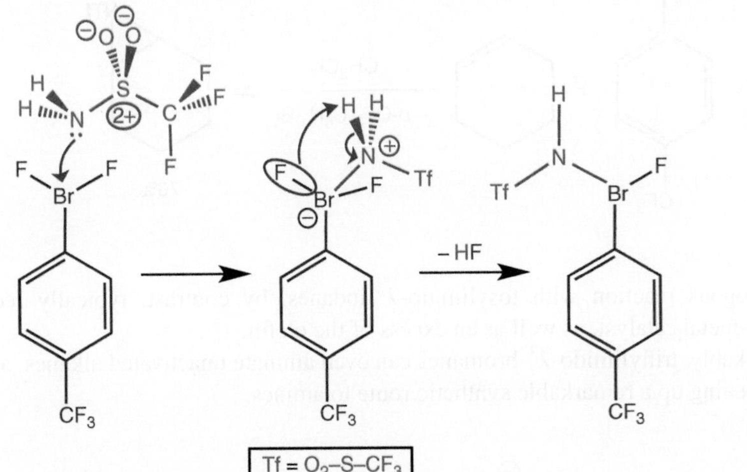

$$\text{Tf} = O_2\text{–S–CF}_3$$

(7.123)

Elimination of a second molecule of HF may be facilitated by coordination of acetonitrile to the bromine, as shown below:

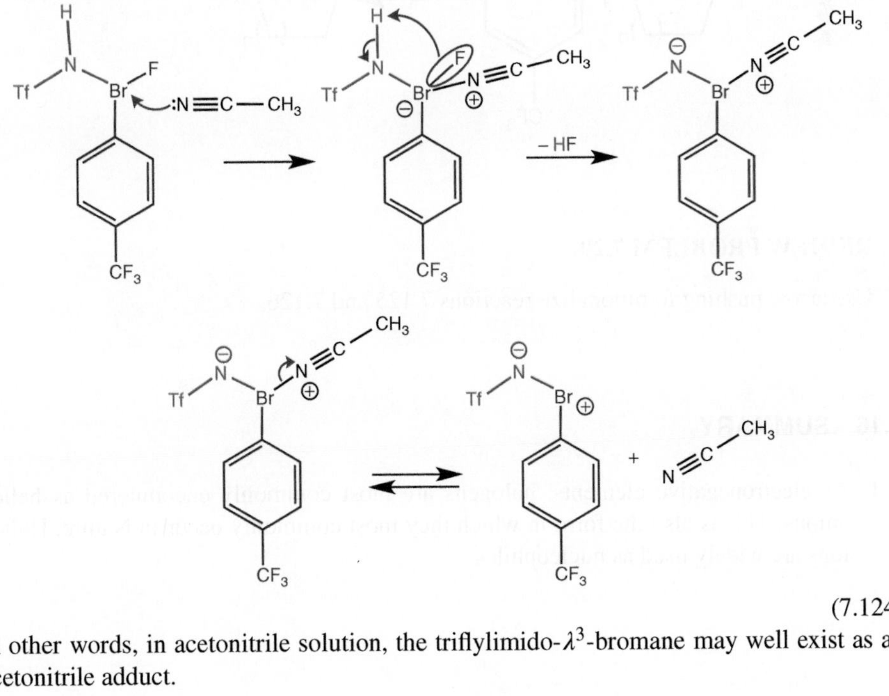

(7.124)

In other words, in acetonitrile solution, the triflylimido-λ³-bromane may well exist as an acetonitrile adduct.

Triflylimido-λ^3-bromanes convert olefins to aziridines in a matter of minutes to about a couple of hours.

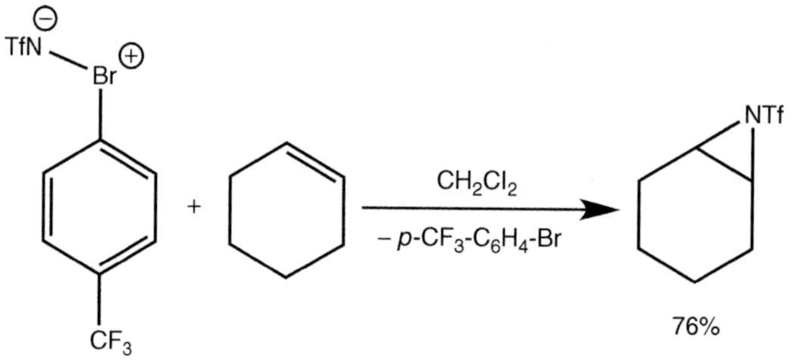

$$(7.125)$$

The analogous reaction with tosylimido-λ^3-iodanes, by contrast, typically requires a transition-metal catalyst, as well as an excess of the olefin.

Remarkably, triflylimido-λ^3-bromanes can even aminate unactivated alkanes, as shown below, opening up a remarkable synthetic route to amines.

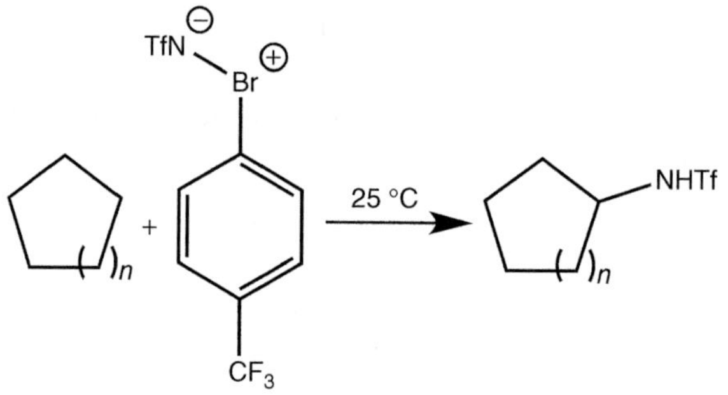

$$(7.126)$$

REVIEW PROBLEM 7.29

Use arrow pushing to rationalize reactions 7.125 and 7.126.

7.16 SUMMARY

1. As electronegative elements, halogens are most commonly encountered as halide anions. This is also the form in which they most commonly occur in Nature. Halide ions are widely used as nucleophiles.

2. The diatomic elements act as oxidants and as electrophiles. Halogen molecules cleave readily via both heterolytic and homolytic pathways:

$$X-X \left\langle \begin{array}{l} \xrightarrow{Nu^{\ominus}} Nu-X + X^{\ominus} \\ \xrightarrow[\text{or } h\nu]{\Delta} X\cdot + X\cdot \end{array} \right. \tag{7.1}$$

3. The disproportionation of molecular halogens in alkaline solution provides an entry into higher oxidation state halogen chemistry. Higher oxidation states of halogens are typically stabilized by strongly electronegative ligands such as fluoride, oxide, and hydroxide.

4. So-called hypofluorous acid, which is not significantly acidic, serves as a useful oxygen atom transfer agent.

5. Numerous interhalogen compounds, as well as cations and anions, are known. Several engage in halide/fluoride ion transfer reactions.

6. By virtue of their tendency to undergo reductive elimination, higher-valent organoiodine compounds serve as important oxidants in organic chemistry.

7. Higher-valent organobromine compounds are emerging as exciting new reagents in organic chemistry. More susceptible to reductive elimination, they are more reactive analogs of related organoiodine compounds.

FURTHER READING

Balard, A. "Sur une substance particulière contenue dans l'eau de la mer (About a Particular Substance Present in Sea Water)," *Annales de Chimie et de Physique, 2nd series,* **1826**, *32*, 337–381.

Weeks, M. E. "The Discovery of the Elements: XVII. The Halogen Family," *J. Chem. Educ.* **1932**, *9*, 1915–1932.

Zhdankin, V. V., ed. *Hypervalent Iodine Chemistry: Preparation, Structure, and Synthetic Applications of Polyvalent Iodine Compounds*; John Wiley & Sons, Inc.: Hoboken, NJ, 2013; 480 pp. *A comprehensive, up-to-date set of reviews.*

Stang, P. J.; Zhdankin, V. V. "Organic Polyvalent Iodine Compounds," *Chem. Rev.* **1996**, *96*, 1123–1178. *An earlier review.*

Rozen, S. "Selective Reactions of Bromine Trifluoride in Organic Chemistry," *Adv. Synth. Catal.* **2010**, *352*, 2691–2707.

Ochiai, M.; Miyamoto, K.; Hayashi, S.; Nakanishi, W. "Hypervalent N-sulfonylimino-λ^3-bromane: Active Nitrenoid Species at Ambient Temperature Under Metal-Free Conditions," *Chem. Comm.* **2010**, *46*, 511–521.

8

The Noble Gases

Because my co-workers at that time (March 23, 1962)
were still not sufficiently experienced to help me with
the glassblowing and the preparation and purification
of PtF₆ [platinum hexafluoride] necessary for the
experiment, I was not ready to carry it out until
about 7 p.m. on that Friday. When I broke the seal
between the red PtF₆ gas and the colorless xenon gas,
there was an immediate interaction, causing an
orange-yellow solid to precipitate. At once I tried to
find someone with whom to share the exciting finding,
but it appeared that everyone had left for dinner!

Neil Bartlett in *Fluorine Chemistry*,
at the Millennium Banks, R. E. *ed.*;
Elsevier: Amsterdam, **2000**, p. 39.

The majority of the noble gases were discovered in the nineteenth century. The discovery that the noble gases are not entirely inert happened in more recent memory, in 1962, and is now part of chemistry's lore. Working alone at the University of British Columbia in Vancouver, Canada, British chemist Neil Bartlett (1932–2008) discovered that the high-valent compound PtF_6 could oxidize molecular oxygen to $O_2^+PtF_6^-$. Since xenon has the same ionization potential as O_2, he reasoned that xenon should also form a similar compound with PtF_6, a prediction that proved essentially correct. Figure 8.1 presents a reproduction of Bartlett's paper, one of the shortest in the history of science for a major discovery. Subsequently, the product was shown to be a bit more complex; it was apparently a mixture of $[XeF][PtF_6]$ and $[XeF][Pt_2F_{11}]$. In the last 50 years since that historic finding, xenon chemistry has grown by leaps and bounds, and a few compounds have also been synthesized for krypton and radon.

Arrow Pushing in Inorganic Chemistry: A Logical Approach to the Chemistry of the Main-Group Elements,
First Edition. Abhik Ghosh and Steffen Berg.
© 2014 John Wiley & Sons, Inc. Published 2014 by John Wiley & Sons, Inc.

Xenon Hexafluoroplatinate(v) Xe⁺[PtF₆]⁻

By Neil Bartlett

(Department of Chemistry, The University of British Columbia,
Vancouver 8, B.C., Canada)

A recent Communication[1] described the compound dioxygenyl hexafluoroplatinate(v), $O_2^+PtF_6^-$, which is formed when molecular oxygen is oxidised by platinum hexafluoride vapour. Since the first ionisation potential of molecular oxygen,[2] 12·2 ev, is comparable with that of xenon,[3] 12·13 ev, it appeared that xenon might also be oxidised by the hexafluoride.

Tensimetric titration of xenon (AIRCO "Reagent Grade") with platinum hexafluoride has proved the existence of a 1:1 compound, XePtF₆. This is an orange-yellow solid, which is insoluble in carbon tetrachloride, and has a negligible vapour pressure at room temperature. It sublimes in a vacuum when heated and the sublimate, when treated with water vapour, rapidly hydrolyses, xenon and oxygen being evolved and hydrated platinum dioxide deposited:

$$2XePtF_6 + 6H_2O \rightarrow 2Xe + O_2 + 2PtO_2 + 12HF$$

The composition of the evolved gas was established by mass-spectrometric analysis.

Although inert-gas clathrates have been described, this compound is believed to be the first xenon charge-transfer compound which is stable at room temperatures. Lattice-energy calculations for the xenon compound, by means of Kapustinskii's equation,[3] give a value ∼ 110 kcal. mole⁻¹, which is only 10 kcal. mole⁻¹ smaller than that calculated for the dioxygenyl compound. These values indicate that if the compounds are ionic the electron affinity of the platinum hexafluoride must have a minimum value of 170 kcal. mole⁻¹.

The author thanks Dr. David Frost for mass spectrometric analyses and the National Research Council, Ottawa, and the Research Corporation for financial support. (*Received, May 4th*, 1962.)

[1] Bartlett and Lohmann, *Proc. Chem. Soc.*, 1962, 115.
[2] Field and Franklin, "Electron Impact Phenomena," Academic Press, Inc., New York, 1957, pp. 114—116.
[3] Kapustinskii, *Quart. Rev.*, 1956, 10, 284.

Figure 8.1 Reproduction of Neil Bartlett's historic paper.

Some general trends in noble gas chemistry are as follows.

- Despite the remarkable progress in xenon chemistry over the last few decades, the noble gases are all exceedingly unreactive, as expected on the basis of their closed valence shells (i.e., their noble gas configurations). Their ionization potentials are so high that all but the most electronegative ligands reduce them back to their elemental state. The ionization potentials of Kr, Xe, and Rn, however, are just low enough that they form a few compounds with fluorine and oxygen, the two most electronegative elements. In addition, xenon forms a few compounds with electronegative groups such as $OTeF_5$, $N(SO_2F)_2$, $N(SO_2CF_3)_2$, CF_3, $C(SO_2CF_3)_3$, and so on. The three uncharged molecular fluorides, XeF_2, XeF_4, and XeF_6, serve as the source of pretty much all xenon compounds.

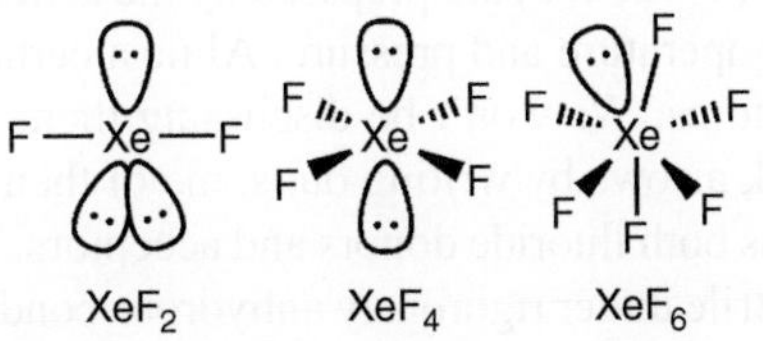

- All noble gas compounds (except for proton adducts such as the gas-phase species HeH^+) are hypervalent; that is, they have more than eight electrons in the noble gas valence shell. In addition, XeF_4 and XeF_6 are susceptible to nucleophilic attack by F^-, which further increases their coordination number. Fluoride ligands are also common leaving or migrating groups in much of xenon chemistry.

- A Xe–F unit may also act as a source of electrophlic F^+. Accordingly, XeF_2 is sometimes used in organic synthesis as an electrophilic fluorinating agent.

- Krypton difluoride (KrF_2) can be synthesized from the elements in a number of different ways, including low temperature photolysis. It is considerably less stable than XeF_2, as shown by the following reaction enthalpies:

$$KrF_2(g) = Kr(g) + F_2(g) \quad \Delta G^0 = -63 \ \mathrm{kJ/mol} \tag{8.1}$$

$$XeF_2(g) = Xe(g) + F_2(g) \quad \Delta G^0 = 105 \ \mathrm{kJ/mol} \tag{8.2}$$

KrF_2 can be stored indefinitely at $-78\,°C$, but decomposes slowly at room temperature. It is an exceedingly powerful oxidant.

- Radon, on the other hand, reacts readily with F_2, yielding radon difluoride. The compound has not been fully characterized. It decomposes on vaporization and is believed to be an ionic compound. The strong radioactivity of all of radon's isotopes has discouraged detailed studies of its chemical properties; the half-life of ^{122}Rn, the longest lived of all radon isotopes, is only 3.82 days.

Most of the discussion below will focus on xenon. We trust that you'll find noble gas chemistry, a term that a bit over 50 years ago would have been an oxymoron, fascinating. Indeed, a few of the mechanisms should be as challenging as anything you have encountered in the rest of this book.

REVIEW PROBLEM 8.1

The first ionization energies of Kr, Xe, and Rn are 1351, 1170, and 1037 kJ/mol, respectively. Use this information to rationalize (i) the instability of KrF_2 relative to XeF_2 and (ii) the likely ionic character of RnF_2.

8.1 THE XENON FLUORIDES: FLUORIDE DONORS AND ACCEPTORS

All three fluorides XeF_n ($n = 2, 4, 6$) are prepared by the action of elemental F_2 on Xe under various conditions of temperature and pressure. Almost certainly, the mechanisms involve radical reactions of F atoms. We won't be discussing them here, but feel free to sharpen your skills with fishhook arrows by writing out some of them.

The compounds act as both fluoride donors and acceptors. Thus, with tetramethylammonium fluoride in acetonitrile under rigorously anhydrous conditions, XeF_4 forms the unique pentagonal-planar (D_{5h}) $[XeF_5]^-$ anion:

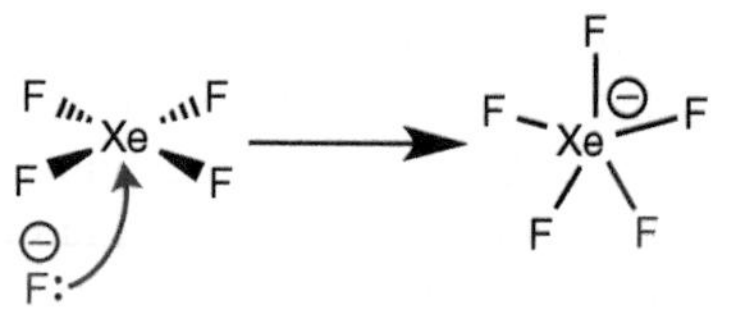

$$\tag{8.3}$$

Stepwise addition of fluoride to XeF_6 leads to the square-antiprismatic octafluoroxenate(VI) ($XeF_8{}^{2-}$) anion, as shown below:

$$XeF_6 + CsF \rightarrow Cs[XeF_7] \tag{8.4}$$

$$Cs[XeF_7] + CsF \rightarrow Cs_2[XeF_8] \tag{8.5}$$

Octafluoroxenate salts are extraordinarily stable, even to temperatures above 400 °C, proof that noble gas compounds are not just laboratory curiosities.

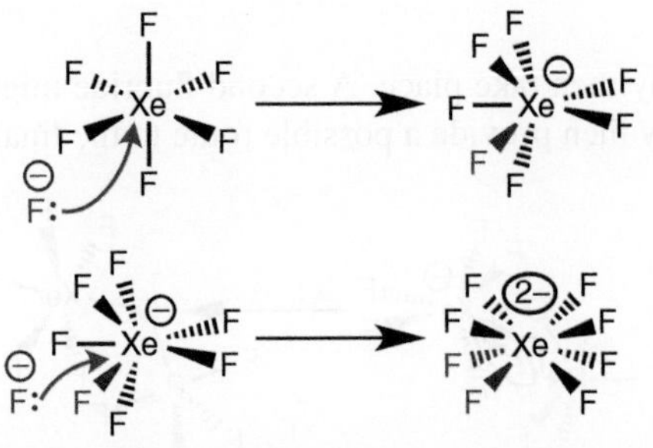

$$\tag{8.6}$$

Thermal decomposition of $[XeF_7]^-$ salts also leads to the $[XeF_8]^{2-}$ anion, in yet another demonstration of the stability of the latter species:

$$2\,Cs[XeF_7] \rightarrow XeF_6 + Cs_2[XeF_8] \tag{8.7}$$

Xenon fluorides also act as fluoride donors toward Lewis acids such as BiF_5 and PtF_5, generating

$$XeF_4 + BiF_5 \rightarrow [XeF_3]^+[BiF_6]^- \tag{8.8}$$

$$XeF_6 + PtF_5 \rightarrow [XeF_5{}^+][PtF_6]^- \tag{8.9}$$

REVIEW PROBLEM 8.2

Is a pentagonal-planar structure for the $[XeF_5]^-$ anion consistent with valence shell electron pair repulsion (VSEPR) theory?

8.2   O/F LIGAND EXCHANGES

By now, you have encountered several examples of ligand exchange reactions in this book (see Section 1.19 for a reminder). XeF_6 undergoes F/O exchange reactions with a variety of nonmetal-oxo species (both neutrals and anions), transferring two fluorides to each molecule of the latter. Thus, with POF_3, XeF_6 yields PF_5 and $XeOF_4$:

$$XeF_6 + POF_3 \rightarrow XeOF_4 + PF_5 \tag{8.10}$$

As in many ligand exchange reactions, the mechanism involves the formation of a bridged intermediate, in this case involving a P–O–Xe bridge:

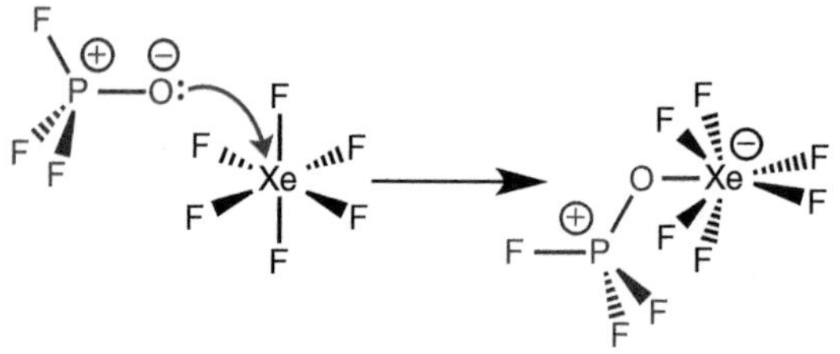

$$(8.11)$$

A fluoride migration may then take place. A second fluoride migration, concomitant with P–O bond cleavage, may then provide a possible route to the final products:

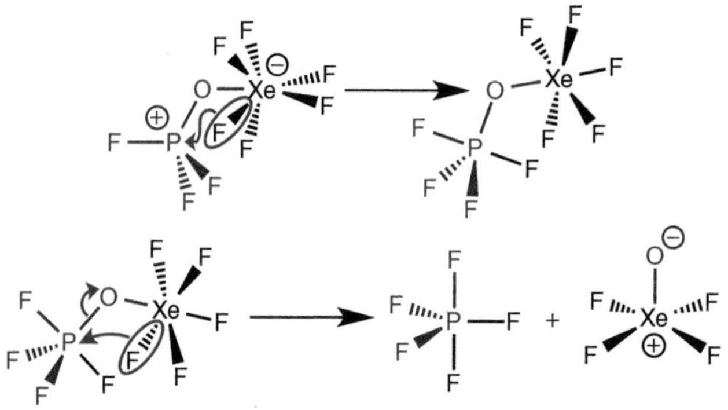

$$(8.12)$$

REVIEW PROBLEM 8.3

XeF_6 reacts with nitrate (NO_3^-) to give nitryl fluoride (NO_2F) and $XeOF_4$:

$$XeF_6 + NO_3^- \rightarrow NO_2F + XeOF_4 + F^-$$

Suggest a mechanism.

8.3   XENON FLUORIDES AS F⁺ DONORS AND OXIDANTS

A perusal of standard reduction potentials (see, e.g., Table 1.6) immediately shows that XeF_2 is a powerful oxidant. Thus, it readily oxidizes HCl to Cl_2:

$$XeF_2 + 2\,HCl \rightarrow Xe + HF + Cl_2 \qquad (8.13)$$

Given that HF is a product and that it has a very high bond dissociation energy (BDE 569 kJ/mol), it's reasonable to start by protonating one of fluorines of XeF_2.

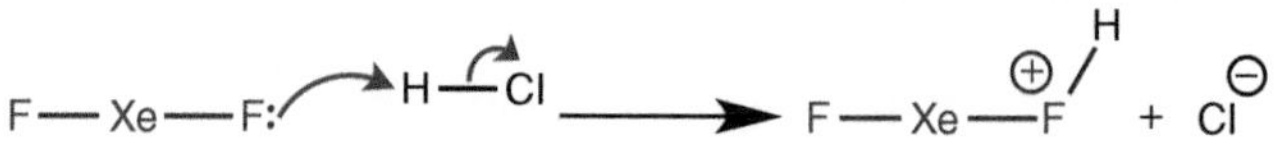

$$(8.14)$$

Chloride might then be expected to attack the unprotonated F, kicking out two excellent leaving groups Xe and HF.

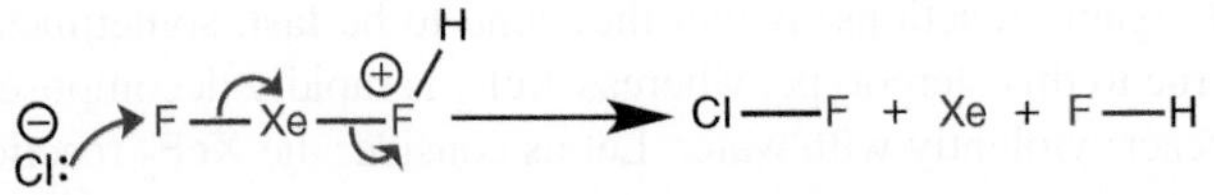

$$Cl^{\ominus} \xrightarrow{} F \xrightarrow{} Xe \xrightarrow{} F \xrightarrow{H}^{\oplus} \longrightarrow Cl \xrightarrow{} F \; + \; Xe \; + \; F \xrightarrow{} H \qquad (8.15)$$

Chloride and ClF then interact to give Cl_2; again, the process is probably facilitated by protonation of the F.

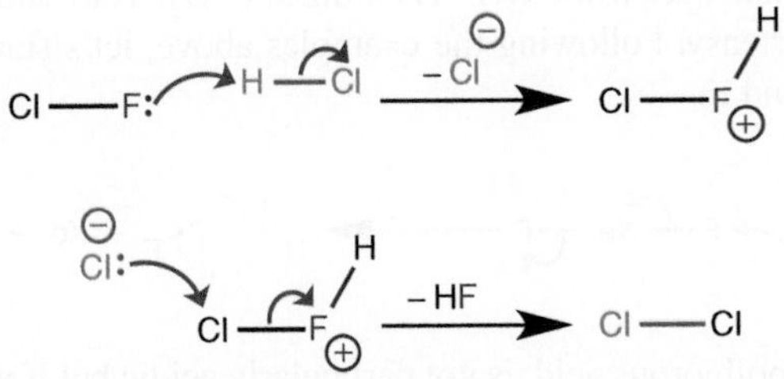

$$Cl \xrightarrow{} F: \xrightarrow{} H \xrightarrow{} Cl \xrightarrow{-Cl^{\ominus}} Cl \xrightarrow{} F \xrightarrow{H}_{\oplus}$$

$$(8.16)$$

$$Cl^{\ominus} \xrightarrow{} Cl \xrightarrow{} F \xrightarrow{H}_{\oplus} \xrightarrow{-HF} Cl \xrightarrow{} Cl$$

XeF_2 is a commercially available reagent, albeit a rather expensive one. Like *N*-fluoro amines and amides (which we discussed in Section 7.6), XeF_2 is used as a F^+ donor in organic synthesis. As in the examples above, the mechanisms typically involve nucleophilic attack on one of the F's of XeF_2.

REVIEW PROBLEM 8.4

XeF_2 has been used to prepare 5-fluorouracil, a cancer chemotherapeutic:

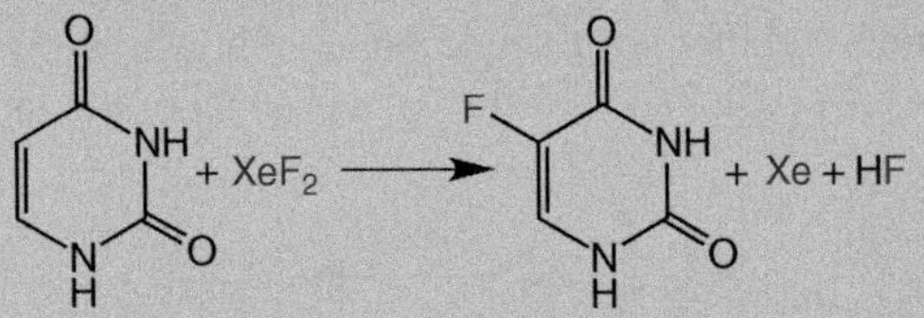

Write a mechanism for the above reaction.

REVIEW PROBLEM 8.5

Write a mechanism for the following reaction:

$$Ph_3TeF + XeF_2 \rightarrow Ph_3TeF_3 + Xe$$

8.4   HYDROLYSIS OF XeF$_2$ AND XeF$_6$

A major reason why main-group reactions have been relatively little studied mechanistically (compared with organic reactions) is that they tend to be fast, sometimes to the point of being violent. True to this stereotype, whereas XeF$_2$ is rapidly decomposed in an alkaline solution, XeF$_6$ reacts violently with water. Let us consider the XeF$_2$ reaction first:

$$2\,\mathrm{XeF_2} + 4\,\mathrm{OH^-} \rightarrow 2\,\mathrm{Xe} + \mathrm{O_2} + 4\,\mathrm{F^-} + 2\,\mathrm{H_2O} \tag{8.17}$$

Observe that some fairly complex redox chemistry must be involved: Divalent xenon is reduced to the elemental state while OH$^-$ is oxidized to O$_2$. That said, obtaining Xe as the final product is rather easy. Following the examples above, let's start off with having an OH$^-$ attack a Xe-bound F:

$$\tag{8.18}$$

The product HOF, hypofluorous acid, is not particularly acidic but if we assume it's deprotonated by hydroxide, the resulting FO$^-$ could attack another molecule of HOF as a nucleophile and produce HOOF, an unstable intermediate that we encountered in Section 7.4, as shown below:

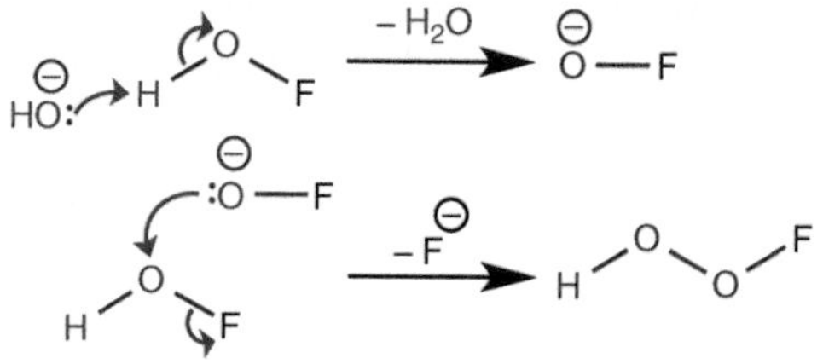

$$\tag{8.19}$$

The instability of HOOF is clearly related to its unique structure with three highly electronegative atoms strung together with single bonds. Shown below is an E2 elimination with a hydroxide ion as the base that leads to O$_2$, a product of the reaction.

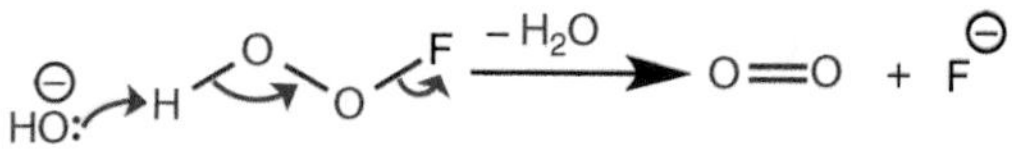

$$\tag{8.20}$$

We have discussed other decomposition pathways for HOOF in Section 7.4.

From an arrow-pushing perspective, the hydrolysis of XeF$_6$ is rather tame (although that's not the adjective one would otherwise apply to such a violent reaction); no fancy redox chemistry is involved. There are two distinct steps, and the reaction can actually be stopped after the first step by limiting the amount of water:

$$\mathrm{XeF_6} + \mathrm{H_2O} \rightarrow \mathrm{XeOF_4} + 2\,\mathrm{HF} \tag{8.21}$$

$$\mathrm{XeOF_4} + 2\,\mathrm{H_2O} \rightarrow \mathrm{XeO_3} + 4\,\mathrm{HF} \tag{8.22}$$

Shown below is a mechanism for the first step. Hydroxide (or water) attacks as a nucleophile; fluoride leaves in the next step, as expected for an S_N2-Si process, followed by deprotonation of the Xe-bound hydroxide.

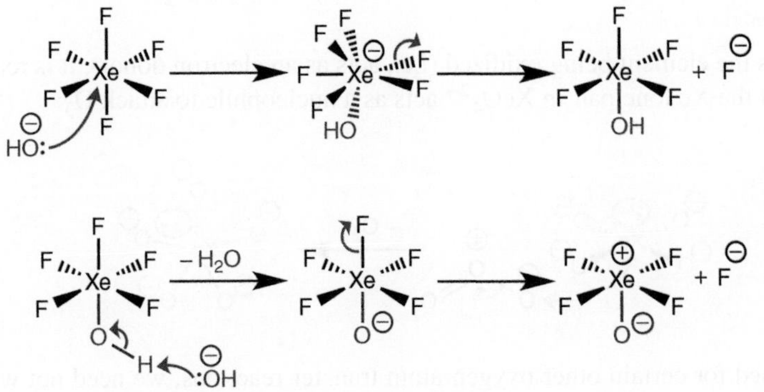

$$(8.23)$$

The $XeOF_4$ so produced may be hydrolyzed further to give XeO_3 (*explosive as a solid!*)

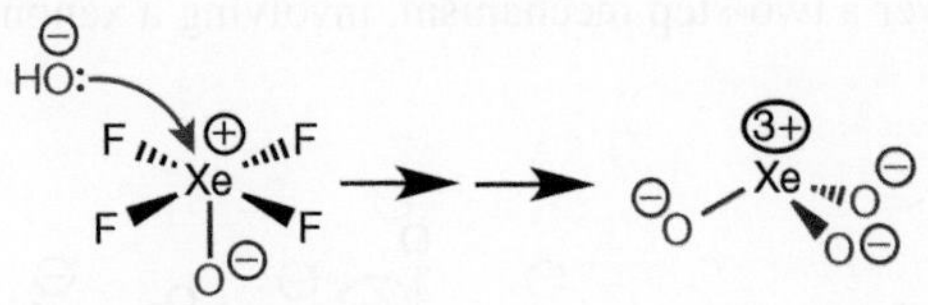

$$(8.24)$$

8.5 XENATE AND PERXENATE

XeO_3 is typically obtained by hydrolysis of XeF_6 (as described above); it's soluble in water to quite high concentrations but does at one point crystallize out as colorless, dangerously explosive crystals. In alkaline solution, XeO_3 forms the anion of xenic acid:

$$XeO_3 + OH^- \rightarrow HXeO_4^- \qquad (8.25)$$

The mechanism is a simple A process (i.e., nucleophile–electrophile association):

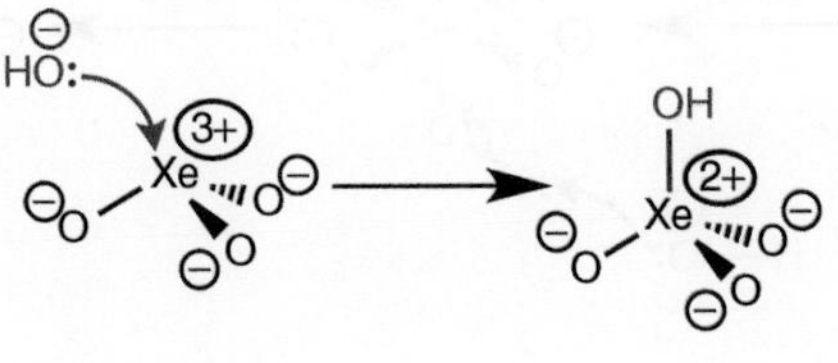

$$(8.26)$$

The xenate anion is unstable with respect to disproportionation, yielding perxenate (XeO_6^{4-}) and elemental Xe:

$$2\, HXeO_4^- + 2\, OH^- \rightarrow XeO_6^{4-} + Xe + O_2 + 2\, H_2O \qquad (8.27)$$

We will discuss the mechanism of this reaction in the next section. An alternative way of obtaining perxenate involves the oxidation of xenate by ozone in alkaline solution:

$$XeO_4^{2-} + O_3 + 2\,OH^- \rightarrow XeO_6^{4-} + O_2 + H_2O \tag{8.28}$$

Since xenon is the element being oxidized (i.e., acts as an electron donor), it is reasonable to suggest that the Xe lone pair in XeO_4^{2-} acts as a nucleophile to attack O_3:

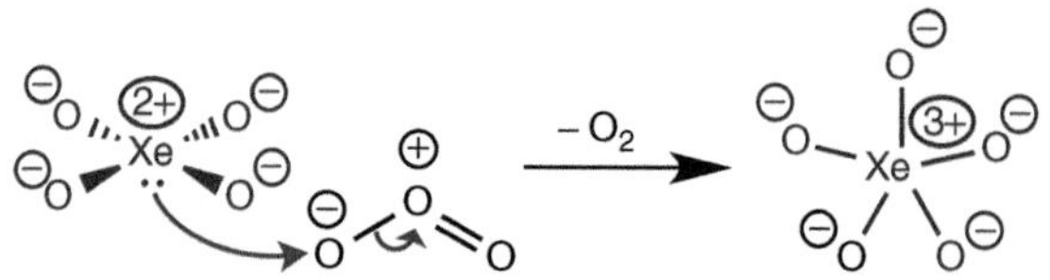

(8.29)

As we discussed for certain other oxygen-atom transfer reactions, we need not worry too much that the electrophilic center is an oxygen atom with a negative formal charge; this is all right, especially when the adjoining oxygen atom has a positive formal charge. Preliminary density functional theory (DFT) studies (by the authors) favor the above S_N2-like direct displacement, over a two-step mechanism, involving a xenon ozonide intermediate, as depicted below:

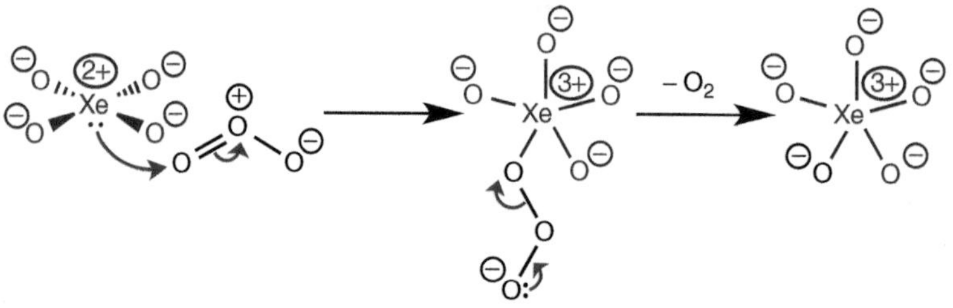

(8.30)

Observe that we have now obtained octavalent Xe (five single bonds and a +3 formal charge), with the same oxidation level as perxenate (XeO_6^{4-}). It only remains to add another hydroxide and deprotonate it:

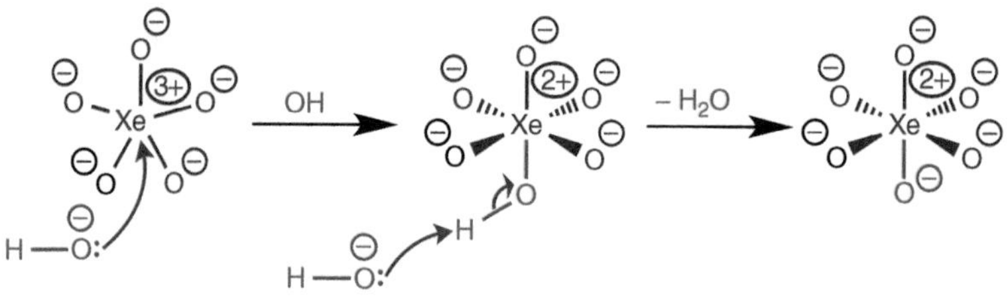

(8.31)

8.6 DISPROPORTIONATION OF XENATE

We promised you challenging mechanisms in our introductory remarks (Section 8.1) and here is a good example. Consider reaction 8.27, which is the disproportionation of xenate to perxenate and elemental xenon in alkaline solution; elemental oxygen is also produced at the same time:

$$2\,HXeO_4^- + 2\,OH^- \rightarrow XeO_6^{4-} + Xe + O_2 + 2\,H_2O \qquad (8.32)$$

By now, we trust you have exercised your arrow-pushing skills a good deal. Nevertheless, as a mechanistic exercise, this reaction may leave you perplexed. How, for example, would one go about stripping all those oxygens from xenate to produce elemental xenon? Before thinking through that question, it might help to focus on the slightly less daunting half of the problem: how does xenate get oxidized to perxenate?

Since oxidation implies electron donation, a reasonable answer to the last question is that we must use the lone pair on the xenon being oxidized to do a nucleophilic attack on one of the oxygens of another xenate, as shown below:

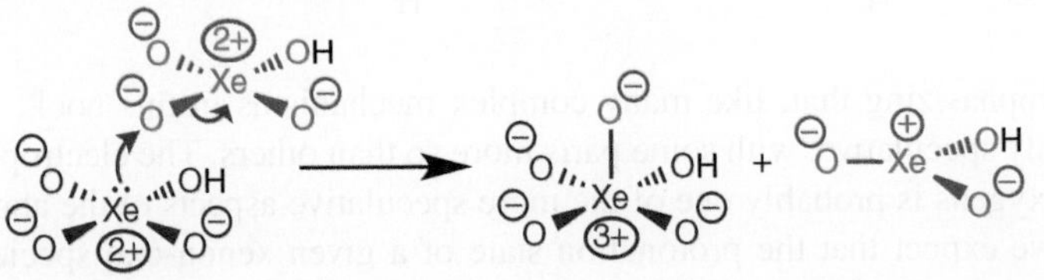

(8.33)

Observe how in one stroke we produced octavalent and tetravalent xenon, or equivalently in this case Xe(VIII) and Xe(IV). The octavalent Xe intermediate readily adds a hydroxide to form perxenate:

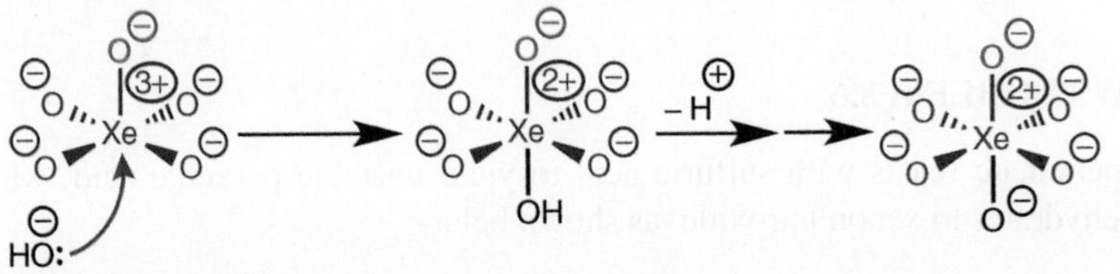

(8.34)

The reactions of the tetravalent xenon intermediate are a bit more complex. If we assume that the oxygens attached to the Xe are electrophilic and susceptible to attack by OH^-, then that would lead to the divalent intermediate $Xe(OH)_2$ and H_2O_2, which is certainly a step in the right direction, namely, toward the final products Xe and O_2:

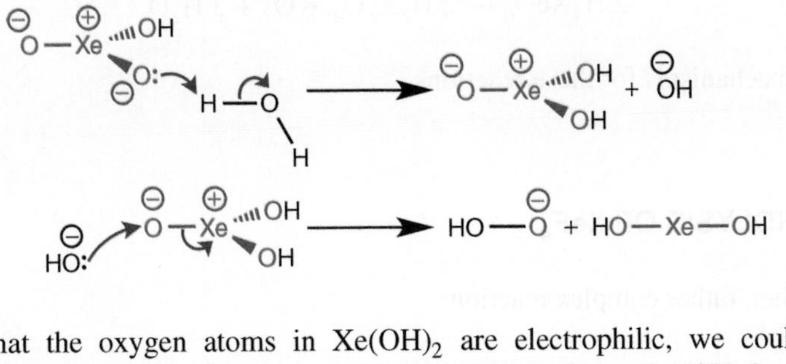

(8.35)

Assuming that the oxygen atoms in $Xe(OH)_2$ are electrophilic, we could have the hydroperoxide anion attack one of them, producing elemental xenon and H_2O_3 (trioxidane), a reactive oxygen species we encountered in Section 6.8.

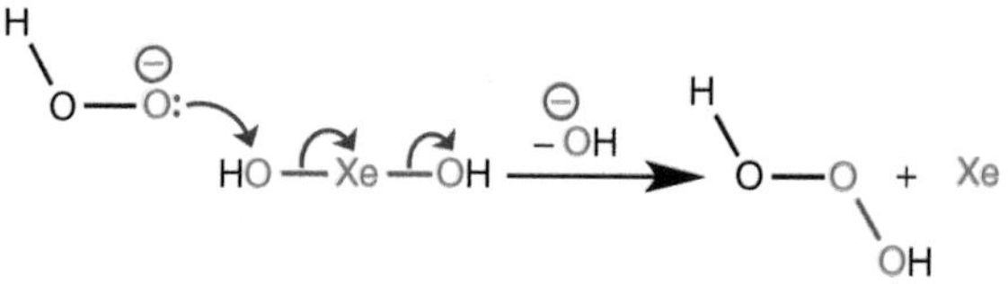

(8.36)

The H_2O_3 then readily falls apart. A cyclic transition state is depicted below; however, a hydroxide-mediated E2 mechanism might also be envisioned for the process.

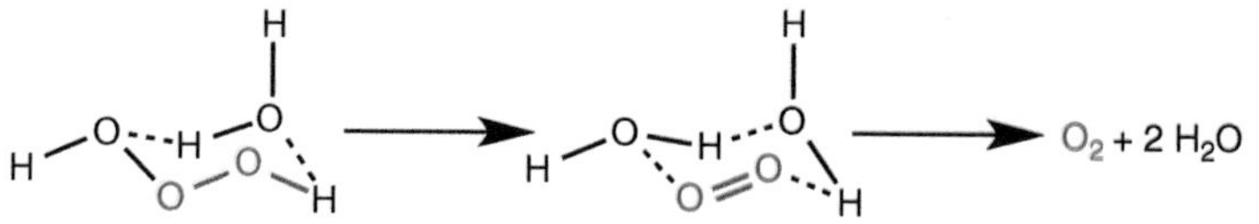

$$O_2 + 2\,H_2O$$

(8.37)

It's worth emphasizing that, like many complex mechanisms in this book, the above pathway is highly speculative, with some parts more so than others. The electrophilicity of xenon-bound oxygens is probably one of the more speculative aspects of the above mechanism. While we expect that the protonation state of a given xenon-oxo species plays a decisive role in determining the electrophilicity of a xenon-bound oxygen, we make no guarantees that the Xe–O species in the above reactions are all depicted in their optimal protonation states. That said, we do believe that there is a good chance that the above mechanism is "correct" in its essentials.

REVIEW PROBLEM 8.6

Barium perxenate reacts with sulfuric acid to yield unstable perxenic acid, which mostly dehydrates to xenon tetroxide, as shown below:

$$Ba_2XeO_6 + 2\,H_2SO_4 \rightarrow 2\,BaSO_4 + H_4XeO_6$$

$$H_4XeO_6 \rightarrow XeO_4 + 2\,H_2O$$

Any remaining perxenic acid decomposes slowly to xenic acid and oxygen:

$$2\,H_4XeO_6 \rightarrow 2\,H_2XeO_4 + O_2 + 2\,H_2O$$

Suggest mechanisms for these reactions.

8.7 HYDROLYSIS OF XeF$_4$

Here is another, rather complex reaction:

$$6\,XeF_4 + 12\,H_2O \rightarrow 2\,XeO_3 + 4\,Xe + 3\,O_2 + 24\,HF$$

(8.38)

With 18 reactant molecules, where would you begin to push arrows? Instead of throwing up your hands at the formidable stoichiometry, it might help to focus on the qualitative aspects of the chemistry. Observe that xenon disproportionates, with concomitant production of molecular oxygen. For tetravalent Xe to be oxidized to the hexavalent state, we need to have a lone pair on the Xe of one XeF_4 attack a F on another XeF_4. Addition of a OH^- should boost the nucleophilicity of the first xenon center, as shown below.

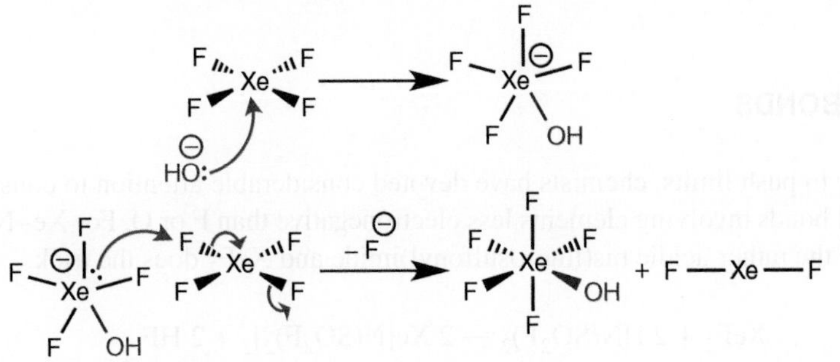

$$(8.39)$$

Observe how in one step we have obtained both hexavalent and divalent Xe. The hexavalent Xe species then hydrolyzes to XeO_3 in much the same way as discussed above for XeF_6 (Section 8.5):

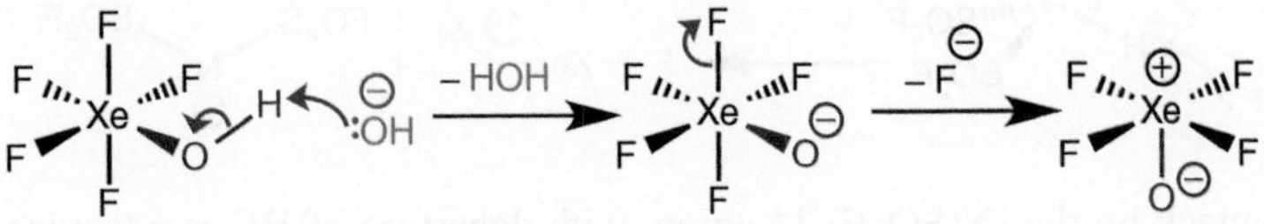

$$(8.40)$$

We have also discussed the hydrolysis of XeF_2 (Section 8.5), which produces Xe, O_2, and HF, thus indirectly accounting for all the observed products.

8.8 OTHER COMPOUNDS CONTAINING Xe–O BONDS

The compound $Xe(OTeF_5)_4$ containing the highly electronegative teflate ligand can be prepared via the interaction of XeF_4 and the Lewis acid $B(OTeF_5)_3$. A fluoride-bridged intermediate presumably forms first:

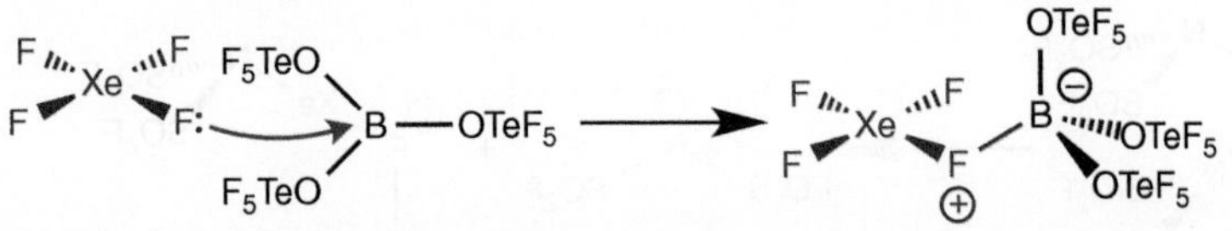

$$(8.41)$$

One of the $OTeF_5$ groups then migrates from the anionic boron center to generate the first Xe–$OTeF_5$ bond:

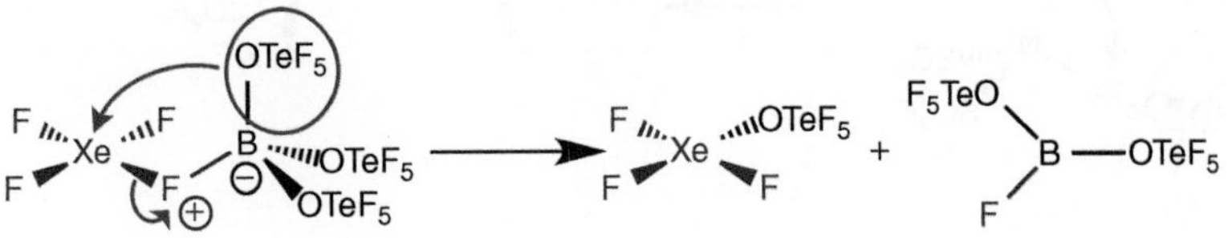

$$(8.42)$$

The process repeats itself until the formation of $Xe(OTeF_5)_4$:

$$(8.43)$$

8.9   Xe–N BONDS

Always eager to push limits, chemists have devoted considerable attention to constructing Xe–nonmetal bonds involving elements less electronegative than F or O. For Xe–N bonds, interaction of the rather acidic bis(fluorosulfonyl)imide and XeF_2 does the trick:

$$XeF_2 + 2\ HN(SO_2F)_2 \rightarrow 2\ Xe[N(SO_2F)_2]_2 + 2\ HF \qquad (8.44)$$

Since HF is a product, protonation of one of the F's of XeF_2 appears to be a plausible first step:

$$(8.45)$$

Nucleophilic attack by the $[N(SO_2F)_2]^-$ anion, with departure of HF as a leaving group, completes the substitution of the first F atom.

$$(8.46)$$

The process repeats itself once more to yield the final product $Xe[N(SO_2F)_2]_2$.

$$(8.47)$$

Note that we have depicted the nucleophilic displacements above as direct S_N2 processes. We have done so for brevity, even though the actual reaction may well proceed via S_N2-Si-type associative processes.

8.10 Xe–C BONDS

Clever use of boron chemistry has allowed the creation of Xe–C bonds, in the form of the $[Xe\text{-}C_6F_5]^+$ cation. The synthesis involves the interaction of XeF_2 and the powerful Lewis acid tris(pentafluorophenyl)borane.

$$XeF_2 + B(C_6F_5)_3 \rightarrow [Xe(C_6F_5)]^+[BF_2(C_6F_5)_2]^- \qquad (8.48)$$

The first step might reasonably be expected to involve one of the F's coordinating to the boron producing an anionic tetracoordinate borate center.

$$(8.49)$$

As elsewhere (particularly Section 3.2), the anionic boron serves as a launchpad for the migration of one of the aryls.

$$(8.50)$$

The two products should then interact in much the same way as the reactants in the first step, setting the stage for the departure of the second F:

$$(8.51)$$

Perhaps not surprisingly, given silicon's strong affinity for fluorine, organosilicon reagents such as $C_6F_5SiMe_3$ (an analog of the Ruppert–Prakash reagent mentioned in Section 4.1) have also been used to construct Xe–C bonds from XeF_2:

$$XeF_2 + C_6F_5SiMe_3 \rightarrow C_6F_5XeF + Me_3SiF \qquad (8.52)$$

The reaction requires an added fluoride ion catalyst; with an excess of the reagent ($C_6F_5SiMe_3$) and of F^-, the second Xe–F bond is also activated, resulting in $Xe(C_6F_5)_2$:

$$C_6F_5XeF + C_6F_5SiMe_3 \rightarrow Xe(C_6F_5)_2 + Me_3SiF \qquad (8.53)$$

Interestingly, no added fluoride was needed for a similar reaction between C_6F_5XeF and $C_6F_5SiF_3$:

$$C_6F_5XeF + C_6F_5SiF_3 \rightarrow Xe(C_6F_5)_2 + SiF_4 \tag{8.54}$$

The compound C_6F_5XeF has proved to be a versatile starting material for other organoxenon compounds, for example, C_6F_5XeCN:

$$C_6F_5XeF + Me_3SiCN \rightarrow C_6F_5XeCN + Me_3SiF \tag{8.55}$$

REVIEW PROBLEM 8.7

Suggest mechanisms for the above silicon-mediated transformations, that is, reactions 8.51 through 8.54.

REVIEW PROBLEM 8.8

Organoxenon compounds involving tetravalent xenon are rare. Frohn and coworkers (*Angew. Chem. Int. Ed.* **2000**, *39*, 391–393) obtained the first example of such a compound via the interaction of XeF_4 and $C_6F_5BF_2$:

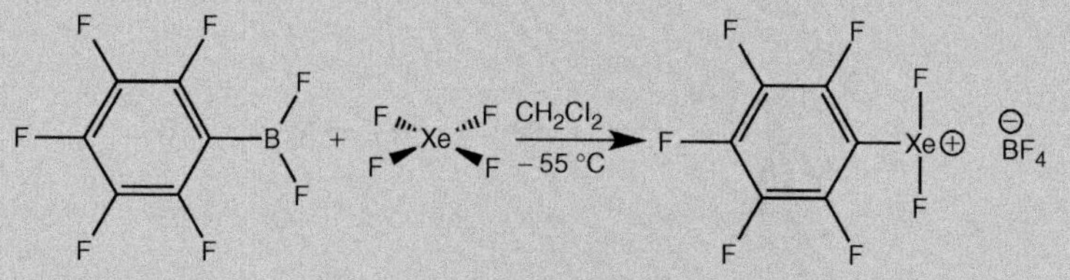

The yellow ionic product precipitated out dichloromethane but proved very soluble in acetonitrile. In the solid state, it decomposed above −20 °C. Suggest a mechanism for this reaction. Also, using VSEPR arguments, comment on the structure of the $C_6F_5XeF_2^+$ cation.

8.11 KRYPTON DIFLUORIDE

As mentioned, krypton difluoride is an exceedingly powerful oxidizing and fluorinating agent, far more potent than XeF_2. Thus, it oxidizes xenon to XeF_6 and gold to the unique Au(V) compound $KrF^+AuF_6^-$:

$$7\, KrF_2 + 2\, Au \rightarrow 2\, KrF^+AuF_6^- + 5\, Kr \tag{8.56}$$

$$3 \, KrF_2 + Xe \rightarrow XeF_6 + 3 \, Kr \qquad (8.57)$$

At 60 °C, the gold-containing salt decomposes to yield the molecular fluoride Au_2F_{10}:

$$2 \, KrF^+AuF_6^- \rightarrow Au_2F_{10} + 2 \, Kr + 2 \, F_2 \qquad (8.58)$$

The molecular structure of Au_2F_{10} is as follows:

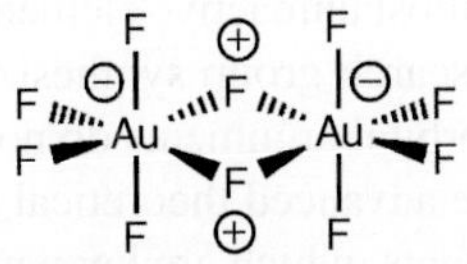

KrF_2 transfers a fluoride ion to strong Lewis acids such as SbF_5, forming the KrF^+ cation:

$$KrF_2 + SbF_5 \rightarrow KrF^+SbF_6^- \qquad (8.59)$$

The KrF^+ cation is one of the strongest oxidants known. As mentioned in Section 7.8, it oxidizes ClF_5 and BrF_5 to ClF_6^+ and BrF_6^+, respectively:

$$XF_5 + KrF^+SbF_6^- \rightarrow XF_6^+SbF_6^- + Kr \quad (X = Cl, Br) \qquad (8.60)$$

Recall that, unlike iodine, chlorine and bromine do not form uncharged XF_7 molecules.

REVIEW PROBLEM 8.9

Suggest a mechanism for the above reaction leading to XF_6^+ salts.

REVIEW PROBLEM 8.10

Use of the teflate ligand permitted the synthesis of the first species containing a krypton–oxygen bond (Sanders, J. C. P.; Schrobilgen, G. J. *J. Chem. Soc., Chem. Comm.* **1989**, 1576–1578). The synthesis involved the interaction of KrF_2 and $B(OTeF_5)_3$ at low temperature (−90 to −112 °C) in SO_2ClF as solvent:

$$3 \, KrF_2 + 2 \, B(OTeF_5)_3 \rightarrow 3 \, Kr(OTeF_5)_2 + 2 \, BF_3$$

Because of thermal instability, the product, krypton "diteflate," could be spectroscopically characterized only in solution. Suggest a mechanism for this reaction.

8.12 PLUS ULTRA

In this final vignette, we'll skip arrow pushing and attempt to present a broader perspective of noble gas chemistry. Not long ago, Seidel and Seppelt at the Free University of Berlin treated AuF_3 in HF/SbF_5 with xenon, obtaining a dark red solution at $-40\,°C$, which yielded crystals of $[AuXe_4](Sb_2F_{11})_2$ at $-78\,°C$ (see *Further Reading* for key references). The researchers' intention was to obtain the simple but elusive compound AuF, using xenon as a very mild reducing agent. Instead, they ended up with the astounding cationic species $[AuXe_4]^{2+}$, made up of two of the most unreactive elements in the periodic table—gold and xenon. Subsequently, the same research group synthesized other Au–Xe complexes such as $[AuXe_2]^{2+}$. Simple molecular orbital arguments do not appear to provide a rationale for the stability of these species. More advanced theoretical studies suggest that their stability owes a great deal to relativistic effects, which are known to be important for gold.

This chemistry is a good reminder that arrow pushing "explains" a lot, particularly the "how" of chemical reactions, but not the "why." Other concepts such as chemical bonding and thermodynamics therefore must not be ignored. The reaction is also an inspiration that truly fundamental discoveries are waiting to be made. The old motto *Plus ultra* (further beyond!) continues to be an appropriate one for inorganic chemistry.

8.13 SUMMARY

The following are some of the highlights of what we discussed in this chapter.

1. Of the three noble gases with a significant chemistry (Kr, Xe, and Rn), the chemistry of Xe is by far the most developed, with the three molecular fluorides XeF_2, XeF_4, and XeF_6 serving as starting materials for most other Xe compounds.
2. The xenon fluorides in general act as both fluoride ion donors and acceptors.
3. Toward certain nucleophiles, xenon fluorides act as a source of electrophilic fluorine, that is, F^+ ion equivalents. Thus, commercially available xenon difluoride is a useful F^+ donor in organic chemistry.
4. The xenon fluorides react vigorously with water. While XeF_6 hydrolyses straightforwardly to XeO_3, XeF_2 and XeF_4 decompose by more complex redox pathways.
5. Xenon–nitrogen and xenon–carbon bonds involving highly electronegative N or C ligands have been synthesized. Among other strategies, organosilicon chemistry has been creatively used for this purpose.
6. KrF_2, one of the strongest oxidants known, has been used to used to effect a number of unique transformations; perhaps the most notable of these is the formation of Au(V)-fluoride compounds.

FURTHER READING

1. Fisher, D. *Much Ado about (Practically) Nothing: A History of the Noble Gases*; Oxford University Press: Oxford, 2010. 288 pp. *A recent popular science book.*
2. Bartlett, N. *Proc. Chem. Soc.* **1962**, 218. *This paper has been reproduced in Figure 8.1.*

3. Hargittai, I. "Neil Bartlett and the First Noble-Gas Compound," *Struct. Chem.* **2009**, *20*, 953–959. *A short, excellent account of the original discovery.*

4. Seidel, S. Seppelt, K. "Xenon as a Complex Ligand: The Tetra Xenono Gold(II) Cation in $AuXe_4^{2+}(Sb_2F_{11}^{-})_2$," *Science* **2000**, *290*, 117–118.

5. Hope, E. G. "Coordination Chemistry of the Noble Gases and Noble Gas Fluorides," *Coord. Chem. Rev.* **2012**, *257*, 902–909. *A review focusing on Xe and XeF_2 as transition metal ligands.*

6. Liebman, J. F.; Deakyne, C. A. "Noble Gas Compounds and Chemistry: A Brief Review of Interrelations and Interactions with Fluorine-Containing Species," *J. Fluor. Chem.* **2003**, *121*, 1–8. *This somewhat specialized review concludes with a poem in honor of Neil Bartlett. We quote four lines, which focus on the elements that we have largely ignored in this chapter: He, Ne, Ar and Rn:*

> *Following Neil's lead, compounds of Rn*
> *And Kr were made, and it seems, Ar's not barren.*
> *But a classical chemical bond, yet remains still beyond*
> *Anyone who studies helium and neon.*

Epilogue

We've reached the end of our journey. We hope you feel you've gained a deeper appreciation of chemical reactivity. In particular, we hope we have helped you look at complicated reaction stoichiometries not as meaningless facts but rather as information that can be understood. That understanding is often speculative, but in general it's experimentally (and computationally) testable, given appropriate resources, and it's far better than not having any insight at all.

If you are planning to become a chemist, you can use this understanding in a creative way to synthesize new molecules and to design new reactions. Indeed that's how many organic chemists view main-group chemistry, as a fount of new reagents and synthetic strategies.

For those of you with other career objectives, we hope you've acquired a new way at looking at the natural world. Whether it's an environmental issue, or one related to health, or something else, you're now in a better position to think about its molecular basis. Once in a while, you may even be able to understand a given issue in mechanistic terms, with actual arrow pushing!

We look forward to hearing from you about your experiences with the approach we have espoused in this book. And we wish you the very best, as you continue to explore the wonderful world of the elements and their compounds!

Arrow Pushing in Inorganic Chemistry: A Logical Approach to the Chemistry of the Main-Group Elements,
First Edition. Abhik Ghosh and Steffen Berg.
© 2014 John Wiley & Sons, Inc. Published 2014 by John Wiley & Sons, Inc.

Appendix A

Inorganic Chemistry Textbooks, with a Descriptive-Inorganic Focus

A.1 INTRODUCTORY TEXTS

1. Housecroft, C.; Sharpe, A. G. *Inorganic Chemistry*, 4th ed., Prentice Hall: Upper Saddle River, NJ, **2012**; 1256 pp.
2. Rayner-Canham, G.; Overton, T. *Descriptive Inorganic Chemistry*, 6th ed., Freeman: New York, **2014**; 768 pp.
3. House, J.; House, K. A. *Descriptive Inorganic Chemistry*, 2nd ed., Academic: Waltham, MA, **2010**; 592 pp.
4. House, J. *Inorganic Chemistry*, 2nd ed., Academic: Waltham, MA, **2012**; 848 pp.

A.2 ADVANCED TEXTS

1. Cotton, F. A.; Wilkinson, G.; Murillo, C. A.; Bochmann, M. *Advanced Inorganic Chemistry*, 6th ed., John Wiley & Sons, Inc.: Hoboken: NJ, **1999**. 1356 pp.
2. Earnshaw, A.; Greenwood, N. *Chemistry of the Elements*, 2nd ed., Butterworth-Heinemann: Oxford, 1997; **1600** pp.

Arrow Pushing in Inorganic Chemistry: A Logical Approach to the Chemistry of the Main-Group Elements,
First Edition. Abhik Ghosh and Steffen Berg.
© 2014 John Wiley & Sons, Inc. Published 2014 by John Wiley & Sons, Inc.

Appendix B

A Short List of Advanced Organic Chemistry Textbooks

Because an elementary knowledge of organic chemistry has been assumed in this book, introductory organic texts are not listed here. The following advanced texts may be worth consulting for a more in-depth treatment of certain topics covered in this book.

1. Carey, F. A.; Sundberg, R. J. *Advanced Organic Chemistry; Part A: Structure and Mechanisms; Part B: Reactions and Synthesis*, 5th ed., Springer: Heidelberg, **2007**; 1200 (Part A) and 1322 pp (Part B).
2. Smith, M. B. *March's Advanced Organic Chemistry: Reactions, Mechanisms, and Structure*, 7th ed., John Wiley & Sons, Inc.: Hoboken, NJ, **2013**; 2080 pp.
3. Anslyn, E. V.; Dougherty, D. A. *Modern Physical Organic Chemistry*, University Science: Mill Valley, CA, **2005**; 1104 pp.

Arrow Pushing in Inorganic Chemistry: A Logical Approach to the Chemistry of the Main-Group Elements, First Edition. Abhik Ghosh and Steffen Berg.
© 2014 John Wiley & Sons, Inc. Published 2014 by John Wiley & Sons, Inc.

Index

Arrow Pushing in Inorganic Chemistry: A Logical Approach to the Chemistry of the Main-Group Elements,
First Edition. Abhik Ghosh and Steffen Berg.
© 2014 John Wiley & Sons, Inc. Published 2014 by John Wiley & Sons, Inc.

Printed and bound by CPI Group (UK) Ltd, Croydon, CR0 4YY

16/04/2025

14658357-0001